Brewing Materials and Processes

Brewing Materials and Processes

A Practical Approach to Beer Excellence

Edited by

Charles W. Bamforth

Department of Food Science and Technology
University of California Davis
Davis, California, USA

AMSTERDAM • BOSTON • HEIDELBERG • LONDON
NEW YORK • OXFORD • PARIS • SAN DIEGO
SAN FRANCISCO • SINGAPORE • SYDNEY • TOKYO

Academic Press is an imprint of Elsevier

Academic Press is an imprint of Elsevier
125 London Wall, London EC2Y 5AS, UK
525 B Street, Suite 1800, San Diego, CA 92101-4495, USA
50 Hampshire Street, 5th Floor, Cambridge, MA 02139, USA
The Boulevard, Langford Lane, Kidlington, Oxford OX5 1GB, UK

British Library Cataloguing-in-Publication Data
A catalogue record for this book is available from the British Library

Library of Congress Cataloging-in-Publication Data
A catalog record for this book is available from the Library of Congress

ISBN: 978-0-12-799954-8

For information on all Academic Press publications
visit our website at https://www.elsevier.com/

Publisher: Nikki Levy
Acquisition Editor: Nancy Maragioglio
Editorial Project Manager: Billie Jean Fernandez
Production Project Manager: Lisa Jones
Designer: Matthew Limbert

Typeset by TNQ Books and Journals
www.tnq.co.in

Contents

List of Contributors

C.W. Bamforth
University of California, Davis, California, CA, United States

C.S. Benedict[†]

R. Biurrun
Research Institute for Beer and Beverage Production (FIBGP), Research and Teaching Institute for Brewing in Berlin (VLB e.V.), Berlin, Germany

N. Davies
Muntons, Stowmarket, United Kingdom

A.J. de Lange
Formerly McLean, Virginia, United States

M. Eumann
EUWA Water Treatment Plants, Gärtringen, Germany

A.E. Hill
Heriot-Watt University, Edinburgh, Scotland, United Kingdom

L.T. Lusk
Formerly Foam and Flavor Development Scientist, Miller Coors LLC

B. Meyer
Research Institute for Beer and Beverage Production (FIBGP), Research and Teaching Institute for Brewing in Berlin (VLB e.V.), Berlin, Germany

R. Pahl
Research Institute for Beer and Beverage Production (FIBGP), Research and Teaching Institute for Brewing in Berlin (VLB e.V.), Berlin, Germany

T.R. Roberts
Steiner Hops Ltd.

I. Russell
Heriot-Watt University, Edinburgh, Scotland

C. Schaeberle
EUWA Water Treatment Plants, Gärtringen, Germany

W.J. Simpson
Cara Technology Limited, Leatherhead, Surrey, United Kingdom

[†]Sadly, our good friend Chaz Benedict died on March 17th 2015.

G. Spedding
Brewing and Distilling Analytical Services, LLC, Lexington, KY, United States

G. Stewart
Heriot Watt University, Edinburgh, Scotland

Preface

I am frequently asked my opinion on the essentials for success for an aspiring new brewing company. I first suggest that they need to be able to get their beer into the marketplace at a price point that customers are prepared to meet but which will allow a satisfactory margin. This is not a straightforward thing by any means, noting that (in the United States) distributorships are still very much influenced by the very large brewing companies and there is only so much space on a supermarket or liquor store shelf. There are literally hundreds of brewing companies at any one time determined to join the thousands already in this increasingly competitive market. Unless they have their own location to serve their beers, it will unfortunately be a struggle. Moreover, even if they do have their own outlet, chances are that the difference between success and failure is the safety net of a food menu and the availability of nonalcoholic beverages.

I also impress upon them the need to brew enough beer to be able to make a livable income. I am well aware of what I call the femtobrewing mentality of some, namely that the only decent brews are those produced on the tiniest possible scale. Beer, however, remains a low-cost product, and the brutal reality is that you need to sell quite a lot of beer to make a living.

Third—and in my view the most critical requirement of all—is the need to deliver a quality proposition. The blunt reality is that (as for any other commodity) beer drinkers will not tolerate inferior products.

In my many years within the brewing industry (approaching 40), I have had the great good fortune to meet many brewers who are committed to quality, in companies large and small. None has surpassed Ken Grossman, founder and President of the Sierra Nevada Brewing Company. His book, *Beyond the Pale* (Wiley), should be compulsory reading for any brewer, but most definitely those who want a realistic take on what is needed to succeed. Although there are other brewers at the upper end (in volume terms) of the brewing industry who turn to mass advertising and trash the products of others to hawk their wares, Ken has always known that developing a strong brand based on the simple ethos of product excellence is the route to sustained success, with the beer speaking for itself. He poured his first six batches of beer away, unprepared to hit the market with a product until he believed it was right and that he was able to deliver that same degree of excellence brew after brew after brew.

That is, for me, fundamentally the most important definition of quality: the meeting of customer expectation and the avoidance of unwanted surprises.

Moreover, that perception is very much a personal thing. This is why I refuse to pontificate about which specific brands are good and which are bad. What you like to drink (or eat, or watch, or read, or believe, etc.) is a matter for your personal preference and delectation. My selections are very likely to differ from yours. So for some in the industry to decry long-standing lager brands as being "yellow fizzy liquid" is to, at a stroke, insult the millions of people who like that style of beer, as well as to display a profound ignorance about the essentials of quality as defined earlier. These beers may not have screaming depths of flavor or the most luscious of foams, but they are astonishingly consistent and pass the test of "fitness for purpose." The brewers of these beers are skilled and knowledgeable individuals—if a grievance might be leveled at some of the larger brewing companies, then it should be targeted at the frequently ruthless business and marketing strategies of such entities.

Armed with this definition of quality, then, we must consider the routes to take to achieve the goal of the right product every time—and one that will remain in a condition to delight the consumer, no matter when the product is purchased and consumed within a realistic life span for that particular beer. I suggest that several needs must be fulfilled:

1. You must have a trained and knowledgeable team, one that is motivated and respected. As a former boss of mine insisted, "there are two things that matter, the people and the quality, and if you look after the first, they will deliver on the second."
2. The facilities must be capable in terms of design to deliver a quality product when operated correctly.
3. Standard operating procedures must be in place, from raw materials-in to products and coproducts-out that are meaningful, up-to-date, and comprehensive, such that if the proverbial bus ran over the brewer, his or her successor could seamlessly step in.
4. Statistically significant methods must be applied judiciously to monitor the process and product, generating information in a timely fashion to satisfy the needs of the operation.
5. Properly maintained instrumentation must be in place to allow those methods to be pursued reliably.

In the Appendix of *Brewing: New Technologies*[1], I discussed much of the salient material pertaining to these items (the recommended methods of the American Society of Brewing Chemists and the fundamentals of statistics). Also addressed were the essential principles of Quality Assurance versus Quality Control, including Quality Standards such as the International Organization for Standardization (ISO) series and Hazard Analysis and Critical Control Points (HACCP). An updated version of that Appendix is included in this new book.

In the first book, *Brewing: New Technologies*, we addressed sequentially the issue of quality classified according to the main attributes of foam, flavor, stability (flavor, colloidal, and microbiological), color, gushing, and impact on health. In this second book, I asked the authors to approach the matter in a somewhat different but complementary manner, with a focus directly on practical aspects. The aim is to

[1]Bamforth, C.W., 2006. Brewing: New Technologies, first ed. Woodhead Publishing, Cambridge, UK.

address in detail the measurement tools and approaches in the context of the nature and significance of the specifications that are applied for a given raw material, process stage, or the final product. The reader will judge if we have succeeded.

Distinguished Professor Charles W. Bamforth PhD, DSc (Hull), DSc (Heriot-Watt), CSci, FIBD, FRSB, FIAFoST
California, United States

CHAPTER

Malts

1

N. Davies
Muntons, Stowmarket, United Kingdom

INTRODUCTION

Love it or loathe it, the traditional malt specification that has been with us for many years is likely to continue in use for the foreseeable future. Notwithstanding the universal acceptance that it is in parts flawed and difficult to interpret, it is still the mainstay of evaluating the suitability of malt for brewing. Over the years many suggestions for novel methods have been researched and suggested as either descriptors of malt quality or indicators of processability. So far, novel methods have had limited acceptance even when these have been extensively peer reviewed. Even with the new methods it can still be difficult to relate the results to brewhouse performance and cause problems when integrating them into a laboratory set up to follow recommended methods which are generally robust and which use recognized equipment and techniques rather than some of the intricacies involved in some newer assays. Familiarity and tradition have bolstered the conventional malt specification and will undoubtedly ensure it remains for some time yet.

What we have to contend with is that the concept of quality is often more perceptive than well defined. In other words, the judgment of the user based on his experience is not always easily and unequivocally traceable back to a specific analytical parameter (analyte). In many instances, one brewer or distiller describes a batch of malt as superb because it runs through his plant like "rocket fuel," whereas another brewer finds the very same batch wholly unacceptable. The dilemma is how to resolve the root cause. Is it malt derived or process derived? Whatever the cause, a problem exists, but resolution may lie not with a fault in supply but rather in what has contributed to arriving at a material apparently in specification that could have been achieved in so many ways.

For consumers the link of malt analysis to product quality is not always appreciated. What malt parameters affect the enjoyment of the final product? Many factors have been found to influence drinkability and many could be attributable to malt (Davies, 2006a,b). We are aware of the influence of malt in beer either consciously or subconsciously and that affects our enjoyment of the product (Fig. 1.1). Precisely how biochemistry in malt manufacture affects some of the more curious aspects of the enjoyment of beer is largely unknown and cannot be

Brewing Materials and Processes. http://dx.doi.org/10.1016/B978-0-12-799954-8.00001-0

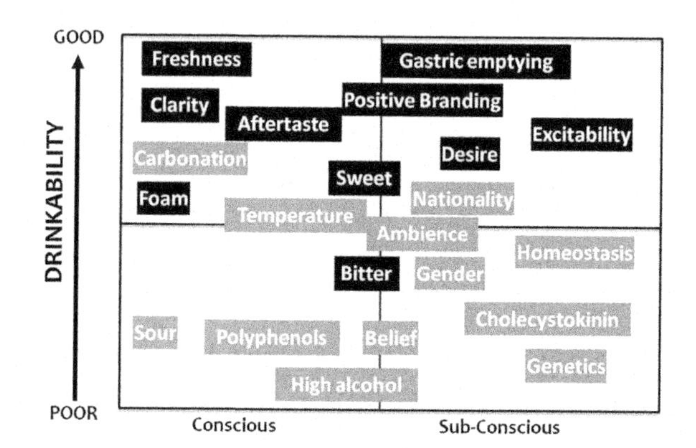

FIGURE 1.1

Factors associated with the drinkability of beer. Some factors we can consciously control and some are subconscious based on biochemistry and social or environmental factors. Arguably, the factors shaded in black are influenced by the malt component. Most of those factors can have a positive impact on the drinkability of beer.

After Davies, N.L., November 2006a. Malt - a vital part of the brewer's palette, Monograph 34: E.B.C. Symposium, "Drinkability," Edinburgh.

reflected in the traditional malt specification. For a small handful of those parameters, however, there are clear links to malt. Freshness can be achieved with a malt that is low in factors that contribute to staling. Clarity can be impacted by malt modification and beta glucan level. Foam quality is affected by malt proteins and protein breakdown. Consumer preference is likely to be more influenced by marketing among the public. If a product is described as full of malt flavor and brewed from the finest malt and hops, it may be perceived as such, whereas a product brewed in the same way but advertised with flavors associated with fruits from esters will not be judged "malty". The brewing and distilling trade has made a concerted effort to promote beer appreciation and wholesomeness of its raw materials, yet specific malt flavors or attributes still take a back seat in most sensory descriptions. In a slightly tongue in cheek attempt to describe how we enjoy beer, it has even been described mathematically. The controlling variables were beer temperature, number of people drinking, days before being back at work, venue mood, and availability of foods or snack, but with no place for the ingredients to contribute to that enjoyment (Fig. 1.2; Mindlab, 2012).

WHAT SHOULD A MALT ANALYSIS TELL YOU?

Of course, this depends who is asking. For the breeder the difference could be as simple as whether the barley is suitable for feed or malting based on broad-brush measures such as hot-water extract, disease resistance, and good agronomy. For

$$E = -(0.62T2 + 39.2W2 + 62.4P2) + (21.8T + 184.4W + 395.4P + 94.5M - 90.25V) + 50(S + F + 6.4)$$

E is a factor describing overall enjoyment.

T is the ambient temperature in degrees Celsius.

W is the number of days until you are required back at work.

P is the number of people with whom you are drinking.

M is related to your mood whilst drinking the pint.

V is related to the volume of the music being played.

S and **F** are related to the availability of snacks and food.

FIGURE 1.2

Formula to describe the enjoyment of beer. Nowhere does this equation allow for the product quality or taste. These parameters seem to be left to the brewer to control and are a given backdrop to enjoying the beverage. Although this is not a serious scientific formula, it illustrates an attitude to how alcoholic beverages are judged by the public.

From Mindlab, 2012. Formula for the Perfect Pint. www.taylor-walker.co.uk/Media/Documents/PDF/news/WPR-Mindlab-The-Perfect-Pint.pdf.

maltsters, a good variety would be one which performed well over a wide range of nitrogen (protein) contents and was easy to germinate and kiln. To a brewer often it is extract yield, color potential, clarity, and flavor. An overarching thread in all these requirements is that the best variety will reduce risk for the whole supply chain. Ideally, the breeder needs a variety that lasts up to 10 years to recoup breeding investment costs. Farmers need the certainty that a crop will grow well and be versatile in different seasonal weather conditions. Maltsters need varieties that can guarantee availability of good quality raw material free from disease and requiring the least processing utilities input. Brewers also need a secure supply of raw material and for the malt not to give rise to any negative processing or taint issues while providing good yeast nutrition and contributing to color and/or flavor. The concept of a risk premium to secure good quality malting barley at the right time is something very real in the modern malting supply chain. Risk can, of course, be managed and reduced for raw material availability, quality, and functional aspects of performance.

For many years, breeders have used their extensive knowledge of plant phenotypes to determine those best for selection from crosses which may number 30,000 individual plants. In recent years, they have been aided through marker-assisted selection or genomics. In some cases it has been possible to link genetic sequences with specific barley or malt attributes and then to select or promote the effect of those sequences in new varieties. Markers based on single-nucleotide polymorphisms (SNPs) have rapidly gained popularity among breeders (Mammadov et al., 2012). There are, however, problems with this type of selection. A major hurdle is the highly repetitive nature of the plant genome (Meyers et al., 2001). The aim is to identify genes or quantitative trait loci (QTL) that contribute to a desired trait in the grain, with agronomy, or even in malt manufacture. Rather than generate a plethora of new QTL for scientific interest only, a recently concluded project sought to use the SNP/QTL libraries available around the world to enable breeding of specific desirable traits for malting barley into existing elite United Kingdom breeding varieties (Thomas et al., 2014; Ramsay et al., 2014). It was notable in the early stages of that research work that the major distinctions between barley varieties were associated with the vernalization gene. Unsurprisingly, we have spring and winter barley varieties as the major barley types. This is just one factor that makes selection of new varieties complex. The complex genetic interrelationships in barley make it difficult to relate process performance to such malt specification parameters as beta glucan, Kolbach index (KI), wort viscosity, fine—coarse extract difference, friability, and free-amino nitrogen (Wentz, 2000; Wentz et al., 2004). What is behind such difficulty in identifying how to breed and improve these characteristics? Many of the genetic traits are carried on different parts of the DNA sequence, on different chromosomes, and some genes exert multiple effects (pleiotropy). The result of this is that one gene or a series of linked genes may control a number of parameters on a malt specification, or that indeed some existing malt analytes are effectively redundant. Similar consequences have been found in attempting to relate diastatic power (DP) to improved fermentability (Evans et al., 2009). In that work, it was made clear that the genetic sequences controlling the three major contributing enzymes to DP (β amylase, α amylase, and limit dextrinase) are all carried on different chromosomes along with other currently unspecified Gibberellin-responsive elements. Selection to improve one of those parameters in isolation is effectively not possible and is an underlying reason why so many malting parameters are interlinked such that changing one inevitably affects a number of others. There is a further layer of uncertainty due to the thermal properties of amylolytic enzymes and the sugar profiles created (Duke and Henson, 2009b; Henson and Duke, 2014). Traits are influenced by genetics and the environment interacting which further complicates breeding selection. It is relatively straightforward to introduce a stable change in a barley variety if the trait is controlled by a heritable gene, but very uncertain and unstable when the impact of the environment on that gene is strong (Kavitha et al., 2012). So much that we think we know is due to the malting process on a current malt specification is influenced much further back in the breeding program and upstream in the mashing process.

As long ago as 1914, a paper in the Journal of the Institute of Brewing by Harold and John Heron on the purchase of malt on the basis of analysis lamented "we feel sure that the majority of brewers fail to fully appreciate the limitations of malt analysis." Taking long leaps through countless brewing research studies over the next 60+ years, we come to the review of Hyde and Brookes (1978) in which links between Hot Water Extract, Gravity and run-off time were reasonably well correlated. They found that just four main analytes (extract, soluble nitrogen ratio, beta glucanase level, and total nitrogen) could account for 85% of the variability in the brewhouse, extract accounting for the lion's share at 70%. However, they too felt that analysis was too skewed toward commercial transactions stating "the information they provide about brewhouse behavior is very limited and that provided about wort or beer quality virtually non-existent." Essentially, a brewer's requirements for malt can be encapsulated as follows: to produce wort as economically as possible that performs well across all brewhouse operations (O'Rourke, 2002).

A traditional malt analysis can be subdivided into five key groups: starch conversion, carbohydrate conversion, carbohydrate extract, color, and enzyme potential. Physical attributes of barley and malt affect all of these, but are not regularly considered important when assessing malt quality. This can, of course, be dangerous. Varieties that differ in grain-size distribution each year will not necessarily mill the same. Quite often, problems with malt processing in breweries can be traced back to an unchanged mill setting. In a year with grains that are wider in diameter (bold) extract yield can be dramatically diminished by generation of dust when mill gap settings are set too small. When size distribution is lower, grains may be only partly milled, giving rise to poorly digestible pieces of endosperm and undigested glucan leading to haze. Mill adjustment is such an important aspect of brewing performance yet is often the source of many hazy beers or reduced brew length. Picking up the problem at the malt specification stage by early discussion in a new season with the maltster could avoid costly volume reduction and filtration time in the brewery.

These days, a malt analysis is very much like a risk management tool. Brewers need consistency of raw materials, yet there will always be seasonal variations and barley malting performance changes throughout the year from harvest onward as the grain matures. Much of that annual change in product is seen only by the maltster who will adjust the process conditions to match the agreed specification. Seasonal changes, in which significant swings arise in the levels of protein or beta glucan or grain-size distribution, can be more of an issue that has to be considered. Grain protein content has been found to correlate with applied nitrogen and beta amylase activity (Yin et al., 2002; Qi et al., 2006). Grain size is known to affect diastatic power and hence can be seasonal (Agu et al., 2007). If the specification requires higher-limit dextrinase activity, for example in distilling or high-adjunct grists, there should be an acknowledgment that this could come at the expense of higher malting loss and reduced extract. That is because this enzyme develops appreciably only after a long germination (Sissons et al., 1993) and it has to be mashed carefully to avoid losses due to heat deactivation during mashing (Stenholm and Home, 1999). A similar trade-off has been found using dehusked barley in which filtration and amylolytic enzyme activity was worse, yet processing and predicted spirit yield improved (Agu et al., 2008). In many ways the malt analysis directs the brewer in how to avoid process difficulties by adjustment of the grist or mashing conditions.

There are many discussion papers on the relevance of laboratory mashes in relationship to brewhouse performance. Are these really an issue? The same is found in small-scale or micromalting in which process conditions that mimic true malting have to be set up quite differently to the main plant to get the best correlation. In the case of micromalt and brewing analyses, the reader is effectively using them as a comparative benchmark. Departures from the norm are just as important as getting an analysis in range.

An issue often arises when improvements in malt analysis are not matched with changes in mash profile. When more-modified malt is specified, it can improve the rate of throughput and the amount of extract and yeast nutrients. However, in many cases the mash profile remains the same as used for malts with higher protein and lower extract levels. The result is that the better malt is overmashed and foam-positive protein can be degraded, extract diminished, and overall produces a poorer product. This illustrates the importance of understanding not just raw-material quality, but the impact of that on subsequent processing.

ANALYTICAL VARIANCE

Variance of analyses between reanalysis in the same laboratory of the same sample and different samples by different laboratories is well documented and mathematically explained. This, of course, adds a degree of controversy at times to the adherence of the maltsters to the expected specification. Whether the variation from specification will actually translate into a real effect on brewing is often difficult to prove. However, it remains important for maltsters to understand the potential variance in a method and somehow to estimate the impact of drifting toward the limits of a specification. Statistically, supplying on the edge of a specification band will inevitably result in some analytical tests being outside the specification. Control of this is often agreed between maltster and brewer by setting blending limits for individual batches which are slightly wider than the blended delivery specification. When laboratories enter results on an electronic enterprise resource planning (ERP) system such as SAP it is often not possible to distinguish the significance of the amount that a particular analytical parameter (analyte) is from the desired final specification. Most systems simply allocate a percentage banding around the analyte value. Other systems are available, though mostly bespoke for each maltster, that grade the significance of each individual analyte with respect to brewing performance. This gives a greater nonconforming score to those analytes that have more significance for brewhouse performance than others. Using the bandings for laboratory variance in analysis (r95) for example, beta glucan being out of specification by one r95 band is more likely to have an impact on brewing than moisture. An effective weighting system for the importance of each parameter and the degree of variance from the specification can be determined through many different mathematical algorithms. This type of system has been around in different guises for a number of years under the general heading of quality scoring or quality number (Aalbers and Van Eerde, 1986). An example of such a systematic approach to malt analysis is shown in Fig. 1.3. Each batch of malt produced can be scored for its adherence to specification firstly for each analyte. In this case, a score of 0 is perfection and within specification. Using banding that approximates to the analytical variance of each analyte,

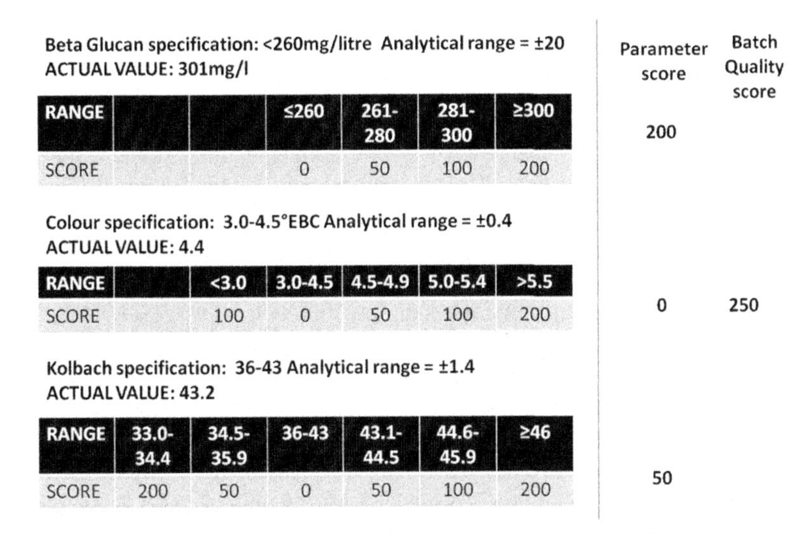

Beta Glucan specification: <260mg/litre Analytical range = ±20
ACTUAL VALUE: 301mg/l

RANGE			≤260	261-280	281-300	≥300
SCORE			0	50	100	200

Colour specification: 3.0-4.5°EBC Analytical range = ±0.4
ACTUAL VALUE: 4.4

RANGE		<3.0	3.0-4.5	4.5-4.9	5.0-5.4	>5.5
SCORE		100	0	50	100	200

Kolbach specification: 36-43 Analytical range = ±1.4
ACTUAL VALUE: 43.2

RANGE	33.0-34.4	34.5-35.9	36-43	43.1-44.5	44.6-45.9	≥46
SCORE	200	50	0	50	100	200

	Parameter score	Batch Quality score
Beta Glucan	200	
Colour	0	250
Kolbach	50	

FIGURE 1.3 Example of a Quality Scoring System for Grading Malt Specifications.

If an analytical result falls outside the specification range, it is assigned a value representing the significance of that to brewing performance. The bands are set approximately matching the analytical variance of the method. Scores can increase more markedly if a parameter is particularly likely to cause a brewing problem over a certain value.

a progressively higher score is assigned to results outside the target range. For some values markedly outside the target range, a jump in quality score can be allocated to indicate this is really not an acceptable batch to blend. A combined score for all the analytes can then be calculated and compared against a grading system for how close to specification is the batch overall (Fig. 1.4). It is also then possible to compare malt production quality for a given period. If a specification is reasonably straightforward

Quality Scoring Bands

Quality Band	A	B	C	D	E	F	G	H
Quality Score	0	2-50	51-99	100-250	251-300	301-400	401-500	>500

FIGURE 1.4 Example of Quality Banding System to Evaluate Batch Performance.

Using this system, the maltster could set an overall performance target of band B. This would mean that, on average, no malt batch was more than one band outside the target specification for just one analyte. This would be a very tightly controlled specification. However, the value would most likely be set for all batches produced in a given period. If the majority of those batches were perfectly in specification and just one was out, the overall performance could still be judged satisfactory.

to produce and the season's barley ideally suited to that malt type, then the majority of malts in the period could be perfect without blending and have a score of 0 (no defects in any parameter). There is then a balance between the number of perfect batches and those falling outside the specification for any reason. Thus one batch that perhaps is dramatically affected by a power outage, for example, does not disproportionately affect the overall plant performance for that period if all other malt batches were well made. This sort of system can also be used to set a target for the adherence of the malting plant to specifications in general by both malt type and by overall performance. In the case illustrated, the quality score for that batch, for just the three parameters selected, is 250 which is just inside band D. A tougher target of band B performance overall could have been set which would equate to just one parameter being outside one quality band in the period. That, of course, may be achievable if only one batch was faulty in the month. For example, if 10 batches of this fictional malt were made in the month and only one was faulty with a score of 250, then the average performance for the month would be 25, which is a band B performance. With any scoring system it is only as good as the parameters being assessed. If there are analyses that potentially could affect brewhouse performance that are not scored, for example LD impact on DP, a key influencer of performance may be ignored because it is not on the specification.

This type of system has also been adopted quite effectively by brewers to evaluate maltsters across a number of important supply-chain performance indicators. Of course, there are merits and demerits of each system and it is often applied as a penalty rather than offering a premium for adherence to very narrow variations in quality. Assigning weighting factors and determining which are the most relevant for maximizing process output entails a large degree of supposition, a reality that is likely to continue. Some comfort can be gained from one single number that represents malt quality or brewhouse performance, but there will likely never be a universally accepted system, and each method will remain an amalgam heavily influenced by mathematical prediction and tradition.

With such a system a judgment has been made as to which parameters are most significant to brewers. It is perfectly possible, however, by understanding the brewer's concerns, to set up quality-scoring systems by customer malt type, not just generically. In this way it builds confidence in the maltster that inevitable batch to batch variations will not have a negative impact on brewing performance.

LABORATORY ANALYSIS AND BREWERY PERFORMANCE

The limitations of direct correlation between laboratory mash conditions and brewery performance are well documented (Axcell et al., 2001; Davies, 2006a,b; Henson and Duke, 2008). Attempts have been made to harmonize isothermal Institute of Brewing (IoB) and EBC Congress mashes to resolve some of these issues and perhaps more accurately reflect brewhouse yield (Evans, 2010; Evans et al., 2011). Why have the old methods prevailed if they are acknowledged as difficult to interpret? Many years of tradition and use undoubtedly have embedded the interpretation of the methods into the minds of analysts, brewers, distillers, and breeders. The use of the specification is to minimize variation rather than predict brewhouse performance. It is a point of reference to compare different suppliers, different crop years, and new varieties and to match raw materials purchasing to customer requirements. As such, it is a well-established benchmark. Choosing the best of the methods for your particular application is important. For example, in Germany it is now common to use just the fine-grind isothermal grist for varietal comparison because the finer grind is closer to commercial practice and is considered a stricter test of barley performance than a temperature-programmed mash.

Essentially, a laboratory analysis is useful if you know what you want out of a raw material and you understand the implications and practical consequences for your process. As such it can be viewed as a safeguard rather than stifling your choice of the best materials available (Axcell et al., 2001). It has to be remembered that malts with apparently the same analytical specification can have widely different performance in brewing and distilling (Axcell, 1998). Table 1.1 indicates why a large degree of interpretation is required in assessing a malt-analysis sheet. It is very important to understand what the possible causes of variation in an analysis could be rather than assume being in specification is sufficient.

A number of malt analyses effectively result in the same impact on the production process, but it is a matter of personal choice which parameters a brewer or distiller selects to ensure that the process will run smoothly. One of the biggest barriers to applying malt analysis results is the limitations of the laboratory mash which can over or under emphasize what will eventually happen during brewing (Davies, 2006a,b). This interaction of parameters has been reflected in the creation of multiparameter equations to predict performance such as fermentability (Evans et al., 2005, 2010).

Attempts to find newer malt analyses that better reflect process performance have been suggested but still not widely adopted, which have included prediction of wort filtration performance. These include TEPRAL (Moll et al., 1989); Stored Waveform Inverse Fourier Transform (SWIFT) (Stewart et al., 2000; Ford et al., 2001); wort osmolyte concentration (Henson and Duke, 2008; Duke and Henson, 2009a); Fourier Transform Infrared (FTIR) spectroscopy (Titze et al., 2009); and medium-infrared analysis to distinguish between extracts that are similar by conventional methods but differ in sugar profiles (Cozzolino et al., 2014). Single-equipment prediction using Near Infrared Spectroscopy is popular and in regular use for measuring malt protein and moisture at grain intake or breeding-line selection. The accuracy and precision of these tests are excellent provided the predictive equations are based on a sufficiently wide range of appropriate samples for the season and geographical location. The limitation remains that they are simply making a faster prediction of an existing parameter and do not necessarily add to the understanding of that value.

There has also been a drive to improve the methods used to analyze existing parameters. Some of the current methods for enzyme analysis are a century old and uprated analyses such as the Megazyme (Ceralpha, Betamyl, and

Table 1.1 What Does the Specification Really Tell You? Does It Really Match What You Need in Your Process?

Malt Analyte	Significance of the Analysis—What to Consider—Does It Protect the Process?
Hot water extract Fine/Coarse difference	Do you need extract as high as possible? Laboratory extract is often much less than brewery due to water: grist ratio hence not an ideal comparison against brewery performance but rather an indication of any potential batch to batch variation in malt. Brewery grind is usually much coarser than laboratory hence there may be little to learn from a fine/coarse difference because the fine result may never be achieved unless hammer milling is used
Total nitrogen	Low is good? not necessarily because although "low" might mean easy modification and milling, it can also mean low enzyme. Why is it low—is it varietal, seasonal, agronomic?
Soluble nitrogen	If modification is low, run off is likely to be slow and turbid and yield reduced If modification is high, beer flavor can be thin and foam diminished. It can also introduce too much color in wort boiling
Viscosity	So it is viscous, but why? Is it β glucan which could also lead to haze? ß glucan is low—but what if you sparge at around 90°C and release the bound glucan, you still have a problem. Is it due to undermodification and presence of small starch granules which can also be a varietal trait? Is β glucan really the issue? Although it can link to viscosity and haze, some varieties differ in their ability to digest β glucan. In some varieties, high levels of β glucan are matched by high levels of beta-glucanase and the malt ends up lower in β glucan than varieties that start out with lower levels of β glucan.
Diastatic enzymes, Diastatic power (DP), Dextrinizing units (DU), Limit dextrinase (LD)	Are you able to utilize the higher enzyme activity by planning a temperature programmed mash? If you mash at 65°C you will lose most of the enzymes in the malt very quickly. Will you mill the malt appropriately so that the enzymes have sufficient substrate to act on? Do you really need DP/DU, or do you need LD which requires longer malting and requires an allowance for greater malting loss?
Kolbach index (KI)/Free amino nitrogen (FAN)	Do you use this to indicate sufficient yeast nutrition, or perhaps that protein is sufficiently modified to generate good quality foam? The spectrum of amino nitrogen is varietally dependent and not controllable simply by increasing KI or FAN.

Table 1.1 What Does the Specification Really Tell You? Does It Really Match What You Need in Your Process?—cont'd

Malt Analyte	Significance of the Analysis—What to Consider—Does It Protect the Process?
Friability	As high as possible? No! This is highly geographically and varietally sensitive and a value of 70–75 in one variety might perform as well as one requiring 90+ in another region of the world.
Homogeneity	If roller milling, this is very important to avoid hard undigestible grist components. If hammer milling and using enzymes, it is likely of much less importance.
Dimethyl sulfide (DMS)	Will getting the DMS level right in the malt give the correct beer flavor? Unlikely as only around 20% of the DMS originates from malt S-methyl methionine (SMM), the remainder being derived from dimethyl sulfoxide (DMSO) degradation by yeast.
Lipoxygenase (LOX)	Driving down malt LOX protects the beer from staling? No it doesn't as only up to 20% of the LOX staling potential can be linked to beer staling. If as a brewer you are concerned that your supplier base of maltsters is generating variable and uncontrollable LOX levels, you may specify one of the new null-LOX barley varieties to control this aspect of quality at source.
Premature yeast flocculation (PYF)	Why has the yeast dropped out of solution? is it really the malt that is at fault? Do all breweries have problems with the same batch. Often, it is found that only certain breweries at certain altitudes have the PYF problem and may not experience it with consecutive brews from the same batch. It is also not always possible to find the putative PYF factor in malt and guarantee that it will cause PYF in the brewery.

For descriptions of the full range of malt analyses most often specified and what they relate to, see Davies (2006b) and O'Rourke (2002).

Limit-Dextrizyme) kits with more reliable substrates are often preferred (Evans, 2008). There have also been numerous adjustments to the recommended fermentability tests to try to match individual brewery conditions, but that in itself inevitably limits their widespread adoption. Changes in the mash-temperature profiles, water: grist ratios, and the use of adjuncts may indeed provide improved discrimination in certain conditions, but are not likely to be transferrable to be reproducible globally as predictors of fermentability (Evans et al., 2007).

What tends to happen with these new tests is that small numbers of brewers may find these tests work well to predict some parts of their process but do not account for the unknowns that can crop up at times. These cause the most problems in the brewery. Tests for Premature Yeast Flocculation (PYF) and Lipoxygenase are a very good case in point. As illustrated in Table 1.1, PYF is difficult to screen for. If it is accepted that a factor is present in malt that can contribute to PYF, then it can only be predictive if it creates the problem every time that factor is detected. Unfortunately, a positive detection of PYF does not always lead to PYF in brewing and may be found at one brewery but not another using the same batch of malt. Lipoxygenase (LOX) is not generated by the malting process in all malting plants, and is a relatively small part of a number of potential contributors to beer staling (Bamforth, 2002). Simply detecting levels of LOX is therefore not a helpful control measure or specification. There are different variants of LOX with LOX-1 present in the barley grain and LOX-2 only detected in germinating barley (Hirota et al., 2006). The activity of both isoenzymes increases during germination and decreases during kilning. Only a small portion of the remaining LOX is reported to be extracted into the mash (De Buck et al., 1998). Several methods have been proposed to control and reduce the beer staling. This has led to development of varieties of barley that can be malted with no possibility of generating LOX (Simpson, 2008). This type of control back at the breeding stage can be the best way of eliminating an issue much in the same way as resistance to fungal attack is effected through resistance in the field.

Too often, specifications are set with antagonistic parameters. For example, low free amino nitrogen (FAN) and a high color and modification; low protein with requirement for good foam; low moisture and low color. Although maltsters can attempt to get as close as possible to these specifications, they can be highly linked to variety and season. Setting such specifications can expose the brewer to uncontrollable seasonal variations in barley and pose barriers to the introduction of new varieties which have been trialed using more generic benchmark specifications for typical brewing malts.

PREVALENCE OF FOOD SAFETY TESTS—DOES IT REALLY REDUCE RISK?

Over the past few years there has been a sustained increase in demand for ever longer lists of chemical residues. A number of countries have introduced requirements to prove that certain chemical residues are not present on the malt. In many cases, it has been a challenge to gain agreement with various foreign food ministries that the majority of the chemicals on their import lists are simply not relevant to barley and malt; for example, they may be veterinary drugs, or applied to wholly different crops, or banned in the country of barley origin. It becomes more difficult when chemicals prohibited in one country may still be used in other parts of the world. When, as in the United Kingdom, grain is bought through assured supply schemes, those scheme auditors when they are on farm will check what chemicals are being used and confirm the correct application rate and will examine chemical storage to ensure only approved chemicals are being bought and used. They can assure that chemicals not permitted to be used on barley are not in the store and not being applied. Certification to an assured scheme is renewed annually and reputable maltsters will not procure high-quality malting barley from nonassured sources. There can be additional safeguards such as an approved list of chemicals that can be applied to malting barley. This list for the United Kingdom (BBPA and Campden BRi, 2013) includes only plant chemical treatments that have no harmful effect on humans as determined through the World Health Organization and other databases, but also do not affect brewing performance. An added safeguard is that the chemicals in testing are applied at twice the recommended rates for farmers. The list of chemicals is thus quite small, but additional confidence can be achieved by then sampling barley and malt to analyze for residue levels. This is when some international legislation drawn up by various ministries of food and health fails to properly recognize risk correctly and malt specifications for supply can require testing of hundreds of chemicals that are simply never going to be present on barley from certain regions. If there is no internationally recognized safe limit for a particular chemical, it is often given a default value. This has led to the anomalous situation in which the import legislation requires maltsters to provide malt with lower levels of a chemical than are naturally present in the grain eg, gibberellic acid.

A number of factors should determine if a chemical residue should be surveyed: eg, if applied at foliar stage before grain emergence or if the chemical is water soluble, then it is not likely to be present on the grain; a fungal-derived mycotoxin may be destroyed in process; the long-standing country risk is exceptionally low. Verification of the former residues by due diligence sampling would be lower frequency than for samples which had received any form of postharvest treatment. Such a sensible scientifically reasoned approach to chemical testing would be ideal. Unfortunately, when government agencies publish lists of banned or restricted chemicals, malt customers tend to want every chemical analyzed even though two layers of protection are already in place: (1) the chemicals are not permitted to be used, (2) external bodies verify farm practice through assured supply certification.

Consistency of food safety controls is also variable on a malt specification. Again, a risk-based approach to ensuring product safety is important. Certain fungi are more prevalent in some countries than others and seasonal conditions and varietal resistance affect the profile of potential fungal species that can be present. If those fungi produce mycotoxins, there is national legislation specifying maximum residue limits. Control of the risk is effected by good farm practice and good storage (HGCA, 2011). Of course, verification that the controls are working is necessary, and malting intake teams will test for certain indicator mycotoxins. For field fungi (*Fusarium* eg, *F. graminearum*, *F. culmorum*) the most common test is for deoxynivalenol (DON) which can in certain conditions be produced from *Fusarium* fungi in the field and includes 3 acetyl DON and 15 acetyl DON, T-2, HT-2 (collectively named "tricothecenes") and Zearalenone. There is no risk-based demand to test for fumonisin (eg, B1) or aflatoxin because they have not been found on barley crops for many years and are extremely unlikely to occur. In some countries, however, the import control for mycotoxins is legislated for as aflatoxin, rather than the more relevant DON; hence, due diligence testing by maltsters often includes aflatoxin or is specified by the customer to reinforce this conclusion. For storage fungi (*Aspergillus ochraceus*, *Penicillium verrucosum*), the test is for Ochratoxin A (OTA). There is a higher frequency of sampling at harvest time than later in the year to establish the seasonal risk. Later in the year, once the barley has been in store for a few months it is more appropriate to test for Ochratoxin A which could develop in storage if the conditions are not controlled. Below 15°C and 15% moisture is the maximum safe point for controlling all fungal species, but to provide a margin for safety, maltsters would generally try to store either much drier (12%) or much cooler (<12°C) or both. Of course, in some countries, it is not possible to store cool, hence drying may be preferable. Stores are temperature monitored to detect the very early stages of insect activity. Even with all these pre- and postharvest checks and verification stages in place, there is still a requirement from many malt customers to have their own schedule of testing and to set their own levels of what is deemed food safe, often much lower than in legislation.

Intake operators are highly trained to look for fungal and other food safety hazards in grain and an effective control point to prevent a problem being brought into the maltings. One of those tests is to look for pink or gray coloration which indicates presence of *Fusarium* mold growth and would trigger a rejection of that load of barley. A musty (damp) smell to the grain would also be a reason to suspect early-stage mold growth or the conditions that would favor molds. It is possible to train operators and screen for their ability to detect mold using sensory analysis with 2,4,6-trichloroanisole as the musty aroma. Operators are required to be able to detect then rank different concentrations of this chemical. Failure to maintain an acceptable sensory ability requires them to be excluded from this specific intake test. Some customers have taken the pink color control and added it to the specification. A maximum number of pink grains per kilogram is allowed. Caution is advised here because the presence of the fungus does not necessarily predict mycotoxin presence (Hook and Williams, 2004). The correlation between the number of pink grains and DON level does appear to relate, albeit with a low 30% percentage correlation (Fig. 1.5). This can be used as an additional specification. If at intake there is a test to control for DON, then the pink grain test is not required. There is, however, a theory that some gushing of beer is attributable to either the *Fusarium* fungus or some secretion from the fungus; hence, the test may have a dual function in which it is found to be critical to operation in certain breweries.

SUSTAINABILITY AS A METRIC FOR MALT SUPPLY

An additional sphere of expectation that is becoming more important as a criterion for raw-material suitability is sustainable sourcing. Sustainability can be defined in many ways with different impacts throughout the malting and brewing supply chain and across different supply chains. This puts pressure on farmers to juggle various approaches for crops in their rotation that currently require certification to a specific named-accreditation scheme. This has led to the creation of a cross supply-chain consortium, the Sustainable Agriculture Initiative Platform (SAI, weblink) with the expressed aim of prescribing one common definition of sustainability. It does

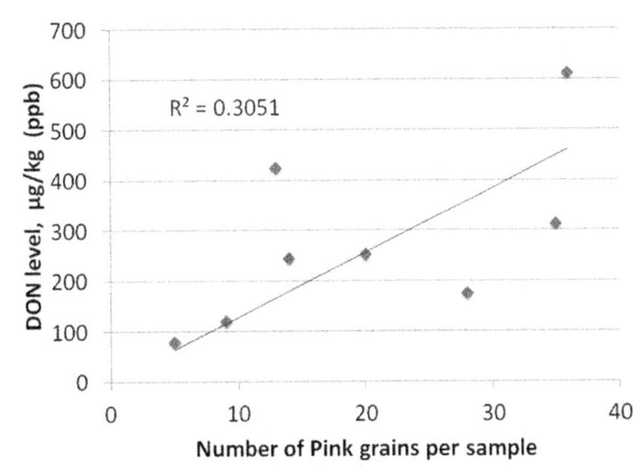

FIGURE 1.5 Using Pink Coloration on Grain to Control Deoxynivalenol (DON) Level and Gushing.

The correlation between the number of pink grains and DON level does appear to relate, albeit with a low percentage correlation (30%). This can be used as an additional specification. If the test is to control for DON, then the pink grain test is not required. There is, however, a theory that some gushing of beer is attributable to either the *Fusarium* fungus or some secretion from the fungus; hence, the test may have a dual function in which it is found critical to operation in certain breweries.

not require farmers to sign up to yet another green sustainability scheme, but rather creates a single benchmark for certification schemes and proprietary codes and removes the need for company-specific sustainable agriculture codes. SAI principles define sustainability in terms of environment, social, and economic measures and assess quality control and ethics. One of the most important tools available to the farmer is a Farm Sustainability Assessment (FSA, 2016) which provides a way to assess farm operational sustainability and gives a basis for improvement plans. When farmers have an existing sustainability system which has been compared with the SAI criteria, it may already be deemed as meeting the criteria at either basic or advanced level. In this way, farmers with existing suitable sustainability programs do not have to make any changes to be compliant with the overarching SAI assessment. A number of international brewers are now setting targets for the proportions of sustainably sourced barley used to make malt and use the Farm Sustainability Assessment as their benchmark to verify sustainable supply. An externally vetted sustainability program is sometimes required to be in place even before contract discussions begin.

The drive to continuously improve sustainable credentials has to be realistic. When a crop such as malting barley already has a low profit margin compared to wheat, we cannot choose to impose unrealistic targets on farmers that might dissuade them from growing the volume and quality required. Nevertheless, there are still many opportunities for the supply chain to become more sustainable and reduce cost by understanding where the opportunities exist. Although it is not just carbon and water reduction that drive sustainability, it is those measures that have come to the fore and driven new metrics primarily for carbon footprinting (Hillier et al., 2011; Davies, 2013a; Whittaker et al., 2013). Each of these has merits in terms of accuracy and user friendliness, but the setting of targets in malt contracts based on carbon footprint will drive an understanding of the relative environmental impacts of the supply chain by using a common currency of carbon dioxide equivalence. Whichever of the many methods are chosen, a baseline for improvement can be ascertained and a capital plan established to reduce the key contributors to supply chain carbon footprint, which the author has found not to vary to any great degree between the various methodologies available.

The implications of sustainability as expressed by carbon footprinting can mislead supply chain partners to think environmental protection is all about carbon. Of course, it is simply a convenient "common currency" to represent use of energy, water, and raw materials and to adjust processing parameters. Carbon footprint analysis enables us to understand the impact of our processing conditions and individual product choice. A result of this is the temptation to reduce malt in the grist and substitute with barley or other adjuncts to avoid the carbon footprint of malting. Is there truly an environmental benefit of switching from malt to barley? At first sight, there is the avoidance of around 40% of the carbon footprint associated with malting which equates to 8–20% of the carbon footprint of beer depending where and how the beer is packaged (Beverage industry Environmental Roundtable, 2012, and Fig. 1.6). Reduction of around 8% of the beer carbon footprint is approximately equivalent to removing the crystal malt component in some ales. That characteristic of the beer is what sells it and what the consumer associates with their preferred beer taste. By implication, the general flavor of malt is something we take out of the grist at our peril. Innovation in the malting supply chain is already progressing toward a dramatic reduction of up to 50% of malting barley carbon footprint which is surely a better way to address the impact of malt than simply removing it from the grist. It does not require us to be scourging about new products that use barley as the raw material for brewing. They are new beverages, not traditional beers to the connoisseur, but appreciated by a different market (Bamforth, 2010).

New products that use barley and external enzyme sources rather than malt are already on the market. Production of nonfermentable malt extracts based on the same principle have been available for many years. Although they can legitimately claim to have a reduced carbon footprint, they are not without problems in making beer. An example is in the use of Ondea Pro (Aastrup, 2010). Barley brews tend toward lower extract and fermentability, and the best brews require the very best malting barley with good levels of natural beta amylase and can produce an equivalent product in terms of processability (Evans et al., 2014a,b). It is most often found that the best result for barley brewing is to use a good proportion of malt (at least 20–25%) with its complement of natural enzymes in the grist (Evans, personal communication, 2014; Davies, unpublished). There is an almost linear relationship of increasing proportions of malt in the grist with extract yield. The sensory impact

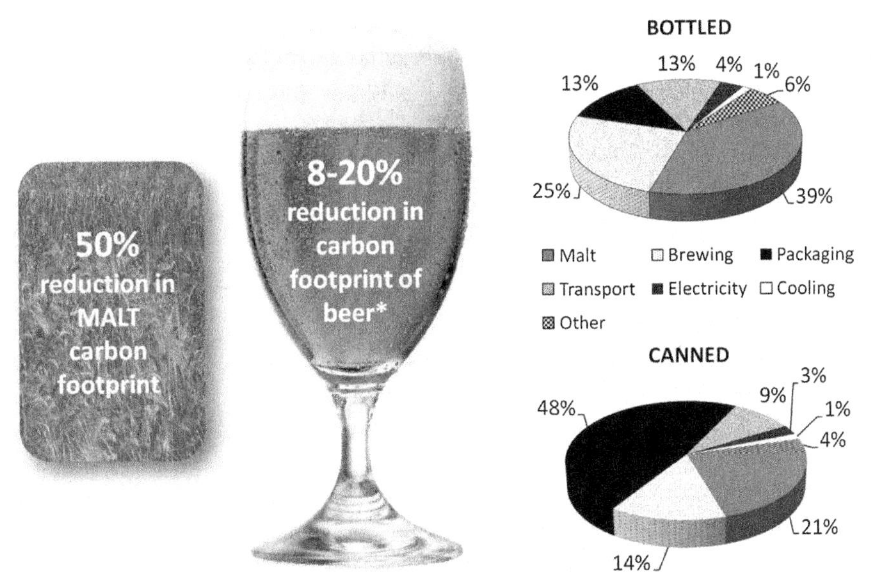

FIGURE 1.6 The Impact of a 50% Reduction in Malting Barley Carbon Footprint on the Carbon Footprint of Beer.

The contribution of malt to the carbon footprint of bottled or canned beer varies appreciably due to grist composition. If the carbon footprint of malt production could be reduced by 50%, the impact to beer would vary between 8% and 20%.

Data from Beverage Industry Environmental Roundtable, 2012. Research on the Carbon Footprint of Beer. http://media.wix.com/ugd/49d7a0_70726e8dc94c456caf8a10771fc31625.pdf; Davies, N.L, April 2013b, Low carbon malt – the journey continues, Brewer and Distiller International, 12–15; Davies, unpublished.

of using malt in the grist is easily perceived. For those who taste the barley-brewed beers for the first time, they believe it is a beer, but those more used to strong malty notes can be disappointed unless the malt proportion is high. Perhaps this will drive the introduction of flavor specifications for malt as discussed next.

FLAVOR SPECIFICATION—USEFUL, A DISTRACTION, OR MEASURE OF THE FUTURE?

It is quite common practice to taste beer and carry out a visual assessment of the product in process or when packaged. This can tell the brewer so many things about product quality based on the interactions of many individual parameters. It is surprising that there is not greater application of this for the malt being used. Some brewers will taste the wort to assess suitability. This has limited ability to discriminate malt flavors because it is predominantly sweet, but in the darker malt containing grists that sweetness is masked and the technique more easily applied (Coghe et al., 2004; Parker, 2012).

Increasingly, when malt is used in food and beverages, the flavor profile and color quality are things that are important for brand differentiation. Flavor profiling of malt is an amazingly simple test that can open up a world of options for product development and quality control.

A system for malt and malted ingredients, flavor profiling has been developed using ground malt made into a "porridge" with just a little water (Murray et al., 1999). It has been possible to identify the best malts in a series of samples all of which are analytically acceptable, to identify malts made in plants with poor hygiene, and to determine the nature of contamination in production, storage, or transit of malt (Chandra et al., 1997; Murray et al., 1999; Davies, 2006a,b, 2010, 2013a). Similar descriptions of malt flavor using different descriptors are also available (Voigt et al., 2013).

Sensory profiling of malt is rarely a feature of a malt specification for brewing or distilling, but is capable of differentiating a spectrum of flavors both positive and negative. It also addresses the conundrum of having two malts identical in conventional specification yet with markedly different flavor impact. This is generally most evident in the specialty malts. Although some would claim that certain barley varieties have an effect on beer flavor, it is the experience of this author that if the varieties are malted to the same specification and milled in the same way the differences in beer flavor are not evident. That is not to say that varieties do not affect flavor profiles. Of course, if a variety generates an amino nitrogen profile that is different from another, there is potential impact on yeast nutrition and the chemical spectrum produced. What we currently lack is a reproducible way of linking sensory variation to specific malt biochemistry and then to process control. That work is ongoing and may improve our abilities in this area for the future.

MALT SENSORY ANALYSIS CAN IMPROVE SELECTION AND SPECIFICATION OF SPECIALTY MALT

Sensory profiling has also proved a useful tool in determining which specialty malt is most appropriate rather than relying on the conventional analysis of color (Davies, 2010). Rather than simply viewing specialty malt as a source of just color, it is now possible to assess the many different hues that contribute to malt color and the associated flavors. When setting a specification for a colored malt, it is not advisable simply to increase the color in the hope that you can then use less in the grist. For example, lower color crystal malts are sweet and fruity, whereas higher color crystal malts are treacly and more bitter. So you have important choices to make: What general color do I need? Do I need to accentuate a particular color hue? What sort of flavor is required? If this possibility for brand differentiation is grasped, it can generate interesting opportunities for developing new products. In the same way that hops can be manufactured into a range of specific essences which can be accurately metered into the process, it is also now possible to have malt extracts in a very liquid and concentrated form. In this form, they can be added to the wort boil or post-fermentation, not just for accurate color adjustment but for flavor introduction or enhancement.

An important note of caution is that specialty malts, or even the white or green malts from which they originate, can be produced in quite different ways. Color and flavor can be stewed in or roasted in and can therefore predominate on the husk or within the endosperm. This can give rise to very different products, and color and flavor extracted in the brewhouse can change appreciably.

There is clearly much to be gained by understanding malt flavor in more detail. The low-tech grinding and wetted-porridge method allows the more obvious differences in flavors to be appreciated by almost anyone with a reasonable ability in sensory analysis and with improved levels of training and description some very elegant profiles and troubleshooting are possible.

MALT COLOR—SURELY ONE OF THE EASIEST PARAMETERS TO SPECIFY?

A casual glance at barley growing in the field might describe the grain as being light brown, golden, or "white for harvesting." Even when barley has been germinated and malted, it still appears brown. Wort is also to a casual observer—brown. Even in a conventional laboratory analysis, the color is often measured as a degree of brownness. As technology advanced some 20 years ago, the ability to describe beer color in terms of other color-contributing components emerged but does not appear to have generated a change in approach to specifying color in the raw material that contributes most: malt. Is there any benefit in trying to describe color in a more sophisticated manner, and what opportunities might that present for product development? The real benefit of a more accurate description of color comes when dealing with specialty malts. So, when a brewer places an order for colored or specialty malt, just what is the certainty that it will match the recipe requirements? Perhaps, you think that is a simple question and if you only ever buy colored malt from one supplier you might never enter the somewhat bewildering world of malt color description that is generated by analytical methods in different countries. The ways of describing and analyzing malt color vary between suppliers, whether local or international. Many disputes and brewhouse corrections could be avoided if the international methods on color were once and for all harmonized. The details of the color issue can be found elsewhere (Davies, 2010), but an example of the 55% difference in color that can be found by specifying color without really appreciating the possible interpretations of that specification is summarized in Table 1.2.

Table 1.2 Example to Show How Different Methods for Measuring Color Can Give Rise to a Product With a 55% Variance in What Might Be Expected

IoB Color Value	Equivalent IoB Value at 450 g	EBC	ASBC[a]
450 g mash 350 EBC	350 EBC	384.8 EBC	145.7[b] 195.3[c]
515 mL mash 350 EBC	406 EBC	446.4 EBC	9.9[a] 6.6[b]

If, on the current specification, the value is an IoB color, it could have been made using one of two mash types: 450 g or 515 mL. Confusion arises because even IoB color value is quoted in EBC units but is very different from EBC color units made using an EBC mash. In this example, the current color specification is 350 EBC analyzed using a 450 g mash. If that sample is analyzed after preparation using a 515 mL mash, the comparative color is 406. There is a significant difference between IoB and EBC mashes already at this point.

[a] *ASBC = American Society of Brewing Chemists.*

[b] *Using the conversion formula: ASBC = (EBC + 1.2)/2.65.*

[c] *Using the conversion formula: ASBC = EBC/1.97.*

IS NOT COLOR MUCH EASIER TO DETERMINE BY EYE?

This is a view held more widely by those using specialty malts for bread making; if the first bag of a new batch does not match by eye the last bag of the previous batch, then it is assumed to be the wrong color. That could be true, but the eye can also play tricks on you.

This introduces the concept of the importance of being able to extract all the colored components from the grain and not be misled by the outer color, which may or may not be the same as the roasted or crystallized product inside. It is possible when roasting to achieve the same color at different points in the roasting process. However, at the later point the ability to extract color is lower, which means that at times it may be necessary to increase the amount of highly colored malt to obtain a given wort color when apparently it is in specification. It is clearly more efficient in terms of energy and color extraction to terminate the roast earlier (Fig. 1.7).

It has also become more important in some circumstances to use a tristimulus (three-color) method to assess malt color. Here, a specialized but relatively low-cost analytical measure is made of the color quality (red, green, blue) expressed as the degree of lightness (L*), the amount of red–green (a*), and the amount of blue–yellow (b*). This technique is also widely used in other industries in which color quality is of prime importance. There are slightly different expressions of these three parameters, and it is important to understand that "L*a*b*" values are different from "Lab" values. The former * values (referred to as star values) are considered more accurate as defined by the Commission Internationale de l'Eclairage under their International Commission on Illumination *L*a*b** (CIELAB) scheme, whereas the "Lab" data uses Hunter calculations to describe color which may misrepresent yellows. Samples can be compared for their color quality using a program that

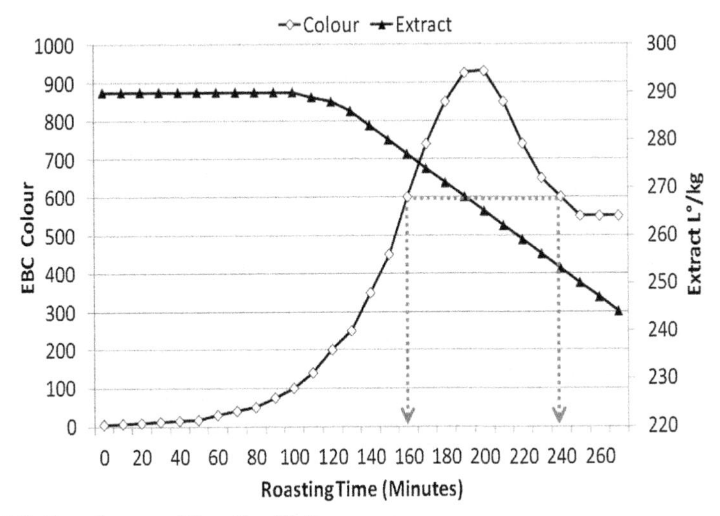

FIGURE 1.7 Time Course of Roasting Malt.

A color of 600 EBC is reached after around 160 min and continues to rise up to 200 min but then starts to decline. This it is possible to roast for 80 min longer than required to reach the color specification. There is a further issue in extended roasting in that the amount of extract in the grain falls in that same period so that not only does the extra roasting time waste energy, but also reduces the effective color that can be extracted from that malt in a mash.

combines all three values to generate one value. A small difference in this single value (ΔE_a) shows the sample to be very close in color quality and vice versa.

It is thus very straightforward to compare batches of roasted material being made or taken in by a customer and to determine the color quality range of batches blended together in that delivery batch if that is important to your application.

Of course, even the tristimulus method has pitfalls to beware of. There can be very good correlation for lightness compared to conventional EBC data, but the method of preparation of the grist can have a marked impact on the reflective properties of the mix and affect color quality detection (Davies, 2010).

SUMMARY

Malt analysis should enable both maltsters and customer to avoid risk and unacceptable variation in process while maximizing quality. Contrariwise, malt is a natural product and subject to a plethora of environmental and processing influences on quality such that a malt specification needs to be understood and interpreted to evaluate that impact. Ultimately, there is a need to match analytical expectation from growing on farm to use in a brewery or distillery to avoid placing undue expectations on the supply chain and to ensure sustainable supply. We are likely to face the dilemma of trying to fit these new demands into old-style specifications for the foreseeable future, but having a clear understanding of the background behind the analyses and their limitations results in them still being suited to controlling our process and driving efficiency and innovation.

REFERENCES

Aalbers, V.J., van Eerde, P., 1986. Centenary review – evaluation of malt quality. Journal of the Institute of Brewing 92 (5), 420–425.

Aastrup, S., 2010. Beer from 100% barley. Scandinavian Brewer's Journal 47 (4), 28–33.

Agu, R.C., Bringhurst, T.A., Brosnan, J.M., 2008. Performance of husked, acid dehusked and hull-less barley and malt in relation to alcohol production. Journal of the Institute of Brewing 114 (1), 62–68.

Agu, R.C., Brosnan, J.M., Bringhurst, T.A., Palmer, G.H., Jack, F.R., 2007. Influence of corn size distribution on the diastatic power of malted barley and its impact on other malt quality parameters. Journal of Agricultural and Food Chemistry 55 (9), 3702–3707.

Axcell, B.C., 1998. Malt analysis – prediction or predicament. Master Brewers Association of the Americas Technical Quarterly 35, 28–30.

Axcell, B.C., Morrall, P., Tulej, R., Murray, J., 2001. Malt quality specifications - a safeguard or restriction on quality. Master Brewers Association of the Americas Technical Quarterly 21, 101–106.

Bamforth, C., 2002. Great Brewing Debates: part 5. Beer freshness: is the maltster to blame? Brewers' Guardian 131 (11), 22–24.

Bamforth, C.W., March/April 2010. What does the future hold for traditional malted ingredients? Brewers' Guardian 43–44.

Beverage Industry Environmental Roundtable, 2012. Research on the Carbon Footprint of Beer. http://media.wix.com/ugd/49d7a0_70726e8dc94c456caf8a10771fc31625.pdf.

BBPA (British Beer and Pub Association) and Campden BRi, 2013. Agrochemicals Accepted by the British Beer and Pub Association and Campden BRI for Use on Cereals. http://www.campdenbri.co.uk/services/downloads/BarleyOct2013.pdf.

Chandra, G.S., Booer, C.D., Davies, N.L., Proudlove, M.O., 1997. Effect of air recirculation during kilning on malt and beer flavour. Proceedings Convention of the Institute of Brewing (Centre of South African Sector), Durban 6, 23–28.

Coghe, S., Martens, E., D'Hollander, H., Dirinck, P.J., Delvaux, F.R., 2004. Sensory and instrumental flavour analysis of wort brewed with dark speciality malts. Journal of the Institute of Brewing 110 (2), 94–103.

Cozzolino, D., Degner, S., Eglinton, J., 2014. A novel approach to monitor the hydrolysis of barley (*Hordeum vulgare* L) malt: a chemometrics approach. Journal of Agricultural and Food Chemistry 62 (48), 11730–11736. http://dx.doi.org/10.1021/jf504116j.

Davies, N.L., November 2006a. Malt - a vital part of the brewer's palette. In: Monograph 34: E.B.C. Symposium, Drinkability, Edinburgh, United Kingdom.

Davies, N.L., 2006b. Malt and malt products. In: Bamforth (Ed.), Brewing New Technologies. Woodhead Publishing, pp. 68–101.

Davies, N.L., 2010. Perception of color and flavor in malt. MBAA Technical Quarterly 47 (4), 3–7.

Davies, N.L., 2013a. Weblink to Poster: Malt, a Vital Part of the Brewer's Palette. http://www.muntons-inc.com/wp-content/uploads/2012/02/brewers-pallette-poster.pdf.

Davies, N.L., April 2013b. Low carbon malt – the journey continues. Brewer and Distiller International 12–15.

De Buck, A., De Rouck, G., Aerts, G., Bonte, S., 1998. Lipoxygenase werking en bierveroudering. Cerevisia 23 (2), 25–37.

Duke, S.H., Henson, C.A., 2009a. A comparison of barley malt amylolytic enzyme activities as indicators of malt sugar concentrations. Journal of the American Society of Brewing Chemists 67 (2), 99–111.

Duke, S.H., Henson, C.A., 2009b. A comparison of barley malt osmolyte concentrations and standard malt quality measurements as indicators of barley malt amylolytic enzyme activities. Journal of the American Society of Brewing Chemists 67 (2), 206–216.

Evans, D.E., 2010. Modernization of malt quality analysis to better inform brewers, maltsters and barley breeders - functional tests and future opportunities. In: 6th Canadian Barley Symposium from July 25–28, 2010, Saskatoon. http://www.canbar6.usask.ca/files/36_Evans.pdf.

Evans, D.E., 2008. A more cost and labour-efficient assay for the combined measurement of the diastatic power enzymes, β-amylase, α-amylase and limit dextrinase. Journal of the American Society of Brewing Chemists 66, 215–222.

Evans, D.E., Collins, H., Wilhelmson, A., Eglington, J., 2005. Assessing the impact of the level of diastatic power enzymes and their thermostability in the hydrolysis of starch during wort production to predict malt fermentability. Journal of the American Society of Brewing Chemists 63, 185–198.

Evans, D.E., Cooper, R.S., Koutoulis, A., 2014a (personal communication). Influence of malt and barley proportions in the grist of mashes made using Ondea Pro®.

Evans, D.E., Dambergs, R., Ratkowsky, D., Li, C., Harasymov, S., Roumeliotis, S., Eglington, J.K., 2010. Refining the prediction of potential malt fermentability by including an assessment of limit dextrinase thermostability and additional measures of malt modification using two different methods for multivariate model development. Journal of the Institute of Brewing 116, 86–97.

Evans, D.E., Goldsmith, M., Dambergs, R., Nischwitz, R., 2011. A comprehensive re-evaluation of the small-scale congress mash protocol parameters for determination of extract and fermentability. Journal of the American Society of Brewing Chemists 69, 13–27.

Evans, D.E., Li, C., Eglinton, J.K., 2007. A superior prediction of malt attenuation. In: Eur. Brew. Conv. Cong. Proc. Venice, vol. 31: presentation #4.

Evans, D.E., Li, C., Haraysmow, S.E., Roumeliotis, S., Eglington, J.K., 2009. Improved prediction of malt fermentability by measurement of the diastatic power of enzymes ß-amylase, α-amylase, and limit dextrinase: II. Impact of barley genetics, growing environment, and gibberellin on levels of α-amylase, and limit dextrinase in malt. Journal of American Society of Brewing Chemistry 67 (1), 14–22.

Evans, D.E., Redd, K., Haraysmow, S.E., Elvig, N., Metz, N., Koutoulis, A., 2014b. The influence of Ondea Pro®, compared by small-scale analysis. Journal of American Society of Brewing Chemistry 72 (3), 192–207.

Farm Sustainability Assessment, 2016. http://www.fsatool.com/.

Ford, C., Barski, M., Logue, S., Stewart, D., Evans, D.E., 2001. Further Development of the SWIFT Filterability Test. Australian Barley Technical Group. www.regional.org.au/au/abts/2001/t4/ford.htm.

Henson, C.A., Duke, S.H., 2008. A comparison of standard and non-standard measures of malt quality,. Journal of the American Society of Brewing Chemists 66 (1), 11–19.

Henson, C.A., Duke, S.H., 2014. A comparison of barley malt amylolytic enzyme thermostabilities and wort sugars produced during mashing. Journal of the American Society of Brewing Chemists 72 (1), 51–65.

Heron, H., Heron, J.M., 1914. The purchase of malt on the basis of analysis,. Journal of the Institute of Brewing 20 (6), 465–487.

HGCA (Home Grown Cereals Authority), 2011. HGCA Grain Storage Guide for Cereals and Oilseeds. http://www.hgca.com/media/178349/g52_grain_storage_guide_3rd_edition.pdf.

Hillier, J.G., Walter, C., Malin, D., Garcia-Suarez, T., Mila-i-Canals, L., Smith, P., 2011. A farm-focused calculator for emissions from crop and livestock production. Journal of Environmental Modelling and Software 26, 1070–1078.

Hirota, N., Kuroda, H., Takoi, K., Kaneko, T., Kaneda, H., Yoshida, I., Takashio, M., Ito, K., Takeda, K., 2006. Brewing performance of malted lipoxygenase-1 null barley and effect on the flavor stability of beer. Cereal Science 83 (3), 250–254.

Hook, S., Williams, R., 2004. Investigating the Possible Relationship between Pink Grains and *Fusarium* mycotoxins in 2004 Harvest Feed Wheat. HGCA Project Report No. 354. http://publications.hgca.com/publications/documents/cropresearch/354_Final_Project_Report.pdf.

Hyde, W.R., Brookes, P.A., 1978. Malt quality in relation to beer quality. Journal of the Institute of Brewing 84, 167–174.

Kavitha, G., Singh, S., Dhindsa, G.S., Sharma, A., Singh, J., 2012. Stability analysis for malt quality parameters in barley (*Hordeum vulgare* L.). Journal of Wheat Research 4 (1), 56–61.

Mammadov, J., Aggarwal, R., Buyyarapu, R., Kumpatla, S., 2012. SNP markers and their impact on plant breeding. International Journal of Plant Genomics 2012, 1–11. http://dx.doi.org/10.1155/2012/728398.

Meyers, B.C., Tingey, S.V., Morgante, M., 2001. Abundance, distribution, and transcriptional activity of repetitive elements in the maize genome. Genome Research 11 (10), 1660–1676.

Mindlab, 2012. Formula for the Perfect Pint. www.taylor-walker.co.uk/Media/Documents/PDF/news/WPR-Mindlab-The-Perfect-Pint.pdf.

Moll, M., Lenoel, M., Flayeux, R., Laperche, S., Leclerc, D., Maulois, G., 1989. The new TEPRAL method for malt extract determination,. Journal of the American Society of Brewing Chemists 47, 14–17.

Murray, J.P., Bennett, S.J.E., Chandra, G.S., Davies, N.L., Pickles, J.L., 1999. Sensory analysis of malt. Technical Quarterly Master Brewers Association of the America 36 (1), 15–19.

O'Rourke, T., 2002. Malt specifications & brewing performance. Brewer International 2 (10), 27–30.

Parker, D.K., 2012. Beer: production, sensory characteristics and sensory analysis. In: Piggott, J. (Ed.), Alcoholic Beverages: Sensory Evaluation and Consumer Research. Woodhead Publishing, pp. 133–158.

Qi, J.-C., Zhang, G.-P., Zhou, M.-X., 2006. Protein and hordein content in barley seeds as affected by nitrogen level and their relationship to beta-amylase activity. Journal of Cereal Science 43 (1), 102–107.

Ramsay, L., Comadran, J., Thomas, W., Marshall, D., Kearsey, M., O'Sullivan, D., Cockram, J., Tapsell, C., Klose, S., Bury, P., Habgood, R., Gymer, P., Christenson, T., Allvin, B., Davies, N.L., Bringhurst, T., Booer, C.J., Waugh, R., 2014. Association Genetics of UK Elite Barley (AGOUEB). http://www.hutton.ac.uk/webfm_send/497.

Simpson, V., 2008. Scandinavian Brewers' Review 65 (5), 25–28.

Sissons, M.J., Lance, R.C.M., Sparrow, D.H.B., 1993. Studies on limit dextrinase in barley. 3. Limit dextrinase in developing kernels. Journal of Ceramic Science 17 (1), 19–24.

Stenholm, K., Home, S., 1999. A new approach to limit dextrinase and its role in mashing. Journal of the Institute of Brewing 195 (4), 205–210.

Stewart, D.C., Freeman, G., Evans, D.E., 2000. Development and assessment of small-scale wort filtration test for the prediction of beer filtration efficiency. Journal of the Institute of Brewing 106, 361–366.

SAI (Sustainable Agriculture Initiative) Platform website. www.saiplatform.org.

Titze, J., Ilberg, V., Friess, A., Jacob, F., 2009. Efficient and quantitative measurement of malt and wort parameters using FTIR spectroscopy. Journal of the American Society of Brewing Chemists 67 (4), 193–199.

Thomas, W., Comadran, J., Ramsay, L., Shaw, P., Marshall, D., Newton, A., O'Sullivan, D., Cockram, J., Mackay, I., Bayles, R., White, J., Kearsey, M., Luo, Z., Wang, M., Tapsell, C., Harrap, D., Werner, P., Klose, S., Bury, P., Wroth, J., Argillier, O., Habgood, R., Glew, M., Bocjard, A.-M., Gymer, P., Vequad, D., Christenson, T., Allvin, B., Davies, N.L., Broadbent, R., Brosnan, J., Bringhurst, T., Booer, C.J., Waugh, R., 2014. Association Genetics of UK Elite Barley (AGOUEB). HGCA Project Report No. 528. www.hgca.com/media/337662/pr528-final-project-report.pdf.

Voigt, J., Richter, A., Krauss-Weyermann, T., 2013. A new approach to malt flavour characterisation – the malt aroma wheel. In: Proc. MBAA Conf. Workshop.

Wentz, M., 2000. Identification of Molecular Markers Associated with Malt Modification in Barley (M.S. thesis). North Dakota State University, Fargo.

Wentz, M.J., Horsley, R.D., Schwarz, P.B., 2004. Relationships among common malt quality and modification parameters. Journal of the American Society of Brewing Chemists 62 (3), 103–107.

Whittaker, C., McManus, M.C., Smith, P., 2013. A comparison of carbon accounting tools for arable crops in the United Kingdom. Journal of Environmental Modelling and Software 46, 228–239.

Yin, C., Zhang, G.-P., Wang, J.-M., Chen, J.-X., 2002. Variation in beta-amylase activity in barley as affected by cultivar and environment and its relation to protein content and grain weight. Journal of Cereal Science 36 (3), 307–312.

CHAPTER

Adjuncts

2

G. Stewart
Heriot Watt University, Edinburgh, Scotland

INTRODUCTION

Since the beginning of the 20th century, there have been increasing brewing developments using high proportions of cereals other than barley malt. This process of barley malt replacement in brewing is accelerating. These materials are called adjuncts (Bamforth, 2012). Today, there are a number of nonbarley beers across the world—details later.

Adjuncts are brewing (and distilling) raw materials and are part of the process's variable costs along with malted cereals, water, hops, processing aids (and additives), energy, cleaning agents, packaging materials, and labor. The cost of yeast is flexible and depends on the production process employed. However, certain raw materials (ingredients) are essential in conventional brewing procedures for the production of beer. Indeed, in many countries, their use is dictated by the appropriate brewing production regulations such as the German Purity Law. These ingredients are: malted cereals, water, hops, and yeast (Narziss, 1984). Adjuncts are not within this dictated category as they are not essential for beer production. However, with certain exceptions (details later), most countries employ adjuncts in the brewing of many beers (Hertrich, 2013).

DEFINITION

Until the second half of the 20th century, brewing adjuncts were of questionable quality and were used with only basic analysis and quality. However, as will be discussed in this chapter, adjuncts are not just starch (or sucrose), and over time their quality has significantly improved. One parameter that has contributed to this improvement is the advent of a number of novel analytical instruments and methods (both on– and offline) (Stewart and Murray, 2011). Particularly notable in this regard are:

- Gas chromatography associated with mass spectroscopy (GC–MS).
- High-performance liquid chromatography associated again with mass spectroscopy (HPLC–MS).
- Confocal imaging for measuring physiological changes in biological systems, including cereal grains and yeast.

Brewing Materials and Processes. http://dx.doi.org/10.1016/B978-0-12-799954-8.00002-2

- Flow cytometric determination of specific enzyme activities in biological systems including cereal grains and yeast.
- Rapid detection and classification of microbial contaminants.

Very important with regard to the use of brewing adjuncts is HPLC. This analytical technique is employed to measure fermentable sugars, such as glucose, maltose, and maltotriose, nonfermentable carbohydrates (dextrins), and amino acids.

The standard definition of an adjunct is: "something joined or added to another thing but not essentially a part of it" (Guralnik and Friend, 1985). Consequently, in the context of beer and brewing, we must remember what is essential for beer production (listed previously) and everything else that finds its way into the process is an adjunct. In the context of this chapter, adjuncts are alternative sources of fermentable extract and are used to replace a proportion of the barley malt which is usually more expensive. Adjuncts will usually be used as less-expensive sources of fermentable extract. In addition, adjuncts can be used to impart elements of beer characteristics and quality such as color, flavor, foam, body, and drinkability. Alternatively, adjuncts are important if there are taxation considerations that make reduced malt use in brewing financially advantageous. For example, Japanese legislation has permitted the development of Happoshu (beer containing 25% malt or less) and Third Category (containing no malt) beer. It should be noted that in Japan, taxation of beer is levied based on the percentage of malt in the grist (Inoue, 1990; Hassell, 2012).

FUNCTIONS IN BREWING

The use of adjuncts in brewing is permitted in most countries. However, a few countries do not allow the use of nonmalted sources of carbohydrates (adjuncts) in brewing. As already mentioned, Germany is the preeminent country that strictly controls the production of its beer. The German Purity Law dates from 1516 (Narziss, 1984) and states that only barley malt, hops, and water (yeast was listed later) can be employed in the brewing process, and the use of adjuncts is forbidden! It is interesting to note that the Purity Law is as strict as the Scotch Whisky Regulations that date from 1909 (Bringhurst and Brosnan, 2014) and were last revised in 2009 (Gray, 2013). Two other European countries have adopted the basis of German Purity Law—Norway and Greece. The adoption of the Purity Law in Norwegian brewing dates from the first decade of the 20th century. In 1905, Norway separated from Sweden and the Norwegian parliament offered the country's monarchy to Prince Carl of Denmark—he took the title of King Haakon VII. King Haakon had a great affection for German beer and insisted that his new kingdom adopt the German Purity Law which meant that the use of adjuncts was not permitted. As a result, Norwegian beers are all malt to this day.

The definition of brewing adjuncts varies from country to country. However, for the purpose of this chapter, a narrow definition will be employed. In the United

Kingdom, the Food Standards Committee defines a brewing adjunct as "any carbohydrate source other than malted barley which contributes sugars to a brewer's wort."

Although a wide range of brewing raw materials fall within this definition, attention will be focused on the use of adjuncts in the following three areas of the brewing process:

- Solid unmalted raw materials usually (but not always) processed within the brewhouse (Schnitzenbaumer and Arendt, 2014).
- Liquid adjuncts (syrups) usually added to the wort kettle (copper) and some specialty liquid products used for priming beer at a later stage in the process.
- Malted cereals, other than barley, such as wheat and sorghum.

Typically, adjuncts (with exceptions that will be discussed later), other than nonbarley malted cereals, only contribute starch, no enzyme activity, and little or no soluble nitrogen and are usually less expensive than barley malt. It is also considered that most adjuncts (not all) do not contribute flavor to the finished product. However, this is not always the case. In general, unmalted barley tends to give beer a stronger, harsher character, particularly stouts (West, 2012), and corn gives beer a full, clean flavor. Wheat imparts beer with a dryness character. Both wheat and barley adjuncts improve beer-head retention (foam stability). Rice will give lager beers a characteristic light flavor with enhanced drinkability. Also included in this chapter is a discussion of barley malt in contrast to malted cereals other than barley.

The overall brewing value of an adjunct can be expressed by the following empirical equation:

Brewing value = (Extract + Contribution to beer quality) − (Brewing costs)

ADJUNCT STARCH

Adjunct starch consists of two types of polymer molecules—the linear amylose and the branched amylopectin (Fig. 2.1). Depending on its source, starch amylose and amylopectin are inherently incompatible molecules with amylose having a lower molecular weight with a relatively extended shape, whereas amylopectin comprises much larger but more compact molecules. Amylose molecules consist of largely simple unbranched chains with 500–2000 α-(1→4)-D-glucose units depending on the plant source (very few α-1→6 branches may be found and they will have little influence on the molecules' overall characteristics). Amylopectin is formed by nonrandom α-1→6 branching of the amylose-type α-(1→4)-D-glucose structure. The branching is determined by branching enzymes that leave each chain with up to 30 glucose residues. The amylopectin molecule contains a million or so glucose residues, about 5% of which form the α-1→6 branch points.

In the United States, current use of unmalted adjuncts in brewing averages about 38% of the total brewing materials employed, excluding hops. The most commonly used adjunct materials are corn (maize) (46% of total adjunct), rice (31%), barley

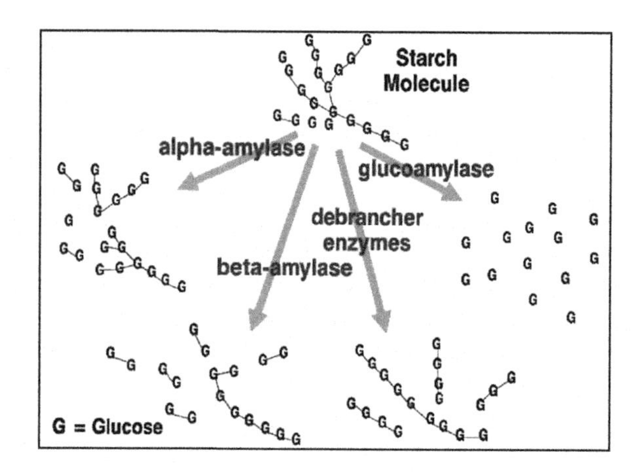

FIGURE 2.1

The enzymatic hydrolysis of starch.

(1%), and sugars and syrups (22%). The following materials have been used as unmalted brewer's adjuncts: yellow corn grits, refined corn starch, rice, sorghum, barley, wheat, wheat starch, cane and beet sugar (sucrose), rye, oats, potatoes, cassava (tapioca), triticale, and even pea starch! In addition, processed adjuncts include corn, wheat, and barley syrups, torrified cereals, cereal flakes, and micronized cereals (details later).

In brewhouses, processing of solid adjuncts is basically similar. However, there are variations to the process. The objective is to hydrolyze the starch into fermentable sugars mainly glucose, maltose, and maltotriose (together with unfermentable dextrins). Processing usually occurs (not always) in a vessel (cereal cooker) separate from the mash mixer (also called a mash tun). Some starches (for example, wheat and barley) have a relatively low gelatinization temperature range (Table 2.1) and can be hydrolyzed mixed with barley malt in the mash mixer.

Table 2.1 Extract (% as is) and Gelatinization Temperature (°C) Range of Starches Employed as Brewing Adjuncts (Briggs et al., 2004)

	Extract (%)	Gelatinization Temperature Range (°C)
Corn (maize)	78	62–77
Barley	70	60–62
Wheat	75	52–66
Rice	84	60–82
Sorghum	82	69–75
Oats	72	52–64
Rye	74	50–62

However, corn and rice starch (the solid adjuncts usually used in North America) have higher gelatinization temperatures (Table 2.1) and consequently must be processed in a cereal cooker. Cereal cookers can be operated simply by raising the temperature of the adjunct slurry gradually to 100°C (under pressure if the brewery is significantly above sea level [for example, Johannesburg; Nairobi; Golden, Colorado; Mexico City]) to ensure liquefaction of the starch. However, in many North American breweries, the milled adjunct is mixed with 10–15% of milled barley malt and the slurry incubated in the cereal cooker for 60 min. The amylases from the malt will partially hydrolyze the starch adjunct. Less frequently, exogenous amylase enzymes (amylases of bacterial and/or fungal origin) are added to the adjunct slurry in the cereal cooker. The slurry temperature is then raised to 100°C.

The slurry (with or without added malt or exogenous enzymes) is then transferred (by use of a pump or gravity) into the mash mixer (which already contains the primary malt charge). After approximately 60 min incubation, the temperature is raised to 100°C (the adjunct slurry from the cereal cooker will have served as an important heat sink) for 5–10 min. At this stage, most of the required starch hydrolysis will be complete, and most (not all) of the enzyme activity in the mash will have been inactivated by heat.

Corn Grits

Corn grits are currently the most widely used adjunct in the United States and Canada. They are produced by dry milling yellow corn outside a brewery. The milling process removes the hull and outer layers of the endosperm along with the oil-rich germ, leaving behind almost 100% endosperm fragments. All of the oil-rich germ must be removed because if it is present in wort and the resulting beer, it will negatively affect its flavor stability and foam stability. The corn oil is used in cooking and is also a key ingredient in some margarines. The endosperm starch fragments are milled further and classified according to brewer's specifications. Corn grits produce a slightly lower extract than other unprocessed adjuncts, such as rice (Table 2.1), due to less dextrin material in the wort after mashing and containing higher levels of protein and fat. The gelatinization temperature range for corn grit starch is 62–77°C. This is slightly lower than that of rice grits starch which is 60–82°C. Flaked grits (Stewart, 2012), flaked micronized grits, and corn grits have been compared with untreated corn grits, and they compared favorably with flaked and micronized barley and wheat adjunct regarding extract yield, fermentability, alpha-amino nitrogen, and wort viscosity. As would be expected, wort protein, peptides, free-amino acids, and nucleic acid derivatives decline in proportion to the adjunct level in the grist. The wort amino acid profile is not affected by any particular adjunct but by its level in the mash compared to the level of malting barley.

Rice

Rice is currently the second most widely used adjunct material in the United States (Holliland, 2012). On an extract basis, it is approximately 25% more expensive than corn grits. Brewer's rice is a byproduct of the edible-rice milling industry. Hulls are removed from paddy rice, and this hulled rice is then dry milled to remove the bran, aleurone layers, and germ. The objective of rice milling is to completely remove these fractions with a minimal amount of damage to the starchy endosperm, resulting in whole kernels for domestic consumption. However, up to 30% of the kernels are fractured in the milling process. The broken pieces (termed "brokens") are considered aesthetically undesirable for domestic use and are sold to brewers at a price considerably less than the whole kernel or mill-run rice price. Rice is preferred by some brewers because of its lower oil content compared to corn grits. Rice has a very neutral aroma and flavor, and when properly converted in the brewhouse it results in a light, clean-tasting drinkable beer.

The quality of brewer's rice can be judged by several factors: cleanliness, gelatinization temperature, mash viscosity, mash aroma, moisture, oil, ash, and protein content. Rice should be free of seeds and extraneous matter. Insect and mold damage cannot be tolerated because these indicate improper storage and/or handling conditions. Rancidity in rice oil can be a problem, but modern storage techniques render this negligible. Nevertheless, laboratory mashes of rice samples (rice tea) should be regularly conducted—they should gelatinize and liquefy in a standard manner and be clean and free from undesirable odors and tastes.

Not all varieties of rice are acceptable for brewing. Rice has a relatively broad gelatinization temperature range (60–82°C) (West, 2012) and is extremely viscous prior to liquefaction in the cereal cooker. Many rice varieties such as Nato will not liquefy properly and are difficult to pump from the cooker to the mash mixer. Other rice varieties, such as short-grain Californian Pearl, Mochi Somi, and Cahose (Holliland, 2012; Taylor et al., 2013), liquefy well in the cooker during a 15 min boil. Both the amylose and liquid content of rice varies with the variety and the cultivation conditions. Consequently, selection of suitable grades is important. Rice liquefies more easily the finer the particle grist, and particles less than 2 mm are considered adequate. Handling of rice is relatively easy, because (unlike corn and wheat starch) the brokens contain little dust and flow easily through standard hopper bottoms and conveying equipment. Unlike corn, which is milled by the cereal processor (offsite), the rice is usually milled in fixed-roller mills inside the brewery. There is no difficulty making the rice mash slurry at 64–76°C, although it is common practice to mash and hold it initially at 36–42°C as a form of protein rest. As with all cereal cooker operations, whatever the starch source, 10–15% of the malt grist may be added to the cereal cooker because the malt enzymes (amylases and proteases) can assist liquefaction to occur so that the cooker mash fluid can be pumped. Atmospheric boiling is required for gelatinization, and some brewers even pressure cook at 112°C. If conversion in the cooker occurs to specification, runoff problems will not be encountered. Rice extract is slightly lower in soluble nitrogen than a similar extract from corn grits (Briggs et al., 2004).

Unmalted Barley

Unmalted barley is used as a brewing adjunct under somewhat specialized circumstances. This raw grain is abrasive and difficult to mill and will yield a high percentage of fine material which causes problems during lautering (West, 2012). Indeed, many brewing operations that include unmalted barley in their grist have, for this reason, installed hammer mills together with mash filters and, as a result, runoff problems have been overcome (Andrews, 2004). In addition, runoff difficulties with a lauter tun can be overcome if the grain is conditioned to 18–20% moisture prior to milling, but this process has not been widely employed in brewing due to operational and quality difficulties!

In the past, barley has normally been partially gelatinized prior to use outside the brewery. Gelatinization occurs either by mild pressure cooking or by steaming at atmospheric pressure followed by passage of the hot grits through rollers held at approximately 85°C. Finally, the moisture content of the flakes is reduced to 8–10%. This process of pregelatinization can also be applied to corn. Pregelatinization of barley affects the ease of β-glucan and pentosan extraction during mashing and consequently the content of these components in wort. Prolonged steaming, prior to rolling the barley, produces a product which yields higher-viscosity sweet worts. This can be controlled by measuring the viscosity of the flaked barley in a cold-water extract which is a good indication of the extent of the steaming process. Barley starch is more readily hydrolyzed than corn or rice starch because of a lower gelatinization temperature (Table 2.1). Barley may be dehusked before use to increase extract yield, but this may lead to runoff difficulties because the husk provides material for filter-bed formation (another reason for the use of a mash filter instead of a lauter tun). If significant proportions of barley are used in the mash, a malt with sufficient enzyme activity is required. The use of unmalted barley as an adjunct leads to a reduction in wort nitrogen content and decreased wort and beer color. No difficulties have been reported in the fermentation of worts containing barley adjunct. However, foam stability is usually improved because foam-enhancing peptides from the barley are not hydrolyzed due to lower levels of proteolysis. A major difficulty associated with using worts with high levels of unmalted barley can be the increase in wort viscosity and runoff times caused by incomplete degradation of β-glucans and arabinoxylans. Mashing at 65°C rapidly deactivates malt β-glucanase activity. Alleviation of these problems can result from pretreatment of the barley with β-glucanase and the use of a temperature-stable β-glucanase in the mash. It is worthy of note that, although a cereal cooker in the brewhouse for the use of corn and rice in the grist is necessary, pregelatinized barley can be added directly to the mash mixer, circumventing the use of a cereal cooker.

Raw (feed) barley can be employed as high as 50% of the grist, and this has been the case in an Australian brewery. Pregelatinized barley is used as a raw material in some well-known stouts at 18–20% of the grist. For reasons already discussed, conventional roller milling cannot be employed; consequently, hammer mills are necessary (Dornbusch, 2012). A high level of malt replacement by unmalted barley usually results in insufficient malt enzymes in the mash for the necessary hydrolysis of the starch, protein, and β-glucans. Consequently, a malt-replacement enzyme system is employed to compensate for the reduced level of malt enzymes (Goode et al., 2005). A number of such enzyme systems have been developed that are usually a mixture of β-amylase, protease, and β-glucanase (and sometimes α-amylase), which are obtained from microbial sources such as *Bacillus subtilis*. Limitations to raw-material choice and processability have been largely overcome in recent years by the use of exogenous enzymes. Traditionally, high proportions of well-modified malted barley are needed in brewing recipes to achieve sufficient extract yield, efficiency, and quality in the brewhouse. Exogenous enzymes have been selected according to cereal-specific substrates, relevant pH, and temperature options. Indeed, processing up to 100% under modified malt, barley, or sorghum, as well as over 60% wheat, rice, and corn are globally well-established procedures today. Raw-material optimization is not only about using more unmalted cereal in the recipes but also about achieving high consistency and efficiency without compromising quality. Many enzyme suppliers to the brewing industry endeavor to address customer needs and enable the industry to enhance their raw-material usage, efficiency, and flexibility.

Barley brewing is the ultimate in high-adjunct brewing without any barley malt being used. The process has been extensively investigated for its economic advantages. In addition, a ban on the import of barley malt was introduced in Nigeria in the late 1980s to conserve valuable foreign currency. This ban forced Nigerian brewers to develop methods for brewing lager beer entirely from locally grown raw materials such as sorghum, corn, and surplus barley malt, although little of this local malt was available at the time.

In barley brewing, it is possible to approximate a typical brewhouse starch hydrolysis profile, and the degree of fermentability, and the wort sugar profile of 100% malt worts. This is possible by substituting malt with unmalted barley at levels of 50% (on an extract basis) and by controlling the main mash schedule (enzyme concentration, time, and temperature). Barley worts have been found to contain less fructose, sucrose, glucose, and maltotriose but more maltose than all-malt worts. No anomalies or difficulties in fermentation, maturation, and beer flavor and stability have been noted. An astringency and harshness of barley beers can be avoided by lowering the wort pH to 4.9 prior to boiling (West, 2012).

Sorghum

Although the use of sorghum (millet) as a brewing adjunct goes back over 50 years, it is only during the last 25 years that real interest by the brewing industry has been shown in this cereal. Sorghum is currently the fifth most-widely grown cereal crop globally. Only wheat, corn, rice, and barley are produced in greater quantities (Taylor et al., 2013). Africa and Central America are major sources of sorghum. However, the absence of a controlled "seed industry" has limited research into a wide range of sorghum genotypes which may be suitable for brewing. As well as its use as an unmalted adjunct, sorghum is also employed as a malt. The development and use of sorghum malt (along with wheat malt) will be discussed in a later section of this chapter.

In Africa, sorghum is a traditional raw material for the production of local top-fermenting beers that are known by various names such as: Bantu beer in South Africa, dolo in Burkina Faso, and billi billi in Chad. These beers are of relatively low-alcohol concentration [approx. 2.5 (v/v)] and are produced without hops. They are slightly sour in taste and are drunk unfiltered (containing "bits"), mainly in rural regions (Holley-Paquette, 2012).

During the Second World War, the US brewing industry employed sorghum as an adjunct because of the scarcity of other raw materials. Unfortunately, the sorghum was of poor quality and was cracked and only partially dehulled and degerminated. As a consequence, brewers obtained poor yields and bitter-tasting beers that lacked drinkability. Because of modern milling techniques and better processing methods, the situation has changed. Today, sorghum brewer's grits are considered of a comparable quality to corn and rice grits (Agu et al., 2011). However, in the United States, because of the bad experience during the 1939–45 war, sorghum brewer's grits are almost never used today. In Africa and Mexico, brewers are currently using an appreciable and increasing percentage of brewer's grits and are producing acceptable beer quality.

In agronomic terms, the advantage of sorghum is its ability to survive under extreme conditions of water stress. This cereal is ideally suited for cultivation in tropical areas including Africa and Central America. The current yield without fertilization is between 2 and 3.8 tons per acre and with the adoption of appropriate enhanced agronomic conditions this yield can be further increased.

In a typical sorghum-brewing process, the dried sorghum is screened to remove extraneous material. The whole grain is then fed into a series of dehullers that produce two product streams. In one of these streams, the husks and embryonic material, constituting some 48% of the original sorghum, are removed, and this fraction, together with the initial screenings, is sold as a byproduct. The second stream, consisting of peeled sorghum together with a small amount of husk, is then passed through an aspirator in which the husk is removed. The purified pearled sorghum, now representing 47% of the original cereal processed, is milled to give 12% of the original material as flour and 35% as sorghum grits. Both of these components are used as brewing adjuncts and can contribute up to 45% of the total wort extract.

Sorghum grain has a very similar starch composition to corn (Table 2.1). Both grains contain starch consisting of 75% amylopectin and 25% amylose. Starch grains are uniform and similar in range, shape, and size. On average, sorghum granules are slightly larger than corn (15 μm compared to 10 μm for corn). Sorghum starch has a higher gelatinization temperature range (69–75°C) than corn starch (62–77°C). In the brewhouse, sorghum brewer's grits perform within acceptable limits. However, the free-amino nitrogen (FAN) concentration of the resulting sorghum wort will be reduced. No special handling or cooking procedures are required. Five percent barley malt in the cereal cooker will provide amylolytic and proteolytic enzymes sufficient to convert the starch and occurs within the mashing time allowed. The beers produced are equivalent in chemical analysis, flavor, and stability to beers produced with other solid adjuncts. Finally, in many areas (for example, Africa and Central America), sorghum offers the lowest-cost source of available fermentable sugar (Ogbonna, 2009).

Refined Corn Starch

Refined corn starch is by far the purest starch available to the brewer (Holliland, 2012). It is a product of the wet-milling industry. Its use is not widespread because its price is higher relative to corn grits and brewer's rice. It can be difficult to handle because the powder is extremely fine. It must be contained in well-grounded lines and tanks to prevent explosions that result from static electricity sparks produced during conveying. The starch easily becomes viscous, and it is nearly impossible to make it flow from tanks unless they have special fluidizing bottoms.

Refined starch can be delivered to a brewery in bags (usually 100 lb) or in bulk which is necessary to reduce starch and labor costs. The starch is removed from a rail car or tanker truck by vacuum and is then blown through an air conveyor to a drop-out cyclone above a bulk storage bin. Dust collection is essential for the air conveyor. Dust can result in explosions and brewery operator lung contamination, causing labor relations problems!

Refined corn starch can be used as the total adjunct or can be mixed with rice or corn grits at the option of the brewer. Gelatinization and liquefaction of the starch proceeds at a lower temperature than rice or corn grists, but it is not sufficiently different to preclude use of a blend with either (Briggs et al., 2004). During use of refined corn starch as a total adjunct care must be taken to prevent sticking on the cooker bottom. Refined starch can be easily liquefied by a similar process to rice. With the exception of trained beer tasters, the resulting beer cannot be organoleptically (or chemically) differentiated from an all-adjunct rice extract. However, brewhouse yield can be increased 1–2% by the use of refined corn starch in place of rice. Assuming an appropriate supply of barley malt, there are no runoff problems with refined corn starch. In addition, beer flavor is unaffected, except that the beer can be slightly thinner because of enhanced wort attenuation limits and lower beer dextrin levels.

The future for the use of refined starch as a brewing adjunct will depend largely on relative pricing. It is unlikely that refined starch will capture any market from dry-milled corn grits. However, a wet-milling starch plant in close proximity to a brewery could provide a slurry starch as an adjunct at a competitive price. Also, the effect of the high-fructose market (details later) and the use of starch slurry in the paper industry could adversely affect starch availability and its price.

Refined Wheat Starch

Refined wheat starch is not presently attractive to most breweries worldwide, particularly in the United States, because of its high price. In the past (not currently), a wheat starch surplus has been used by Canadian breweries (this was also true of pea starch—which enjoyed a brief period of popularity in the 1970s). Chemically, wheat starch is similar to corn starch. Its gelatinization temperature is similar to that of malt gelatinization and, consequently, can be added directly to the malt mash (Table 2.1). However, 10% higher brewhouse yields can be obtained by cooking the wheat starch separately in a conventional adjunct cereal cooker. Lautering times with wheat starch can be 10% longer than with a similar grist containing corn grits.

Wheat starch has similar conveying and handling problems to refined corn starch. Slurrying should take place below 52°C to prevent lumping. The cooker temperature should not exceed 98°C because the starch foams excessively upon boiling. Wheat starch is higher in pentosans, and it is suggested that the cooker mash, with 10% of malt added, should stand at 48°C for 30 min prior to the 98°C rise to give the β-glucanase time to break down the β-glucans at its optimal temperature. This procedure will result in little or no runoff problems. The resulting beer is comparable to beer brewed with corn grits in analysis, flavor, and drinkability. If wheat starch is available at prices competitive to other adjuncts, it would be a perfectly suitable adjunct.

Torrified Cereals

Torrefaction of cereals (Stewart, 2012) is a process by which cereal grains are subjected to heat at 260°C and expanded or "popped" (popcorn process). This process renders the starch pregelatinized and thereby eliminates the brewhouse cooking step. Torrefaction also denatures a major portion of the protein in the kernel with the result that the water-soluble protein is only 10% of the total. Barley and wheat are potential candidates for torrefaction as is their use as torrefied adjuncts. The chemical analyses are quite similar for each. They both contain approximately 1.4% wort-soluble protein and, therefore, could permit the use of lower-protein malts or high-adjunct levels and still maintain soluble-protein levels similar to worts produced with lower soluble-protein adjuncts.

No handling or dust problems are associated with the use of torrefied cereals. It is possible to blend these torrefied products with malt. They can then be ground simultaneously and mashed together in a mash mixer. However, if they are used and the torrefied product cooked at 71–77°C prior to addition to malt, higher extract yields will result. Torrefied cereals lead to increased lauter grain-bed depth and to slight runoff penalties. Particle size and mill settings are critical with large particle size leading to poor yield with too fine a grind resulting in runoff problems. The beer flavor produced with torrefied cereals is reported (Stewart, 2012) unchanged, and if torrefied cereals became economically competitive with other adjuncts they could become employed as an alternative source.

Liquid Adjuncts

Liquid adjuncts, as already briefly discussed, are usually added to the brew at the wort-boiling stage. The major sugars are glucose syrups, cane-sugar syrups, and invert syrups. Although these syrups differ in details, the essential similarity is that they are all largely concentrated fermentable solutions of carbohydrates. The term glucose can be misleading. Although glucose is the commonly used name for dextrose glucose syrups used in brewing, they are solutions of a large range of sugars and will contain, in varying proportions, depending upon the hydrolysis method employed: glucose (dextrose), maltose, maltotriose, maltotetraose, and larger dextrins (Helstad, 2013).

Cane-sugar syrups contain sucrose derived from sugar cane and sugar beet, and sometimes, depending upon the grade, small quantities of invert sugar. Invert syrups, as the name suggests, are solutions of invert sugar—a mixture of glucose and fructose. Invert sugar is produced in nature and commercially, by the hydrolysis of sucrose which, together with glucose and fructose, occurs abundantly in nature. Commercially, sucrose is extracted from sugar cane or sugar beet and is used extensively in the food and candy industries. Candy syrups tend to be dark in color due to extensive malanoidin formation and is only used for specialized dark beers—usually ales. Glucose syrups are manufactured from starch which is usually derived from corn and wheat grains.

Glucose syrups have been commercially available since the mid-1950s. They were originally produced by direct acid conversion of starch to a 64–48 dextrose equivalent (DE) range. The degree of starch conversion is usually expressed as DE. This is a measure of the reducing power of the solution, expressed as dextrose in dry solids. For example, pure starch would have a DE of 0 and pure dextrose solid a DE of 100. In the mid-1960s, new developments in enzyme technology, in addition to the poor quality of straight and converted syrup, led to acid conversion resulting in a 42 DE product followed by a 64 DE product with enzyme conversion. Most syrups were refined by activated carbon filtration. Table 2.2 shows the sugar profile of these syrups in comparison to a typical all-malt wort. It can be seen that the only similarity was the content of the higher saccharides or nonfermentables. At the time of this development, a brewery's main concern was that the apparent extract of the finished beer did not change with the addition of the liquid adjunct—the syrup had to be approximately 22% nonfermentable (Dutton, 1996).

The use of liquid adjuncts continued in the 1980s, and it became apparent that they had a number of quality shortfalls. The high level of glucose in these syrups particularly became a concern (Helstad and Friedrich, 2002). Conversion of starch

Table 2.2 Sugar Spectra (%) of Brewing Syrups Compared to an All-Malt Wort

	Acid	Acid/ Enzyme	Enzyme/ Enzyme MS	Enzyme/ Enzyme VHMS	All-Malt Worts
Glucose	65	40	15	5	8
Maltose	10	28	55	70	54
Maltotriose	5	12	10	10	15
Dextrins	20	20	20	15	23

MS, *maltose syrup;* VHMS, *very high maltose syrup.*

with the aid of a mineral acid produces predominantly glucose as the hydrolysis product. When brewer's yeast is exposed to high concentrations of glucose, a phenomenon referred to as the "glucose effect" may be experienced with poor-quality yeast which can result in sluggish and "hung" wort fermentations (Hardwick and Skinner, 1990).

Brewers expect consistency, and the production of acid and acid/enzyme syrups depends on the termination of a reaction by chemical or mechanical means. An inconsistent syrup will prejudice beer quality. It is difficult to attain proper production consistency when parameters such as temperature, time, pH, and concentration are collectively critical. In addition, there has been concern regarding the physical properties of acid and acid/enzyme syrups. Large-volume users and breweries located long distances from the supplier experience inconsistent syrup color when stored for lengthy periods at elevated temperatures. This is the "browning reaction" (melanoidin formation) of sugars and particularly high levels of glucose in syrup will darken (melanoidin reactions), catalyzed by the presence of metal ions and protein (Hornsey, 2012). To inhibit this reaction, wet millers add sulfites. However, sulfites have been shown to cause allergic reactions with some people and, consequently, food and drug agencies look upon their use with disfavor. In addition, "darkened" syrups can adversely affect beer flavor stability and leave beer with a bitter taste, stringent overtones, and characteristic "corn" flavors. Finally, there has been concern regarding the chemical composition of acid/enzyme carbon-refined syrups. Syrup production involves exposure to low pH conditions followed by neutralizing with a sodium base. Consequently, residual compounds will form including high levels of sodium hydroxymethylfurfural (HMF). Also, high overall beer sodium levels can have negative health implications for some consumers.

Although this acid/enzyme carbon-refined process had disadvantages, it was accepted by many brewers until the late 1980s when there was a need for changes. At that time, many changes in the brewing and wet-milling industries were occurring, and this led to the development of a new series of liquid adjuncts. The two most important brewing production issues at that time were an increase in the adjunct levels of North American beer (up to 50% of the wort content) and increasing wort gravities as part of high-gravity brewing (HGB) procedures; the latter procedures were employed to ease capacity constraints and to improve brewhouse efficiencies. Indeed, the need for capital investment has been largely alleviated by increasing output by as much as 50% through the practice of fermenting wort at gravities of 16°Plato or higher. Details of the use of brewing syrups during the use of HGB procedures has recently been discussed in detail in a separate publication (Stewart, 2014).

With the advent of new enzyme liquefaction technologies and downstream multistage enzyme hydrolysis, production of water-white corn syrups with virtually any required carbohydrate (sugar) profile is possible. This new enzyme-liquefaction technology is an immobilized enzyme/enzyme system employing a number of amylase enzymes of bacterial/fungal origin (Helstad, 2013).

A major stimulator of change in the syrup manufacturing industry was the advent of high-fructose syrup (HFS) (Vega, 2012). This is a process in which glucose is isomerized by immobilized glucose isomerase to fructose. This reaction typically produces an HFS containing 42% fructose and 58% glucose. HFS is extensively employed in the soft drink and candy industries as a replacement for cane and beet sugar because fructose is 1.6 times sweeter on the human palate than sucrose on a molecule-for-molecule basis. Although HFS does not have a direct application in brewing, it has been used as a sweetener in the production of "beer coolers"—a blend of beer and fruit flavors plus sweetener.

Depending on the sugar spectrum required following the hydrolysis of cornstarch, the following matrix of high-quality enzymes can be employed: glucoamylase, α-amylase, β-amylase, glucose isomerase, and pullulanase (debranching enzyme). The specificity and efficiency of these processes have been enhanced by using thermotolerant enzymes, some developed as a result of genetic manipulation (GM) techniques (Stewart, 2013). Beccause of the application of this technology, some, but not all, breweries are reluctant to use syrups produced with GM techniques. Table 2.2 shows the carbohydrate profile of a "new generation" high-maltose corn syrup (enzyme/enzyme) (MS) compared to a typical all-malt wort. The profiles are almost identical. This novel enzyme/enzyme manufacturing process for "tailor made" syrups permits brewers to introduce liquid adjuncts at any level into the brewing process without changing the carbohydrate profile of the wort. Brewing with these syrups as adjuncts is now routine, and no difficulties in either the brewhouse or fermentation cellar have been reported. The author can confirm these findings from his own experience! In addition, syrup (and consequently beer) sodium levels decreased by as much as 60% [no need to employ sodium hydroxide (or sodium carbonate) to neutralize the starch hydrolysate], in comparison to beer produced with acid/enzyme syrups as an adjunct. This development has been looked upon with favor by the US Food and Drug Administration (FDA) and similar agencies from other countries.

USE OF LIQUID ADJUNCTS

Although these newer-generation syrups fulfilled the ongoing needs of the brewing industry, demands were made for syrups with higher fermentabilities, lower glucose, and higher maltose (and maltotriose) concentrations. These studies have led to the production of syrups containing higher levels of HFS (Vega, 2012), very high levels of maltose (VHMS), and syrups containing a spectrum of different molecular-weight dextrins (Helstad and Friedrich, 2002) (Table 2.2).

PRIMING SUGARS

Priming sugars and carbohydrates are used as wort supplements, and syrups are used as priming sugars. A priming sugar (or sugars) is any sugar added to a fermented beer with the principal purpose being to induce a secondary fermentation in a tank, cask, bottle, or, more rarely, a keg. The end result is natural carbonation and, usually, flavor development (Parkes, 2012). The sugar may be added as a solid, but is more often added in liquid (syrup) form, just prior to racking the beer into a conditioning tank or final container.

Yeast, which may be added at the same time as the priming sugar(s), takes up these sugars and produces carbon dioxide. This process carbonates the beer in the tank, keg, or package. Priming sugars are usually highly fermentable, with the most common being sucrose, glucose, fructose, maltose (rarely), and dextrins (unfermentable). Dextrins are usually added to give a beer body and are often added to low-alcohol products.

MALTED CEREALS

Malted cereals other than barley are classed by many jurisdictions as adjuncts. They are produced in relatively small amounts. Some are used in brewing European-style beers, others are used in the manufacture of opaque beers, distilled products, foodstuffs, and confectionary. These malts are made from temperate-zone cereals (wheat, rye, triticale, and oats) and from tropical cereals [sorghum, corn (maize), and rice]. Only barley and oat grains in the temperate-zone cereals are husked. Wheat, rye, and triticale are huskless, and this can give rise to malthouse problems. Due to the small production quantities normally produced, these malts can often be of variable quality. Only wheat, rye, and sorghum malts will be discussed here.

WHEAT MALTS

Wheat malts have been prepared on the industrial scale for many years. Most wheat malts used in brewing are pale and lightly kilned (Delvaux et al., 2004). In the United Kingdom, wheat malts may be used in the grist for beers and stouts (at a rate of 3 to 10%) to improve beer-head formation and head retention and possibly to enhance the diastatic activity of the grist and the supply of yeast nutrients. Wheat malt is also being used by some brewers in the United States to produce wheat beers which are part of the craft brewing sector. Wheat malt lacks husk, and this contributes to the way that it is prepared in the grist and the wort is separated in the brewhouse. In contrast to barley malts, wheat malts contain very little tannin material. In Europe, much larger proportions of wheat malt are used in particular beers. For example, in traditional German and Belgian wheat beers [Weizenbier or Weissbier (white beers)] 75–80% of the grist may be wheat malt (Delvaux et al., 2004). Although, unlike barley, wheats are not widely developed and selected by breeders for their malting qualities, but regular reports appear on the malting characteristics of German wheats (Kattein and Narziss, 2013).

Table 2.3 Analysis of Barley, Wheat, and Sorghum Malts (Briggs, 1998)

	Barley	Wheat	Sorghum
Extract (% dry matter)	81.0	85.7	80.8
Moisture	4.3	8.1	6.5
Color (EBC units)	3.5	8.0	Variable
α-Amylase (units dry matter)	35	55	40
Diastatic power (DP)	66	106	23
FAN (% dry matter)	84	143	70

EBC, *European Brewery Convention;* FAN, *free-amino nitrogen.*

The physical and chemical differences between wheat and barley render the malting of wheat a different challenge to barley. The lack of a husk means that it absorbs water more quickly than barley. Consequently, steeping times will be shorter. In addition, the wheat grain achieves modification more quickly than barley, but wheat malt is relatively less well modified than barley malt. Kilning is conducted at a lower temperature (35–40°C air temperature) compared to 70–80°C air temperature for pale barley malt. Because wheat contains a higher protein content than typical malting barley, this results in a slightly darker color, even in the palest wheat malt (Table 2.3).

Wheat malting is hastened and α-amylase levels are enhanced by the application of gibberellic acid to the germination. Malt extracts can be increased from 82.5 to 87.6% (Agu, 2012) with the application of gibberellic acid. However, currently the use of gibberellic acid during malting is not encouraged. As well as having a high extract, wheat malts are often rich in FAN, which assists fermentation (Table 2.3). However, mashed-wheat malts tend to give slow wort separations, probably due to the pentosans that are present. This will often give rise to viscous worts. A wide range of malts can be made from wheat including smoked, roasted, and pale malts.

RYE

Rye grains are hull-less; consequently, malting rye has all the problems (milling, ease of damage, etc.) and all the advantages (rapid water uptake, high potential extract) of wheat. Therefore, rye grains are relatively small and thin. Little rye grain is malted for brewing, but rye is malted for use in the production of rye whisky in Canada. However, even in these circumstances, rye malt is only 5–10% of the grist. Nevertheless, rye malts can have very high levels of starch-degrading enzymes and this makes them attractive to distillers using mashes rich in unmalted cereals in which fermentable extract yield is critical!

SORGHUM

Sorghum has been malted for the production of opaque Bantu beers and related products for a long time. For centuries, sorghum has been one of the third most important cereal crops grown in the United States and the fifth most-important cereal crop grown in the world. Nigeria is the world's largest producer of grain sorghum, followed by the United States and India. In developed countries, and increasingly in developing countries such as India, the predominant use of sorghum is fodder for poultry and cattle. Leading exporters in 2010 were the United States, Australia, and Argentina—Mexico was the largest importer of sorghum.

Sorghum grains are huskless. The grains are variously pigmented having red, white, brown, yellow, or pink coloration. The internal endogenous structure of sorghum grain is similar to maize. It comprises a single layer of aleurone cells, an outer steely area, and an inner mealy area. Like corn, the embryo of sorghum is large (approx. 10% of the grain's dry weight) and it contains most of the lipids.

Sorghum germinates rapidly at between 20 and 30°C. Unlike barley, in which gibberellic acid will induce the aleurone cells to produce quantities of α-amylase, gibberellic acid in sorghum does not increase α-amylase development. Some cultivars of sorghum have moderate levels of β-amylase whereas many others contain low β-amylase levels. Limit dextrinase (debranching enzyme) (Fig. 2.1) and protease activities are found mainly in the endosperm, whereas carboxypeptidase activities are present mainly in the embryo.

During steeping, formaldehyde (approx. 0–1%) is sometimes added to suppress fungal growth, but this practice is not permitted in most jurisdictions globally (Nso et al., 2006). When the grain has been steeped at 25°C for 24 h, it usually contains 33–35% moisture. Water is sprayed onto the grain during germination to maintain embryo growth and enzyme modification of the endosperm. During germination extensive amylolytic digestion of starch granules can occur, especially in mealy endosperm areas.

Although starch breakdown in mealy areas of sorghum endosperm can be extensive, cell-wall breakdown is limited. This contrasts greatly with cell-wall breakdown in the malted barley endosperm. Beer filtration problems in the brewing of sorghum malt have been observed, and this may reflect the possibility that during malting localized breakdown of endosperm cell walls may expose portals through which proteases and amylases migrate. During malting, troublesome quantities of β-glucans, pentosans, and proteins may be insufficiently broken down and may be extracted into the wort. Nitrogen solubilization, starch extract development, wort separation, and beer filtration are more limited when beer is produced from sorghum malt rather than from barley malt. Despite the low levels of β-glucans found in sorghum grain and malted sorghum, worts of low viscosity can contain troublesome quantities of β-glucan.

Irrespective of commercial enzyme addition, sorghum malts, mashed at 65°C, produce low extracts. It has been suggested that increased α-amylase stability can result from calcium ion addition, and enhanced extract development in sorghum mashes will occur. Although wort viscosities from sorghum malts are lower than those of barley, such low viscosities may not indicate trouble-free beer filtration or beer physical stability. In contrast to barley malt worts, sorghum worts usually contain high percentages of dextrins. The lower wort-maltose concentration (approx. 15%) of some sorghum worts reflects the reduced β-amylase activity (diastatic power) and fermentation extract of some sorghum malts. Higher concentrations of high molecular weight α- and β-glucans may aggravate filtration problems with sorghum beers.

Although barley malt is the preferred cereal malt for brewing beer, other malted cereals can be employed for making beer. Sorghum-malt and wheat-malt beers have been marketed as both traditional and new types of beers. The response of the customer to such products will indicate whether acceptable beers can be produced with minimal quantities of barley malt. An intensive review on the use of unmalted oats and unmalted sorghum in brewing has recently been published (Schnitzenbaumer and Arendt, 2014).

SUMMARY

The primary raw material to provide fermentable sugars for producing beer (and whisk(e)y) is malted barley. However, there are a number of other fermentable substrates that can be used in association with it. These materials are defined as adjuncts, and they can be employed at various stages of the brewing process including the brewhouse, the kettle, or blended as a fermented unmalted grain "neutral base" with the fermented barley malt extract.

REFERENCES

Agu, R.C., 2012. Effect of short steeping regime on hull-less barley and wheat malt quality. The Master Brewers Association of the Americas. Technical Quarterly 49, 82–86.

Agu, R.C., Goodfellow, V., Bryce, J.H., 2011. Effect of mashing regime on fermentability of malted sorghum. The Master Brewers Association of the Americas. Technical Quarterly 48, 60–66.

Andrews, J., 2004. A review of progress in mash separation technology. The Master Brewers Association of the Americas. Technical Quarterly 42, 45–49.

Bamforth, C.W., 2012. Adjuncts. The Oxford Companion to Beer. Oxford University Press, New York, pp. 11–13.

Briggs, D.E., 1998. Malts and Malting. Blackie Academic Professional, London, pp. 722–741.

Briggs, D.E., Boulton, C.A., Brookes, P.A., Stevens, R., 2004. Brewing Science and Practice. Woodhead Publishing Limited, Cambridge, England, pp. 34–44.

Bringhurst, T., Brosnan, J., 2014. Scotch whisky: raw material selection and processing. In: Russell, I., Stewart, G.G. (Eds.), Whisky: Technology, Production and Marketing. Elsevier, Amsterdam, pp. 49–112.

Delvaux, F., Combes, F.J., Delvaux, R., 2004. The effect of wheat malting on the colloidal haze of white beers. The Master Brewers Association of the Americas. Technical Quarterly 42, 27–32.

Dornbusch, H., 2012. Wheat Malt. The Oxford Companion to Beer. Oxford University Press, New York, p. 839.

Dutton, I., 1996. Corn syrup brewing adjuncts – their manufacture and use. The Master Brewers Association of the Americas. Technical Quarterly 33, 47–53.

Goode, D.L., Wijngaard, H.H., Arendt, E.K., 2005. Mashing with unmalted barley – impact of malted barley and commercial enzymes (*Bacillus* spp) additions. The Master Brewers Association of the Americas. Technical Quarterly 42, 184–198.

Gray, A.S., 2013. The Scotch Whisky Industry Review, 36th ed. Sutherlands, Edinburgh, Scotland.

Guralnik, D.B., Friend, J.H., 1985. Adjuncts. In: Webster's New World Dictionary of the American Language, p. 18.

Hardwick, B.C., Skinner, K.E., 1990. Fermentation conditions associated with incomplete glucose and fructose utilization. The Master Brewers Association of the Americas. Technical Quarterly 27, 117–121.

Hassell, B., 2012. Japan. The Oxford Companion to Beer. Oxford University Press, New York, pp. 503–504.

Helstad, S., 2013. Topics in brewing: brewing adjuncts – liquid. The Master Brewers Association of the Americas. Technical Quarterly 50, 99–104.

Helstad, S., Friedrich, J., 2002. Flavour characteristics of liquid adjuncts derived from corn. The Master Brewers Association of the Americas. Technical Quarterly 39, 183–190.

Hertrich, J., 2013. Topics in brewing: brewing adjuncts. The Master Brewers Association of the Americas. Technical Quarterly 50, 72–81.

Holley-Paquette, M., 2012. Sorghum. The Oxford Companion to Beer. Oxford University Press, New York, p. 743.

Holliland, C., 2012. Sucrose. The Oxford Companion to Beer. Oxford University Press, New York, p. 775.

Hornsey, I., 2012. Melanoidins. The Oxford Companion to Beer. Oxford University Press, New York, p. 582.

Inoue, T., 1990. What is the state of brewing science in Japan? The Master Brewers Association of the Americas. Technical Quarterly 27, 38–41.

Kattein, U., Narziss, L., 2013. 35 years of malting and brewing in Germany – experience with improvements in quality characteristics of raw materials and changes in technical in maltings and brewhouse. The Master Brewers Association of the Americas. Technical Quarterly 50, 93–96.

Narziss, L., 1984. The German beer law. Journal of the Institute of Brewing 90, 351–358.

Nso, E.J., Nanadum, M., Palmer, G.H., 2006. The effects of formaldehyde on enzyme development of sorghum malts. The Master Brewers Association of the Americas. Technical Quarterly 43, 177–182.

Ogbonna, A.C., 2009. Proteolytic enzymes and protein modification in malting sorghum – a review. The Master Brewers Association of the Americas. Technical Quarterly, 46-3-0714-01.

Parkes, S., 2012. Carbonation. The Oxford Companion to Beer. Oxford University Press, New York, p. 221.

Schnitzenbaumer, B., Arendt, E.K., 2014. Brewing with up to 40% unmalted oats (*Avena sativa*) and sorghum (*Sorghum bicolor*) – a review. Journal of the Institute of Brewing 120, 315–330.

Stewart, G.G., 2012. Flaked Barley. The Oxford Companion to Beer. Oxford University Press, New York, pp. 357–358.

Stewart, G.G., 2013. Biochemistry of brewing. In: Eskin, N.A.M., Shahidi, F. (Eds.), Biochemistry of Foods. Elsevier, Amsterdam, pp. 291–318.

Stewart, G.G., 2014. Brewing Intensification. American Society of Brewing Chemists, St. Paul, Minn.

Stewart, G.G., Murray, J., 2011. Using brewing science to make good beer. The Master Brewers Association of the Americas. Technical Quarterly 48, 13–19.

Taylor, J.R.N., Dlamini, B.C., Kruger, J., 2013. 125th Anniversary Review: the science of the tropical cereals sorghum, maize and rice in relation to lager beer brewing. Journal of the Institute of Brewing 119, 1–14.

Vega, R., 2012. Fructose. The Oxford Companion to Beer. Oxford University Press, New York, p. 377.

West, C.J., 2012. Barley. The Oxford Companion to Beer. Oxford University Press, New York, pp. 85–90.

CHAPTER

Hops

3

T.R. Roberts
Steiner Hops Ltd.

INTRODUCTION

The hop plant is described as being dioecious indicating that it has separate male and female plants. However, only the female plants produce the hop cones within which reside the all-important yellow lupulin glands. Within these lupulin glands, the compounds important to the brewer and beer quality are found.

In 1967, Stevens, in his comprehensive review on hop chemistry, showed a typical analysis of dried cone hops as follows:

Resins	15%
Proteins	15%
Monosaccharides	2%
Tannins (polyphenols)	4%
Pectins	2%
Steam volatile oils	0.5%
Ash	8%
Moisture	10%
Cellulose, etc.	43%

With the introduction of the newer superhigh-alpha varieties, the percentage of resins and oils today may be proportionately higher. In practice, only the resins, oils, and polyphenols significantly contribute to beer quality.

IMPACT OF HOPS ON BEER QUALITY

To fully understand the practicalities and problems of using hops and hop products in the brewing process and the control strategies required to ensure product quality, it is essential that the role of hops in the process be understood. Although thoroughly explored in many other texts (for example Roberts and Wilson, 2006), it is worth briefly reviewing here the impact of hops within the brewing process and on final beer quality.

Brewing Materials and Processes. http://dx.doi.org/10.1016/B978-0-12-799954-8.00003-4

FLAVOR

Hop Resins

The most obvious contribution of hops to beer flavor is that of the so-called soft resins, principally the alpha acids (also known as humulones), which are ultimately responsible for the characteristic bitter taste. There are three main homologs of alpha acid—humulone, cohumulone, and adhumulone—he ratios of each varying from variety to variety. The alpha acids themselves are nonbitter, and it is only when converted to their isomerized forms (iso-alpha acids) during the boiling process that they become bitter to taste. There are differences in the solubility of each of these iso-alpha-acid homologs with the isocohumulone being the most soluble and chemically active. Although shown to be less bitter than the other homologs (Hughes and Simpson, 1996), some believe that isocohumulone gives a more lingering bitterness on the tongue and therefore a less pleasant, harsher flavor.

Nonisomerized alpha acids can sometimes be found in beer in small amounts although values as high as 4 ppm have been found in heavily dry-hopped American India Pale Ales (IPA) (Smith et al., 2004). These do not contribute to perceived bitterness but will be measured as bitterness units when using the Bittering Unit (BU) method [European Brewery Convention (EBC) 9.8 or American Society of Brewing Chemists (ASBC) Beer-23].

The other significant part of the soft resins is the beta-acid fraction. The amount of beta acids is normally lower than alpha acids and the ratio of alpha to beta acids is variety dependent. Beta acids are also known as lupulones and, similarly to the alpha acids, exist as three homologs (lupulone, colupulone, and adlupulone). They are largely insoluble and are normally lost during boiling (leaving a maximum of 1 ppm in beer). Isomerization of beta acids is very difficult even under extreme boiling conditions; however, a small amount of the beta acids is oxidized to humulones in the hops or hop product and, being more soluble, are more likely to end up in the beer. These humulones are bitter to taste and therefore will make some small contribution to beer bitterness. A long-established "rule of thumb" to assess the bittering potential of alpha and beta acids is:

$$\text{Bittering potential} = \text{alpha} - \text{acids} + \frac{\text{Beta} - \text{acids}}{9}$$

Hop Oils

Of equal importance in terms of beer flavor are the tastes and aromas produced by the hop oils. Gas chromatographic analysis of hops shows a large number of oil compounds (more than 300 chemical species) with the individual chromatographic patterns being distinctive for each variety and forming the basis of one method of variety identification (Peacock and McCarty, 1992; Walsh, 1998).

The oil compounds present in hops are normally classified into three distinct groups:

- Terpenes:
  Accounting for up to 90% of the total oils present, these compounds are all categorized by their high volatility, and therefore they tend to disappear rapidly up the kettle chimney during boiling or be flushed out by the evolution of carbon dioxide during fermentation. The principal terpenes present in hops are the sesquiterpenes—myrcene, α-humulene, and β-caryophyllene—which collectively amount to 80% of the total. These compounds are also almost insoluble but small flavor-active amounts may be found in final beer, particularly in the case of dry-hopped beers or when "green hop" beers are produced using late additions of fresh, unkilned hops.
  Although in themselves not normally significant to final beer quality, the sesquiterpenes can be oxidized within hops and hop products to compounds described as the oxygenated fraction (see following) which do play a significant role in beer quality.
- Oxygenated Fraction:
  Hoppy flavor in beer is usually associated with relatively small amounts of oxygenated derivatives of the terpenes, usually categorized into groups such as alcohols, aldehydes, ketones, esters, etc. (Deinzer and Yang, 1994). Although more soluble than the terpenes, very little of these compounds survives into the final beer and, therefore, to be flavor active, they must have very low flavor thresholds measured in parts per billion (ppb). Often present in concentrations below the flavor threshold, there is some evidence to suggest the presence of these compounds may be synergistic in contributing to the "hoppiness" of beer (Siebert, 1994).
  Maximum retention of these oxygenated compounds is usually achieved by late kettle addition but there is some evidence to suggest that so-called first-wort hopping also can result in significant hop character (Preis and Mitter, 1995). In this practice, oxygenated materials are dissolved into the wort during kettle filling but before the vigorous boil starts.
- Sulfur Fraction:
  A few sulfur compounds have also been found in hop oil. These tend to have negative impact on beer flavor and therefore should be avoided as far as possible. The spraying of elemental sulfur to act as a fungicide during the growing season has been linked to the formation of these compounds (Peppard, 1981) as well as the, now largely defunct, practice of adding sulfur during kilning to improve hop color and presentation (Pickett et al., 1976).

Hops also contain a group of compounds known as glycosides that are formed by various materials, including hop oils, covalently bonding with sugar moieties. These compounds can be split by hydrolysis releasing the flavor-active components into the beer producing a detectable hop flavor (Goldstein et al., 1999).

Polyphenols

Polyphenols can also have positive effects on beer flavor as well as on flavor stability. Thus, certain hop polyphenols are thought to contribute to beer flavor by adding bitterness and body as well as the less well-defined characteristic "mouthfeel" (Forster et al., 1999). However, polyphenols can produce astringency, considered by some detrimental to beer quality. There is evidence to suggest that lower molecular weight hop polyphenols provide antioxidant activity in beer, increasing the reducing power of the beers with resultant improvements in flavor stability (Buggey, 2001).

Negative Influences on Flavor

It is also worth noting that both alpha acids and iso-alpha acids can be involved in the production of off-flavors and in beer staling:

a. In aged hops, the alpha acids can be oxidized to various fatty acids which are characterized by strong, cheesy flavors and which can find their way into finished beer.
b. Iso-alpha acids will slowly oxidize in beer, particularly at elevated storage temperatures, to nonbitter humulinic acid resulting in loss of bitterness.
c. Iso-alpha acids can also be converted to carbonyls via oxidation and condensation reactions to contribute to the unpleasant staling flavors in beer typically described as "cardboard" or "papery" flavors.
d. Iso-alpha acids can react with sulfur compounds naturally present in beer to produce a very unpleasant material called 3-methyl-2-butene-thiol (MBT). The process requires light in the wavelength range 350–500 nm for activation as well as the presence of the cofactor riboflavin. The flavor threshold of MBT is variously described as between 2 and 20 parts per trillion (ppt) and the character contributed is often described as "skunky." Traditionally, the formation of MBT was alleviated by the use of dark-glass bottles preventing access to the beer by ultraviolet (UV) light. However, today, hop products are available in which the reactive side chain of the iso-alpha acid is stabilized by the reduction of the carbonyl group or by hydrogenation of the C]C double bond.

MICROBIOLOGICAL STABILITY

Hop resins and their isomerized forms provide varying degrees of antimicrobial resistance particularly against Gram-positive bacteria such as *Acetobacter, Lactobacillus*, and *Pediococcus*. Traditionally, it was considered that a minimum of 18 mg/L was required in wort and beer to have any effect but more recently levels as low as 10 mg/L have been found to show some antibacterial activity. Unhopped worts can be particularly susceptible to infection. When no kettle-added hop products are used, and bittering is achieved by 100% addition of light-stable hop extracts added postfermentation, significant infection problems can occur.

PHYSICAL STABILITY

It is well known that polyphenols play an important part in the physical stability of beer through their ability to bind with proteins forming a coagulum that readily comes out of solution either during boiling, fermentation, or cold storage. However, some polyphenol/protein complexes can also be deposited in the beer in the final package resulting in unwanted haze. Hops contain more polyphenols per unit weight than does malt but obviously the contribution to the total polyphenolic content in wort is greater from malt than hops owing to the proportionately larger proportions of malt used in the brewing process. Despite this, it is thought that hop polyphenols do play an important part in the removal of proteins and the physical stability of beer. This may be particularly apparent when large amounts of low-alpha, aroma hops are used to achieve high bitterness and pronounced aroma, the quantity of polyphenols being proportionately higher than bittering to the same value by using high-alpha hops.

The amount and influence of hop polyphenols can be significantly affected by choice of hop product, because, in some products, polyphenols may be reduced or removed completely.

FOAM FORMATION AND STABILITY

It is well documented that hops acids make a major contribution to beer foam, both in terms of formation and stability (Hughes, 2000). These foam-positive effects are based on:

- The hydrophobicity of hop acids
- Their ability to bind polypeptides
- Their tendency to reduce surface tension in liquids
- Their ability to increase surface viscosity.

Although soluble to a limited extent, the iso-alpha acids readily come out of solution, migrate to the foam in which they combine with polypeptides present in the beer. However, as we have previously seen, isocohumulone is more soluble than either of the other homologs and therefore is found to be less foam active than either isohumulone or isoadhumulone.

As well as being light stable, reduced and hydrogenated forms of iso-alpha acids, in particular tetrahydroiso-alpha acid and hexahydroiso-alpha acid, are also significantly more foam active than the unreduced material (Kunimune and Shellhammer, 2008). This is partly due to the fact that both these forms of iso-alpha acid are less soluble and migrate even more readily to the beer foam. As with isocohumulone, the co-forms of both of these hydrogenated hop acids are less foam active than the other variants.

It is also worth noting that alpha acids are also foam positive if they can be initially dissolved in the beer. A recent patent application describes the foam-positive effect of a stable aqueous solution of alpha acids when added to beer.

GUSHING

In the early days of producing postfermentation isomerized hop extracts, poor manufacturing practices (eg, contamination with iron, presence of oxygen, etc.) resulted in the use by the brewer of oxidized and damaged iso-alpha acid extracts. These extracts containing these oxidation materials were often found associated with beer gushing. Fortunately, the various postfermentation additions today are produced to a much higher standard and are rarely associated with gushing.

On the positive side, some hop acids and oils (particularly linalool and β-caryophyllene) have been shown to be gushing inhibitors (Hanke et al., 2009).

VARIABILITY

Brewers will fully appreciate that it is difficult enough to produce consistent beer without the added problems derived from the potential variability of raw materials. Cone hops are a naturally grown raw material and are therefore subject to several sources of variability and inconsistency typically caused by:

1. Structure of the hop cone—the hop cone itself is not in any way a homogeneous structure and separation of the lupulin glands during picking, processing, and packing readily occurs.
2. Position on the hop bine—leading to variation in ripeness or development of the cones (see Table 3.1).
3. Influence of geographical location—the same variety grown in different areas may vary considerably; even within a relatively small hop garden of a few hectares, significant variations in quality can be found depending on orientation, slope, degree of exposure, etc.
4. Seasonal variation—like most other crops hops are affected by seasonal variations in weather and growing conditions. Hop growers will consistently claim that there has been too much or too little sun, heat, cold, rain, frost, etc. Table 3.2 demonstrates the variation in alpha contents for three varieties over a 10-year period from 2004 to 2013.

Table 3.1 Typical Variation in Alpha % and Total Oil in Magnum Throughout the Height of the Bine

Position on the Bine	Alpha Acid (%)	Total Oil (ml/100 g)
Head	14.2	2.92
Top	13.8	2.77
Middle	12.1	2.43
Bottom	10.9	1.39

Table 3.2 Seasonal Variation in Average Alphas for Three Varieties in Crop Years 2004–2013

Crop Year	Hallertau Hersbrucker	Styrian Goldings	CTZ#
2004	3.0	4.5	15.0
2005	3.0	4.0	15.2
2006	2.2	2.9	15.5
2007	2.5	3.0	14.5
2008	2.9	4.5	15.0
2009	3.4	4.2	15.2
2010	3.5	4.0	15.0
2011	4.5	4.1	15.5
2012	3.0	3.2	15.5
2013	1.9	2.2	15.8
10-year Average	3.0	3.7	15.2
Range	1.9–4.5	2.2–4.5	14.5–15.8

 # Columbus, Tomahawk, and Zeus.

5. Pests and Disease—hops are susceptible to attack from many different pests and diseases which can impact on appearance but more significantly on the development of the essential compounds within the lupulin glands.
6. Storage—even when stored at $<5°C$, leaf hops can lose over 20% of the alpha acids over 24 months (see Fig. 3.1).

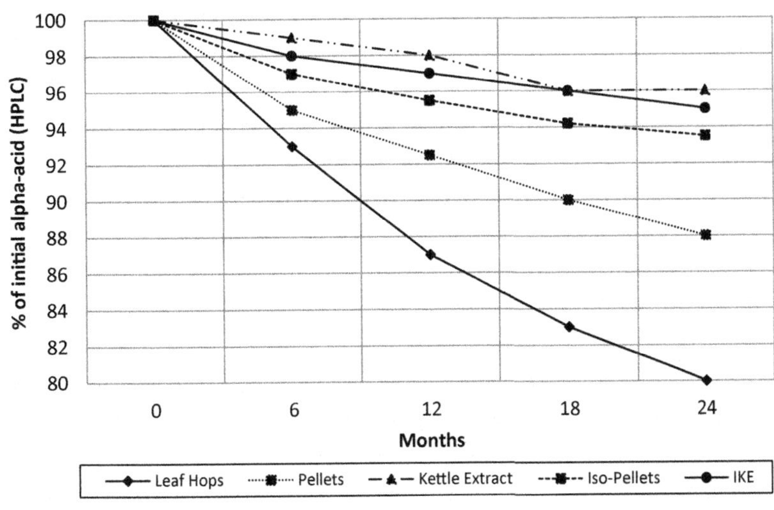

FIGURE 3.1

Stability of leaf hops and kettle-added hop products measured by loss of (iso-) alpha acid during storage at their recommended storage temperatures.

HOW BEST TO MANAGE HOPS TO ACHIEVE BEER EXCELLENCE

Below are discussed the choices and decisions involved in choosing hops for use in the brewery. The range of varieties and hop products has never been greater, but this in itself brings with it problems that need to be considered.

CHOICE OF SUPPLIER

As with all raw materials the choice of hop supplier is one of the most important purchasing decisions. Although price is important, consideration should also be given to the following:

- Product quality
- Quality of service
- Reliability and ability to deliver on time, every time
- Integrity
- Quality systems and safe methods of working

It is essential that good relationships are built with "approved" suppliers based on understanding, clarity of needs, and mutual trust.

STATUTORY ISSUES

Product Safety

Brewers have a duty of care to ensure that their products are safe for the consumer to drink. Therefore, it is essential that the brewer has confidence that his raw materials fully meet product safety requirements.

Very few hop-growing countries are able to successfully grow hops without recourse to the use of chemical applications for prevention of attack by pests and diseases.

Each country will have regulations covering which chemicals are approved for use specifically for hops, the frequency and amount allowed to be applied, and the maximum residue levels permitted to be present in the picked and dried crop.

International harmonization on the use of chemicals and their residues has been progressing for many years and should remain a priority to ensure the free movement of hops between countries. This is particularly important in the current beer market with brewers' desire for new and exotic flavors being satisfied from a seemingly ever-increasing range of hops from all parts of the global hop market.

Most hop farmers work hard at ensuring their crops fully meet the needs of the customer. However, there have been occasional instances of hops not complying with regulations either due to negligence by the grower, poor agricultural practices, or accidental contamination from neighboring crops caused by spray drift. It is, therefore, important that hop buyers, when possible, carry out the following:

- Farm Audits:
  Not all customers have the resources to carry out full supplier audits. However, when possible, it is beneficial to visit growers to review quality issues and develop an understanding of mutual problems and concerns. Some growers, particularly those who also grow other crops for supermarkets, will have some formal, independently audited, quality accreditation with corresponding documented procedures for each stage of hop production. When this is not the case, the grower should still be able to demonstrate good growing practice (particularly the documented application of chemicals), handling, processing, and packing of hops.
- Supplier Information:
  As a minimum requirement, brewers should insist that all records relating to the quantity and timing of spray applications are fully completed and either supplied by their merchant or grower with each consignment of hops or at least available for inspection as required.
  Suppliers should also be able to demonstrate due diligence by providing analytical evidence that statistically significant, random samples of their annual harvest meets the appropriate regulations in terms of chemical residues.

Traceability

In case of quality issues, it is essential that suppliers must be able to clearly trace the origin and integrity of the hops used in the production of any hop product right back to the farm where the hops were originally grown. In the event of a public safety concern, the source of the problem should be identifiable and any other potentially contaminated products isolated or recalled.

Certification

Most hop-producing countries have formal certification procedures for hops and hop products aimed at ensuring a common, minimum standard of product (Biendl et al., 2012/2013). These regulations may vary slightly from country to country, but as a minimum they should ensure:

- Confirmation of variety and traceability of harvested hops
- Levels of moisture content and extraneous matter (often referred to as "leaf and stem").
- Rules regarding the blending of hops
- Minimum requirements for packaging providing rules on sealing and labeling which identify—crop year, variety, name, and place of production.
- Provision of a signed certificate from an authorized source confirming relevant details.

This information should provide the minimum basis for any commercial transaction.

ANALYTICAL ISSUES

Over the course of the last century, a variety of analytical methods have been developed for hops and hop products. Primarily developed for the measurement of bitterness, these methods are regularly updated to include new methods for newly introduced products as well as improving the accuracy and relevance of existing methods. Following extensive collaborative testing, these methods are agreed upon and published as internationally accepted recommended methods. It is important that the methods selected for routine use should be quick, capable of automation (when possible), repeatable, and, above all, meaningful for purpose.

Currently, there are three widely used analytical compilations:

- Analytica-EBC (incorporating the Methods of Analysis of the Institute of Brewing & Distilling since 2004)
- ASBC Methods of Analysis
- Mebak Methods (Brautechnisce Analysenmethoden)

Care must be taken that, when commercial transactions are based on analytical data, the details of the precise method are clearly stated and agreed upon.

Typically, the recommended methods cover the following:

Sampling of hops and hops products:

It has previously been discussed that leaf hops, in particular, show considerable variability. Despite these reservations, sampling methods are normally specified for unpressed and pressed hops, hop pellets, extracts, and isomerized pellets and extracts. Obviously, greater confidence can be placed on the sampling integrity of processed products.

Moisture content of hops and hop products:

Oven-drying methods for green hops, kilned leaf hops, hop powders, and pellets provide a reasonably accurate measure of the weight loss due to water evaporation expressed as a % (m/m). These methods do not necessarily give the true moisture content as the loss of some oils also occurs.

Seed content of hops:

Applicable to leaf hops only, methods for measuring seed content involve drying a sample, physically separating, and sieving. Results are expressed as % (m/m). In the ASBC methods, procedures for measuring leaf and stem involve the manual separation of the extraneous material with forceps with the results expressed again as % (m/m). These results can influence commercial transactions as well as certification.

Alpha-acid, hop resins or "bitter substances" in hops and hop products:

The recommended methods for the routine analyses of hop resins in hops and hop products are generally based on one of the following, all of which involve (1) extraction into a solvent and (2) measurement of concentration in the solvent:

- Lead conductance value (LCV or CLV) methods for alpha acids:
  This method involves the extraction of hops in a nonpolar solvent or solvent mixture followed by partial dilution in methanol. The resultant mixture is titrated with a methanolic solution of lead acetate (PbAc) and the conductivity monitored. As the alpha acids are precipitated as lead salts, the conductivity remains unchanged but, as soon as all the alpha acids have been consumed, the conductivity increases rapidly (see Fig. 3.2). The point at which this increase occurs is the "end-point" from which the concentration of alpha acids can be determined by use of a simple equation. The results are expressed as % LCV which provides a good measure of the alpha acid content of fresh hops but, as the hops deteriorate, more of the alpha and beta acids are converted into compounds that may also be extracted and react with lead acetate. Hence, the rate at which the % LCV may decline is often less than the true alpha-acid decline. However, as some of these degradation compounds are bitter, the % LCV is often considered a better representation of bittering value than more specific alpha-acid analysis. Consequently, many brewers continue to use this value as the basis for hop buying and usage, instead of the more-specific analytical values.
- Spectrophotometric methods for alpha acids, beta acids, and hop storage index (HSI):
  This method, most commonly used in the United States, is based on measuring the absorbance of a dilute, alkaline solution in methanol of organic solvent-extracted

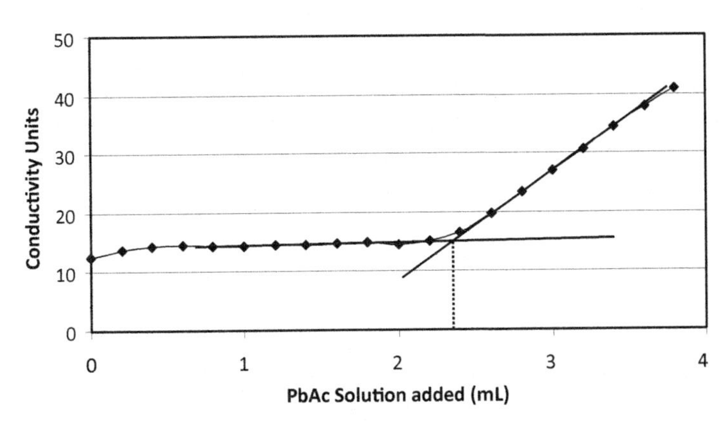

FIGURE 3.2

Lead conductance analysis titration curve.

hops. The absorbance is measured at three UV wavelengths—275, 325, and 355 nm—and the results inserted into two regression equations from which the % alpha acids and % beta acids can be calculated.

As the majority of the hop degradation products absorb more strongly at 275 nm than either of the alpha and beta acids, it is possible to obtain a measure of the degradation by using the ratio of absorbance at 275:325 nm. This figure is known as the HSI and the lower the figure the less degradation has occurred.

- High-pressure liquid chromatography methods

  High-pressure liquid chromatography methods (also known as high performance liquid chromatography or HPLC) for alpha acids, beta acids, and iso-alpha acids were developed in the 1970s and 1980s and have been successfully fine-tuned since then, particularly to include the reduced, light-stable variants of iso-alpha acids. Such procedures have revolutionized hop resin analysis and provide the most accurate and specific measurement of these compounds. The methods involve the following steps:

  a. Sample Preparation:

  It is often convenient to use the same primary methods developed for sample preparation in lead conductance or spectrophotometric methods. The sample is normally greatly diluted with methanol before use.

  b. Separation of Components:

  The separation of hops components depends on utilizing the differences in the relative affinity of solutes for absorption onto the "solid phase" packed within the HPLC column, which, for hop analysis, is normally a silica-based material combined with long-chained hydrocarbon molecules. The "mobile phase," usually a mixture of organic solvents and water, is pumped through the column at a steady rate.

  c. Detection

  The hop acids are normally detected by passing a beam of UV light through a small inline cell. As in a spectrophotometer, the light absorbance is dependent on the spectral characteristics of the solute and its concentration. The results from the detector are electronically integrated, typically drawing peaks on a screen with the concentration of individual compounds being relative to the area under the peaks.

  d. Calibration

  It is essential that the HPLC columns are calibrated using known concentrations of pure standards. These standards are now internationally agreed, produced, and available for purchase by analytical laboratories. There is one International Calibration Extract (ICE) for measuring alpha and beta acids and four International Calibration Standards (ICS) for measuring iso-alpha acid and the three forms of commercially available reduced isomerized extracts. It is essential that, when specifying HPLC analysis or comparing analyses from different sources, the correct most up-to-date standard is used.

Hop Oils:

Analysis of hop oils normally consists of two elements:

- Measurement of total oil
  Hops, pellets, or extract are weighed into a distillation flask and mixed with a defined amount of water. The mixture is then heated for 3–4 h and the vapors continually condensed with the water being returned to the flask. At the end of distillation, the volume of oil is measured and expressed as milliliters of oil per 100 g of sample.
  Because of the high temperatures used, some of the compounds within the total oil are effectively artifacts being converted from other compounds by the conditions of the process. This could have impact on the component analysis described in the following.
- Hop oil components
  Compositional analysis is usually determined using gas chromatography (GC). In modern GC machines long (c.50 m) coils of capillary glass with a suitable coating acting as the stationary phase through which the hydrodistilled oil, mixed in a carrier organic solvent, is passed using a stream of hydrogen or helium. The temperature of the coil is very carefully controlled, and compounds are detected in the exiting gas using a suitable detector. Today, there are a number of detectors each used for a different group of compounds—hydrocarbons, sulfur compounds, nitrogen compounds, and organochlorine compounds.
  Peak identification, particularly for some of the minor constituents, is not easy and can require further analysis using gas chromatography mass spectroscopy (GC–MS) techniques or calibration with pure compounds. The individual results are usually expressed as a % of the total oil content.

Polyphenols:

Although polyphenols are considered to play an important role in beer quality, no internationally agreed methods of analysis exist. Tannins as a group can be measured but detailed knowledge of the constituent polyphenols is not easily established. HPLC methods for the separation of polyphenolic materials do exist but are not commonly employed by hop suppliers or breweries.

Details of all of the above analytical methods are available by reference to the appropriate recommended methods of analysis.

CHOICE OF HOPPING METHOD OR PRODUCT

When using processed hop products, there are several advantages compared to leaf hops:

- Improved homogeneity
- Improved shelf life and stability (see Fig. 3.1)
- Increased density and reduced bulk (see Table 3.3)
- Potentially lower wort losses

Table 3.3 Comparison of Typical Bulk Densities of Hop Products

Product	Typical Bulk Density (kg/m^3)
Bale hops	100–150
Pellets	450–550
Kettle extracts	960–1020
Postfermentation extracts	1010–1030

- Reduction in chemical and heavy metal residues
- Improved utilization (particularly the preisomerized products)
- Easier handling in the brewery with the potential for automated dosing

Additionally, some of the more recently introduced products demonstrate further quality benefits such as light and/or foam stability.

LEAF HOPS

Although the use of leaf hops is now exclusively confined to some breweries within the regional, craft, or microbrewery sectors (estimated to account for only around 2% of the total annual world hop production), it is often the case that many brewers still insist on selecting the leaf hops from which their hop products will be produced. Leaf hops are the ultimate starting point from which all other products are manufactured, and it is therefore important that buyers understand the quality issues and acceptance criteria for leaf hops.

Packing

Leaf hops are compressed and packed in hessian or polypropylene-covered bales following kilning and after a period of cool conditioning when the residual moisture is allowed to evenly disperse throughout the hop cone. In Europe, 60-kg bales have become the standard whereas, in the United States, bales can be typically 200 pounds or 90 kg. The compression of the bale is important in that it helps to exclude air and therefore on the one hand slows oxidation but, on the other, overcompression can damage the lupulin glands, thus, conversely, exposing the resins and oils to possible oxidative damage.

Other smaller packs are available eg, 5, 10, or 20 kg, and these are usually supplied vacuum packed in triple-layer aluminum foils packed in cartons. These smaller, more stable packs are particularly useful for smaller breweries in which the usage is low and storage is at ambient temperatures.

Physical or Hand Evaluation

Physical evaluation of hops is not included in all recommended methods and, when it is, the procedure is simple, concentrating mainly on seed, leaf, and stem (ASBC

Hops-2). However, when buying leaf hops either for direct use or for processing into hop products, many brewers still insist on carrying out detailed hand evaluation of the offered hops based on the samples supplied by the seller, supported by analytical data as appropriate. This procedure is thought particularly important when buying aroma hops. When hops are being bought for their alpha-acid content only, this practice is sometimes deemed unnecessary and decisions are made purely on analytical data.

Samples for hand evaluation can be submitted as either a "roll" sample of loose hops, taken from the conditioning floor, as a bore sample from a bale, or as a "cut" sample in which a rectangular sample of hops is literally cut, using a sharp knife, from the side of a bale. Both provide perfectly acceptable samples for the physical assessment procedure. For aroma hops, one in 10 bales is normally sampled, whereas for alpha varieties the sampling is often reduced to one in 30 bales. It is important for the buyer to establish that the hops are reasonably consistent within a defined lot or batch.

The evaluation process can be cursory or very specific, and often the precise detail is developed by the experienced buyer over many years. Generally, the procedure will involve the following steps:

a. Visual examination of the loose hops or cut sample looking for:

- Color and wholeness of cones
- Cone brittleness and degree of shatter
- Contamination with extraneous matter (including leaf and stem)
- Seed content
- Quantity and color of lupulin glands
- Cone damage arising from wind, spray burn, fungal attack, and pest damage

These assessments can give the buyer indications of possible growing and handling problems, which may affect brewing quality, as well as the degree of physical contamination which affects the monetary value of the hops.

b. Assessment of moisture or degree of drying by pressing down on the cut sample or squeezing a handful of hops to assess "springiness." Overdried hops lack spring and easily break up. Excessive drying can lead to damage of the key hop components and negatively affect brewing value.

c. Assessment of aroma by rubbing a sample of hops between the hands. Often this process is split into an initial "light" rub followed by a more vigorous "big" rub and finishing with a "hold." At each stage, the rubbed hops are sniffed to evaluate aroma quality, trueness to type, and also to detect any undesirable traits.

Usually, each aspect of the process is scored and the resultant information used as a basis for acceptance or rejection. It should be understood that early-picked hops normally have a better appearance than their later-picked counterparts. Later-picked hops are often riper, however, and have a more developed aroma which may be particularly desirable when buying aroma hops. It is the brewers' choice as to whether they select on appearance, aroma or, as is more usual, a combination of both.

Analyses

Assuming satisfactory hand evaluation and confirmation of compliance with pesticide residue limits together with provision of appropriate certification, the analytical data supplied with leaf hops usually include:

a. Moisture (EBC 7.2 or ASBC Hops-4)
The level of moisture in leaf hops is important as wet hops can encourage microbiological activity leading to composting of the hops and deterioration of key components. Of equal importance is the danger of heat buildup within the bale which can lead to spontaneous combustion and, in extreme circumstances, cause major hop-store fires. Because of this, hops should always be probed for moisture levels at various points within the bale over a few days to ensure no heat buildup prior to safely placing in store.
Moisture levels are quoted as a percentage of the total weight and should be between 8% and 10%.

b. Alpha-acid (EBC 7.4 or ASBC Hops-6)
Obviously, because alpha-acid levels are subject to yearly variation, it is important that, to calculate the quantity of hops to be used to achieve the desired bitterness, the alpha-acid levels of the individual batches of hops are measured and provided at the time of purchase. However, because of the stability issues, it may be desirable to recheck alpha levels once hops have been in store for a long period.
In many instances, beta acids are not routinely measured for each batch of leaf hops; however, as previously described, beta acids can albeit indirectly contribute to bitter flavor in beer. By use of any LCV method, some of these bitter oxidation products from both alpha and beta acids will be measured and included in the "alpha" % or more accurately the "total bittering" value.
Other useful analysis sometimes specified for leaf hops are:

c. Total Oil (EBC 7.10 or ASBC Hops-13)
As with hop resin contents, the oil content of hops will vary year by year but not necessarily exactly in line with alpha acids. As any annual or batch adjustment to the hop, grist is usually made on the basis of alpha acid, it is apparent that most brewers do not adjust for varying oil content even for aroma hops. This could have a significant effect on beer flavor.
Some brewers insist on total oil values (ml/100 g) being supplied for each lot of aroma hops and use this value as a basis for adjusting the amount of late-kettle hops to be added. Although oil retention is variable, this adjustment attempts to ensure consistency of hop oil addition while ignoring the limited impact of the alpha-acid content on bitterness owing to the normally poor utilization of late-addition hops.

d. Hop Storage Index (ASBC Hops-12)
As mentioned previously, a useful measure of hop degradation is the so-called hop storage index or HSI. It is normally a standard analysis in the United States and can be calculated from the ASBC spectrophotometric method for measurement

of hop resins. Although somewhat variety dependent, HSI values of <0.35 are considered good, whereas values >0.40 would indicate the start of degradation possibly due to poor harvesting and kilning.

e. Cohumulone (EBC 7.7) or *ASBC Hops-7*
Not normally considered a routine analysis, cohumulone (CoH) levels may be specifically requested for the following reasons:
- Flavor—as described, some brewers consider that CoH produces a harsh bitterness when compared to the other alpha-acid homologs.
- Production of downstream products—isocohumulone is more soluble than the other homologs, and therefore yields can often be better when using high-CoH hops.
- Beer Foam—as isohumulone and isoadhumulone are less soluble than isocohumulone, these former two homologs are considered better for foam retention. Therefore, in the production of reduced isoproducts for foam enhancement, high-alpha hop varieties with lower CoH levels may be preferred.

PROCESSED PRODUCTS

HOP PELLETS

Hop pellets are the most commonly used processed hop product accounting for almost 60% of the world's total hop usage. Pellets are produced from leaf hops in three different forms:

Standard or Type 90 Pellets

The term Type 90 (T90) refers to the weight yield (ie, 90%) that was typically expected after the removal of excess moisture and extraneous matter prior to processing into pellets. The production of T90 pellets involves a sequence of physical steps in which hops are blended, redried to <10% (if required), extraneous matter removed, milled in a hammer mill, before the resultant hop powder is forced through c.6 mm holes in a bronze or steel die. The pellets are then packed in a triple-ply, metalized laminate bag which is either evacuated before sealing ("hard" packs) or gas flushed with a nitrogen/carbon dioxide mixture ("soft" or "pillow" packs) before heat sealing and packing into cartons.

As the production of T90 pellets involves very little change to the original dried hop, these pellets most closely resemble the use of leaf hops.

Concentrated or Type 45 Pellets

In the production of Type 45 pellets (T45) some of the "leaf" material is removed. This separation is achieved after milling and at temperatures of −20°C by passing the cooled powder through a series of vibrating screens which separate the powder into two fractions—the lupulin and leaf fractions. A calculated amount of the leaf fraction can then be added back to the lupulin fraction. This renders the sticky lupulin fraction easier to handle and pack as well as allowing the alpha-acid value in the pellets to be standardized to an agreed level, higher than the harvest alpha. T45 pellets are pelletized, cooled, and packed in metalized laminate bags as described earlier. The additional advantages of T45 pellets over T90 are therefore:

- Reduction in weight and volume
- Reduction in chemical and heavy-metal residues by removal of some of the leaf fraction
- Standardization of the alpha value removing any potential problems caused by the yearly variation of the harvest alpha acid %

However, a possible disadvantage is that much of the polyphenolic material is removed which could have implications on beer quality and stability.

Pre-isomerized or Iso-Pellets

Isopellets are hop pellets in which the alpha acids are converted into iso-alpha acids as part of the processing operation. This is achieved by adding a calculated amount of magnesium oxide (approximately 1–3% by weight) to the milled hops, mixing, and then pelletizing and packing as normal. The cartons of hop pellets are then warm conditioned at temperatures between 45 and 55°C for 8–14 days during which time the isomerization proceeds catalyzed by the presence of magnesium ions. Although there are small changes to the hop oils, practical experience has shown that no flavor differences can be detected in comparable beers brewed with both T90 and Isopellets (Taylor et al., 2000). Because the alpha acids are already isomerized, dissolution in the wort is rapid and more efficient than standard pellets. Typically, utilizations (in final beer) are increased to at least 50–60% compared to 30–35% for T90 and T45 pellets.

Quality Issues—Pellets

a. Packing:
   - It is essential that the integrity of the laminate bag remains intact to prevent ingress of air allowing oxidative degradation of the important hop compounds. Problems can arise from pinhole puncture damage to the laminate itself or poor sealing at the bag closure. With hard packs, any pack failure is clearly visible, whereas any leaks in soft packs are almost impossible to detect without analysis of the gas mixture in the bag. A maximum rate of bag failure should be established as part of the processing specification, and the supplier should be able to demonstrate adequate checks within their quality control system to give the customer statistical evidence that pack integrity meets the agreed specification. For practical processing reasons, it is necessary to use only soft packs when producing isopellets.
   - Carton strength is important when the cartons are packed onto pallets, and storage space dictates that double or even treble stacking is required. Inadequate carton thickness and strength can cause unsafe stacking resulting as a minimum in ruptured bags and, at worst, injury to brewery staff.

- Labeling requirements should be specified by the hop buyer at the time of placing the order or processing contract. Typically, labels should adhere to country certification standards including as a minimum—variety, crop year, alpha %, net weight, and batch number (essential for traceability). Other information such as country of origin, total oil content, and sequential carton numbering can also be requested.

b. Physical Examination:

- Pellets should be smooth and olive green (but not too dark or glassy) with no unpleasant aromas and measure approximately 6 mm × 10–15 mm. The pellets should not be too hard and the pack should break apart easily upon opening to aid dispersion in the kettle.
- If the pellets are dark in color, long, and tapered (rat-tailed), it could indicate that they have been overheated during passage through the pellet die. It is recommended that customers specify a maximum exit die temperature which typically should be between 50 and 55°C. This can be achieved by changing the die design or cooling at the exit with nitrogen or CO_2.

c. Analyses:

- Moisture (EBC 7.2 or ASBC Hops-4):
  The moisture levels should be <10% and preferably around 8%. It is difficult to achieve a complete vacuum or gas flush within the pellet packs, and it is possible for microbiological activity to cause slow degradation at higher moisture levels even when only traces of oxygen are present.
- Alpha acid or Iso-alpha acid:
  Alpha-acid measurement for T90 and T45 pellets is normally by Lead Conductance (EBC 7.5) or Spectrophotometric (ASBC Hops-6) methods, whereas for Iso-pellets the iso-alpha-acid value is measured by HPLC methods (EBC 7.11 or ASBC Hops-15). In iso-pellets the residual alpha-acid value is also measured by HPLC and the results reported as an iso-alpha + alpha value.
- Total Oil (EBC 7.10 or ASBC Hops-13):
  As with leaf hops, a total oil value may be important for control of late-addition aroma hops.

d. Processing yields:

If the brewer is buying leaf hops to be subsequently processed into pellets, it is important that the appropriate yields are specified before processing. The most important of these are:

- Alpha-acid recovery—expressed as a % of the total weight of alpha contained in the finished pellets compared to the amount in the original hops (T90 and T45 pellets). Normally, >96% is specified.
- Weight yield—expressed as a % of total weight of pellets produced compared to original weight of hops. Typically, a minimum of 98% is specified.
  For isopellets:
- Iso-alpha acid + residual alpha acid—expressed as a % of the original weight of alpha. Typically, >90% should be achievable.

- Conversion % of original alpha acid into iso-alpha acid. Typically, a minimum of 95% should be achieved.

In the event of redrying during processing, it may be necessary to adjust these calculations to take account of changes in dry matter.

KETTLE EXTRACTS

Hop extracts are the second most commonly used processed hop product. Historically, kettle extracts were produced using a variety of organic solvents, but today only two solvents are used. Other added-value kettle extracts are also available.

CO_2 Extract

Depending on process conditions, hop pellets, or coarsely remilled pellets, are packed into a series of high-pressure extractors. Liquid CO_2 is heated and pressurized to a specific temperature and pressure before being pumped through the bed of hops within a series of extractors. The nonpolar hop resins and oils dissolve in the CO_2 which is passed continuously through the extractors until virtually all have been extracted. The dissolved material is then separated from the CO_2 by reducing the pressure and passing through an evaporator at a temperature of 45–50°C. The extract is collected and the CO_2 pumped through a condenser, liquefied, and returned to a storage tank for reuse.

CO_2 extracts are commonly referred to as either Liquid or Supercritical extracts referring to their process conditions differing only in the temperatures and pressures of the CO_2 used. Under the more extreme, supercritical conditions, the CO_2 behaves as a supercritical fluid extracting some of the polar degradation products and hard resins. As the extraction efficiency is higher under supercritical conditions, commercial plants usually operate under these conditions. CO_2 extracts are the starting material for production of virtually all postfermentation added extracts.

As a solvent, CO_2 extracts very little polar polyphenolic material, and there is some evidence to suggest that beers produced with extracts are less flavor stable than equivalent beers brewed with whole hops or pellets.

Ethanol

In the production of ethanol extracts, whole hops are gently crushed in a wet mill and mixed with 90% ethanol. The resultant slurry is dropped into a continuous, revolving, two-deck, segmented extractor in which ethanol is percolated through the hops in a countercurrent direction. The extraction takes place at ambient pressure and temperatures of between 40 and 60°C. The mixture of ethanol and extracted hop material is centrifuged and sent to a multistage vacuum evaporator to remove the ethanol which, along with the material recovered from the spent hops, is sent to a rectification plant for concentration back to 90% for reuse. The hop material is split into two fractions—the resin and the aqueous fractions—by centrifugation. The aqueous fraction contains a high concentration of undesirable material such as nitrates, pesticides, etc. and is normally discarded.

Ethanol extracts contain a more complete mixture of hop components (both polar and nonpolar) and therefore are thought by some more representative of leaf hops. However, because they are less "pure" than CO_2 extracts, they are not suitable for the production of downstream products.

Isomerized Kettle Extract (IKE) & Potassium-Form Isomerized Kettle Extract (PIKE)

Similar to the production of Isopellets, a magnesium salt (in this case, magnesium sulfate) can be used to catalyze the isomerization of alpha acids in CO_2 extract. The magnesium sulfate is added to the extract along with deionized water, warmed to 75–85°C, and the pH adjusted to 8.5–9.5. The mixture is stirred for several hours under a blanket of nitrogen during which time the alpha acids are isomerized to iso-alpha acids. The pH of the mixture is reduced to around 2 by the addition of mineral acid which causes the hop resins and oils to separate from the aqueous phase. After settling the phases can be separated.

The IKE can be processed further by the addition of a stoichiometric quantity of potassium hydroxide which reacts with the free-acid forms of iso-alpha acids to produce their neutral, potassium salts.

Both forms of isomerized extract provide much improved utilizations when compared to normal kettle extracts (usually a minimum increase of 50%). However, there is some small loss of hop oils during the production of both forms of extract.

Light-stable Kettle Extract (LSKE)

One of the problems associated with the use of postfermentation added, light-stable products (see later) is that these aqueous solutions are pure solutions of reduced iso-alpha acids in which the hop oils have been removed during processing. To achieve hop character in light-stable beers, light-stable kettle extracts have been developed. In these extracts the alpha is separated from the beta and oils, isomerized, reduced to the *rho*-iso-alpha-acid form by a sodium borohydride reduction, and then added back to the cleaned (removal of any remaining alpha acids) beta and oil mixture. In the absence of any unreduced iso-alpha acids, this product can successfully be used for producing light-stable beers with, depending on time addition, some hop character.

Quality Issues—Kettle Extracts

a. Packing:

- Extracts can be packed into a variety of containers ranging from lacquer-coated steel cans, polypropylene pails, and bulk steel, or polypropylene drums with sizes varying from 0.5 kg to 200 kg. In all types of pack, it is preferable to flood the surface of the filled container with an inert gas prior to sealing to remove as much air as possible. In cans or pails of extract, it is possible to fill each container with a fixed quantity of alpha acid (eg, 300 g in a 1-kg can) so that fixed multiples of containers can be added to the kettle reducing the possibility of dosing errors. The previously common practice of standardizing the quantity of alpha in a container by the addition of a calculated amount of glucose syrup is nowadays seldom used as much of the world's supply of glucose syrup is produced from genetically modified crops.

- The pH of extracts varies considerably with typical values shown in Table 3.4. Although polypropylene containers are inert to the pH of hop extracts, steel cans and drums rely on the integrity of the lacquer coating (or polyethylene liner in drums) to protect the metal from corrosion. This can be a particular problem with the low pH of both CO_2 extract and IKE. If the lacquer is damaged or incomplete, most likely at the seams of the containers, corrosion can readily occur and the product within contaminated. Another potential concern is the sealing of can lids which, if not correct, can cause leakage, ingress of air, and corrosion damage. The supplier should be able to demonstrate that can lining and seal tests are routinely carried out before use and during filling to identify and prevent any problems being released to the customer.
- Kettle extracts lend themselves to bulk storage and handling in the brewhouse. Owing to the pH issues described, the materials of construction of tanks, pipes, and pumps needs to be considered carefully. Flushing of the system between use and exclusion of air in the bulk tanks (particularly with isomerized extracts) should be considered in the design.
- As described for pellets, labeling requirements should be specified by the hop buyer at the time of placing the order or processing contract. Typically, labels should meet minimum certification standards.

b. Physical Examination:

Physical examination of extracts is not particularly significant as the appearance will vary according to process conditions. However, typically the extracts should have the following appearance:

CO_2 Extract—a semifluid syrup; golden, amber, or green
Ethanol Extract—a viscous, dark-green syrup
IKE—golden, pale-amber mobile syrup
PIKE—golden, pale-brown thick syrup or semisolid paste
LSKE—golden, amber syrup

Table 3.4 Typical pH Values for Kettle-added Extracts

Product	pH
CO_2 Extract	c.4.0
Ethanol Extract	c.6.2
IKE	c.2.8
PIKE	c.6.5
LSKE	c.7.8

c. Analyses:

- Alpha acid (EBC 7.7 or ASBC Hops-14), Iso-alpha acid (EBC 7.8 or ASBC Hops-16) and *rho*-iso-alpha acid (EBC 7.9) as appropriate:
  For extracts, HPLC analysis are usually preferred but with ethanol extract, which can contain 0.5–2.0% iso-alpha acids, a hybrid figure called the conductometric bitter value (CBV) is usually quoted. This figure is derived by measuring the alpha content by an LCV analysis (EBC 7.6) + 50% of the iso-alpha acids by HPLC (EBC 7.8).
- Beta acids (EBC 7.7 or ASBC Hops-14):
  Extract users may request beta-acid analysis, particularly when buying extract off the shelf, as high beta-acid content will result in higher viscosity, and therefore the extract may be difficult to handle in the brewhouse and not be suitable for bulk handling. Similarly, buyers of extracts destined for further processing into downstream products normally prefer extracts with low beta-acid contents.
- Solvent residues:
  In CO_2 extracts, all of the CO_2 is removed. With ethanol extracts, a limit of <0.3% ethanol is normally specified.

d. Processing Yields:

When leaf hops are bought for subsequent conversion into kettle extracts, it is advisable to specify and agree on appropriate yields. These typically would be:

- CO_2 Extract: >95% (by HPLC)
- Ethanol Extract: >95% (by CBV)

POSTFERMENTATION EXTRACTS

The earliest postfermentation products were introduced in the 1960s. These products were developed to provide a more efficient, cost-effective form of bittering and to allow late adjustment of bitterness to bring beer back into specification. As previously explained, many of these early extracts caused quality problems, most commonly either haze or gushing. With improvements to CO_2 extraction technology and the development of HPLC analyses to assist in better process control, the quality of these extracts improved dramatically.

Modern extracts are normally clear solutions of pure, potassium salt forms of iso-alpha acids and, apart from bitterness, make no other contributions to hop flavor or character. Usually added just prior to final filtration, iso-extracts can achieve very high utilizations (up to c.90% for iso-alpha acid) although the reduced forms typically achieve less (c.60–80%).

Isomerized Hop Extract, Iso-Extract or Postfermentation Bittering (PFB)

The starting material for the production of iso-extract is CO_2 extract. The process involves addition of magnesium salts under alkaline conditions and at elevated temperatures during which the alpha acids are converted to iso-alpha acids. The beta acids and hop oils are separated from the iso-alpha acids by pH adjustment using a mineral acid. Typically, two separations take place to ensure that the levels of any contaminating beta acids are very low. Potassium hydroxide is then added to the free iso acid to form the potassium salt which is standardized in pH-adjusted, deionized water to a concentration of either 20% or 30% prior to sale.

Reduced Iso-Extracts

Produced from iso-alpha acids, either by a reduction process or by hydrogenation (or a combination of both), reduced iso-extracts were introduced in the late 1990s. Depending on the precise location of the reduction process within the iso-alpha-acid molecule, these products provide varying degrees of perceived bitterness as well as either light stability and/or improvements to beer foam (see Fig. 3.3). Sometimes these extracts are added into the kettle, in particular *rho*-iso extracts, and less frequently Tetra extracts, but more normally they are added postfermentation. To achieve light stability, it is essential that all traces of unreduced iso-alpha acids are removed from the surface of vessels, pitching yeast, etc. as even very small amounts can be converted to MBT resulting in a detectable "skunky" flavor.

Rho-Iso-extract (Rho)

Normally produced from the reduction of purified iso-alpha acids in the presence of sodium borohydride at elevated temperatures and at alkaline pH, Rho is normally sold as the potassium salt solution standardized to either 35% or 10% (by spectrophotometric analysis). The 35% material is opaque in appearance and can be added either into the kettle or postfermentation, whereas the 10% product is a clear solution which is normally only added postfermentation.

Used in isolation, or with another light-stable product, Rho should produce a light-stable beer. It is described as having a perceived bitterness of only 0.7–0.8 times that of unreduced iso-alpha acid. Although still involved in foam formation and stabilization, it is considered similar, or only slightly better, than normal iso-extract in foam enhancement.

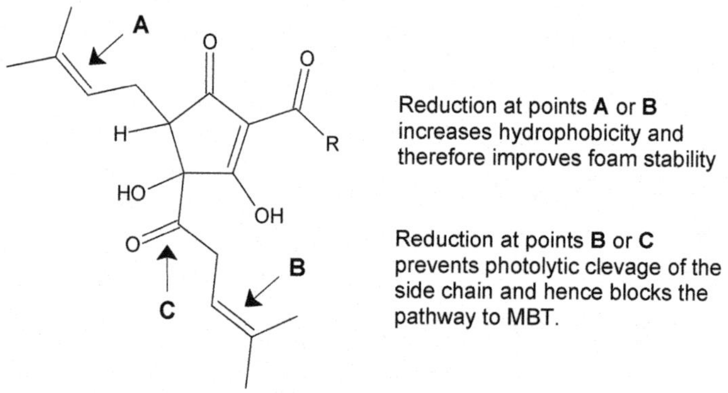

A
O
O
H
R
HO
OH
O
B
C
Reduction at points **A** or **B** increases hydrophobicity and therefore improves foam stability

Reduction at points **B** or **C** prevents photolytic clevage of the side chain and hence blocks the pathway to MBT.

FIGURE 3.3

How chemical reduction alters the properties of iso-alpha acids.

Tetrahydro-Iso-Alpha-Acid (Tetra)

Tetra can be produced from a number of different, patented processes using either alpha or beta acids. More commonly, pure iso-alpha acids, either in an aqueous or solvent solution, are hydrogenated using gaseous hydrogen in the presence of a palladium/carbon catalyst. The resultant tetrahydro-iso-alpha acids are separated from the catalyst by filtration. The process for Tetra production from beta acids is more complex and involves three steps—hydrogenation, oxidation using peracetic acid (or air), and finally isomerization. In both cases, the product is a clear, aqueous 10% solution of Tetra in the potassium salt form.

When added to beer, Tetra (without any contamination from unreduced hop acids) produces light-stable beer. In concentrations as low as 3 mg/liter, Tetra will significantly enhance beer foam formation and stability, and in many beers addition of Tetra has now become the norm particularly in draft lagers in which presentation in the glass is very important.

Although broadly similar, Tetra produced from beta acids will have much higher levels of tetrahydro-iso-co-humulone present (often >60%), and therefore foam performance could be inferior when compared to the alpha-derived product.

Tetra is arguably described as having a perceived bitterness of 1.7 to 1.8 times that of unreduced iso-alpha acid, and the flavor is often described as harsh. This certainly appears the case when tasted in water but in more complex, flavored liquids, such as beer, this apparent increase appears much less intense. Brewers have variously reported figures ranging from 1.2 to 1.7.

Hexahydro-Iso-Alpha-Acid (Hexa)

In the production of Hexa, hydrogen is introduced into the iso-alpha-acid molecules by a combination of both borohydride reduction and hydrogenation using the palladium/carbon catalyst described earlier. The starting point can either be Rho which is subsequently hydrogenated or Tetra which is subjected to borohydride reduction.

Hexa is sold in the potassium salt form as a clear aqueous 10% solution or a 20% solution stabilized by the presence of propylene glycol or ethanol. Hexa is light stable and produces excellent foam. However, some brewers maintain that Hexa produces foam that is too "stiff" and that looks unnatural. The relative bitterness of Hexa is considered slightly more bitter than unreduced iso-alpha acid (×1.2) but the flavor is described as "soft," and therefore it is very suitable to be used in combination with the more harsh flavor of Tetra.

Quality Issues—Postfermentation Extracts

a. Packing:
Typically, 20-kg polypropylene containers are used but, for large volume users, 1000-kg bulk tanks (totes) can be provided. Unlike most other hop products, these products should be stored at ambient temperatures to avoid precipitation.

b. Physical Examination:
Products should be clear on receipt but occasionally a haze or precipitate can appear. This will normally go back into solution with gentle warming.

c. Analyses:

- Reduced iso-alpha-acid content:
  The reduced iso-alpha content is normally measured by HPLC methods (EBC 7.9) using the relevant ICS standard. Buyers should be aware that, when spectrophotometric methods are used, the equivalent figure by HPLC could be 9–10% lower.
- Iso-alpha acid:
  As contamination with even small quantities of nonreduced iso-alpha acids will result in "skunky" flavors, some buyers specify a maximum level of iso-alpha acid present. Typically, this would be <0.05% w/w.
- Haze:
  It is also common for brewers to specify a maximum haze—typically, <3 EBC.
- Stability:
  Because of physical stability issues, stability limits may be specified, particularly with Tetra and Hexa. These may specify (1) the total absence of any precipitate at 20°C after 4 weeks and/or (2) more stringent tests involving refrigeration, storage for 7 days, and then warming back to room temperature for 1 day before evaluating any remaining precipitate.
- Impurity Ratio:
  As previously discussed, contaminating materials present in downstream products can lead to gushing problems when added to beer. It is therefore quite common for some form of measurement of purity or impurity to be included in the specification. One such specification requires that the ratio of the total area of the three largest non-iso-alpha peaks to the total area of the iso-alpha peaks is less than 0.12.

d. Processing yields:
Processing yields can vary enormously depending on the process used and the starting material. Consequently, when having products toll processed from their own source material, customers must agree appropriate yields with their processor.

e. Dosing:
Care must be taken when using these reduced iso-alpha products to ensure that addition systems provide very precise control of addition rates. Best results are achieved when product is added over at least 70% of the beer run from tank to filter. These products can usually be dosed at sales strength but, if the product needs to be diluted before dosing (say to a concentration of 2–3%), demineralized, pH-adjusted water (to pH 10–11) should be used. Wherever possible, air should be excluded from any containers, dosing tanks, etc. and dosing systems should be flushed with water between uses.

AROMA PRODUCTS

A number of aroma products are available to add to beer usually postfermentation but some can be added into fermenter. The oldest and most traditional of these is:

Dry Hop "Plugs" ("Whole Hop" Pellets, Type 100 Pellets)

These are most commonly used in the United Kingdom with a long tradition of unfiltered, cask-conditioned beer. More recently, dry hopping has also become popular among craft brewers worldwide.

The production of hop plugs merely involves the breaking up of baled hops, pushing into a die producing a loose, large pellet usually weighing either approximately 14 or 28 g (0.5 and 1.0 ounces), and packing into vacuum packs to preserve the oil content.

The hop plugs are added into unfiltered beer either in racking tank or more usually into the cask or container in which the loose pellet breaks up and over time the hop oil constituents dissolve into the beer including, as discussed previously, some of the terpene fraction. Brewers should be aware that the development of hop character will depend on contact time, and therefore it is essential to treat all beer batches consistently in terms of storage temperature and time prior to dispatch.

Often the main objective of using postfermentation added aroma products is to improve the consistency of, or accentuate, hop character in a way that late kettle hopping and dry hopping often fails to reliably achieve. However, attempts to duplicate the subtle flavors resulting from late kettle additions have had limited success. This is partly due to the influence of yeast during fermentation which can modify certain hop compounds by enzymic means and enzyme-mediated hydrolysis.

Despite these reservations more refined hop oil products have been developed and used:

Pure Hop Oil

Whole hop oil products are usually produced from whole hops by steam distillation or from pure resin extracts (both CO_2 and ethanol) by either steam distillation, low-temperature vacuum distillation, or molecular distillation.

Fractionated Products

The whole hop oil can be further fractionated into a number of fractions variously described as "spicy," "floral," "estery," "herbal," or "citrusy."

The resultant oils (either whole or fractionated) are normally either dissolved in ethanol, propylene glycol or mixed with a food-grade emulsifier and standardized to a 1% solution. Addition to beer is normally before final filtration to avoid any potential problems with haze, and recovery in beer can be as high as 90%.

Quality Issues—Hop Oils

a. Packing:

One percent dilutions of hop oil products are normally packed in aluminum bottles containing either 0.5 or 1 kilo. Pure oil products are supplied in glass or aluminum bottles and, owing to the small quantities required, in sizes usually measured in grams rather than kilograms.

b. Analyses:

- Obviously, the precise composition of oil products is dependent on the hop variety, starting material, and resultant fractionation.
  Individual hop oil components can be measured by gas chromatography techniques according to methods such as EBC 7.12.
- Maximum levels of "contaminating" hop resins can be specified—usually at <0.1%.

PRODUCT SHELF LIFE

Table 3.5 shows the typical, recommended storage temperatures and "best before dates" for the range of hop products discussed. These will vary from supplier to supplier depending on product specification, the nature of the production process, packing, etc.

Products over their recommended shelf life are not necessarily unfit to use. However, some deterioration may have occurred and the respective brewing value may have declined.

Table 3.5 Typical Recommended Storage Conditions and Shelf Life for Hop Products

Product	Storage Temperature °C	Best Before Date (years)
Leaf hops	<5	1
Pellets (T90 & T45)	<5	3
Iso-pellets	<5	4
Pure resin extracts (CO_2 & ethanol)	<10	6
IKE & PIKE	<10	2
LSKE	<10	4
Iso-extract	5–15	2
Rho	10–25	2
Tetra	10–15	1
Hexa	10–25	1
Hop oils	<10	1

REFERENCES

Biendl, M., Brunner, W., Hormansperger, L., Schmidt, R., 2012/2013. Three steps for best quality hops made in Germany. Hopfen-rundschau International 8–18.

Buggey, L., 2001. A review of polyphenolic antioxidants in hops, brewing and beer. The Brewer International 1, 21–25.

Deinzer, M., Yang, X., 1994. Hop aroma: character impact compounds found in beer, methods of formation of individual compounds. In: Eur. Brew. Conv. Monograph XXII, Symposium on Hops, Zouterwoude, pp. 181–197.

Forster, A., Beck, B., Schmidt, R., 1999. Hop polyphenols – do more than just cause turbidity in beer. Hopfen-rundschau International 68–74.

Goldstein, H., Ting, P., Navarro, A., Ryder, D., 1999. Water soluble hop flavor precursors and their role in beer flavor. In: Proc. 27th Cong. Eur. Brew. Conv., Cannes, pp. 53–62.

Hanke, S., Kern, M., Herrmann, M., Back, W., Becker, Th, Krottenthaler, M., 2009. Suppression of gushing by hop constituents. Brewing Science 62, 181–186.

Hughes, P., Simpson, W.J., 1996. Bitterness of congeners and stereoisomers of hop-derived bitter acids found in beer. Journal of the American Society of Brewing Chemists 54, 234–237.

Hughes, P., 2000. The significance of iso-α-acids for beer quality – Cambridge prize paper. Journal of the Institute of Brewing 106, 271–276.

Kunimune, T., Shellhammer, T.H., 2008. Foam-stabilizing effects and cling formation patterns of iso-α-acids and reduced iso-α-acids in lager beer. Journal of Agricultural and Food Chemistry 56, 8629–8634.

Peacock, V.E., McCarty, P., 1992. Varietal identification of hops and hop pellets. Technical Quarterly – the Master Brewers Association of Americas 29, 81–85.

Peppard, T.L., 1981. Volatile sulphur compounds in hops and hop oils; a review. Journal of the Institute of Brewing 87, 376–385.

Pickett, J.A., Peppard, T.L., Sharpe, F.R., 1976. Effect of sulphuring on hop oil composition. Journal of the Institute of Brewing 82, 288–289.

Preis, F., Mitter, W., 1995. The rediscovery of first wort hopping. BRAUWELT International 13 (308), 310–311, 313–315.

Roberts, T.R., Wilson, R.J.H., 2006. Handbook of Brewing, second ed., pp. 177–279.

Siebert, K.J., 1994. Sensory analysis of hop oil-derived compounds. In: Quality Control, Eur. Brew. Conv. Monograph XXII, Symposium on Hops, Zouterwoude, pp. 198–215.

Smith, R.J., Wilson, D.D., Wilson, R.J.W., 2004. The Influence of Naturally Occurring Hop Acids on the BU Analyses of Dry-hopped Beer. World Brew. Cong. 2004 (poster 54).

Stevens, R., 1967. The chemistry of hop constituents. Chemical Reviews American Society 67, 19–71.

Taylor, D., Humphrey, P.M., Yorston, B., Wilson, R.J.H., Roberts, T.R., Biendl, M., 2000. A guide to the use of pre-isomerized hop pellets, including aroma varieties. Technical Quarterly – the Master Brewers Association of Americas 37, 225–231.

Walsh, A., 1998. The mark of nobility. An investigation into the purity of noble hop lineage. Brewing Techniques 60–69. Mar/Apr.

CHAPTER

4 Yeast

I. Russell
Heriot-Watt University, Edinburgh, Scotland

INTRODUCTION

One of the most difficult raw materials to manage in the production of beer, in terms of monitoring quality, is the yeast culture, because it consists of living organisms in which health and ability to ferment can be affected by a multitude of factors. It is difficult to define exactly "what is a good brewing yeast?" The yeast culture should healthy and free of contaminating bacteria and wild yeast. The yeast strain should have good resistance to numerous stress factors (eg, temperature, ethanol, pH, osmotic pressure) with no overt weaknesses in terms of nutrition (oxygen and carbohydrate uptake, nitrogen, mineral requirements, etc.). The yeast strain should consistently develop the desired flavor profile in the beer, have the required wort attenuation ability, be easily separated from the wort at the end of fermentation for reuse either by its flocculation ability or by centrifugation, and should have good long-term storage ability between fermentations.

YEAST LIFE SPAN

Yeast has a finite replicative life span that is a function of the number of divisions (generations) that the yeast has undertaken. Once it reaches its preset limit, it enters senescence and eventually dies. Under optimal conditions, a typical yeast cell will divide to produce a new daughter cell every 90–120 min (depending on the strain, environmental conditions, and temperature). Polyploid brewing yeast has a life span varying from a high of 21.7 ± 7.5 divisions to a low of 10.3 ± 4.7 divisions (Smart, 2000). It is possible to approximate the age of a yeast cell by counting the number of bud scars. The bud scars are rich in chitin and can be stained with calcofluor white and counted with a fluorescent microscope. Birth scars are low in chitin and fluoresce weakly and there is just one such scar per cell.

The following signs of aging can be seen when examining yeast cells using a light microscope: an increase in bud scar number, cell size, surface wrinkles, granularity, and retention of daughter cells (a form of chain formation). As the mother cell ages, protein damage occurs within the cell. The mother cell tends to retain the problems and sends the daughter cell (bud) on its way in a healthy condition.

Brewing Materials and Processes. http://dx.doi.org/10.1016/B978-0-12-799954-8.00004-6

The newly formed yeast cell transports the damaged and aged proteins back into the mother cell, thus guaranteeing that the new daughter cell is born healthier than the parent cell (Liu et al., 2010). Aging causes a progressive impairment of cellular mechanisms, which eventually results in a yeast cell with a reduced capacity to adapt to stress and to ferment.

WORT CARBOHYDRATES AND YEAST

Yeast prefers simple sugars (glucose and fructose) and will always take these up first as these two sugars can enter the cell by facilitated (passive) diffusion and go directly into the various biosynthetic pathways (Fig. 4.1). Sucrose (which consists of a linked glucose and fructose molecule) cannot cross the plasma membrane. It is hydrolyzed into its two components by the invertase enzyme present in the intracellular thin space between the cell wall and the plasma membrane, and these two simple sugars can then easily enter the cell. It is more difficult for the largest concentration wort sugar maltose (which consists of two linked glucose units) to cross the yeast cell plasma membrane. This requires both energy and the strain must have the correct genetics in place (maltose permease genes for transport). Maltose uptake does not commence until most of the glucose has been taken up from the wort. The enzyme maltase (α-glucosidase) is synthesized by the cell to hydrolyze the maltose inside the cell into single glucose units, which the cell can then utilize. Maltotriose (which consists of three linked glucose units) is even more difficult for the cell to

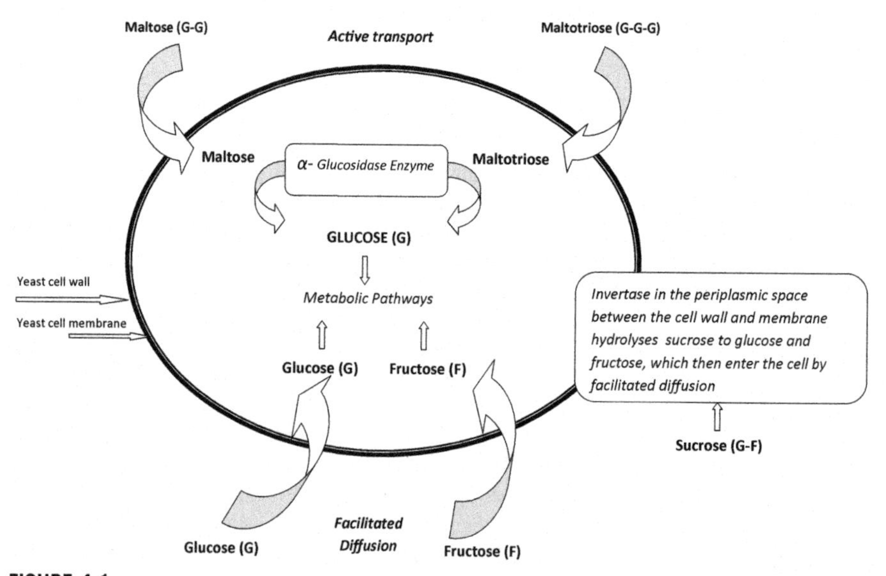

FIGURE 4.1

Wort sugar uptake by the yeast cell.

utilize. When the yeast is stressed, it will often not take up the sugar maltotriose, resulting in this sugar being left in the wort.

Dextrins are larger molecules (many linked and branched glucose units). Dextrins do not enter the cell and remain in the fermented beer and give the product body (calories) and mouthfeel. Some yeast strains (not typical brewer's yeast) have the ability to break down some of this dextrin material resulting in an atypically low final Plato and a beer that is thin on the palate and no longer within specification. This is usually due to a wild yeast infection that has gone undetected, with the wild yeast having the ability to metabolize some of the dextrin material.

HIGH-GRAVITY BREWING

There is no doubt that high gravity brewing exerts extra stress on the yeast both in terms of the osmotic pressure encountered early in the fermentation and the ethanol effects on the cell membrane later in the fermentation. With high-gravity brewing, it is even more important that the pitching yeast be very healthy because it is entering into such a stressful environment (Stewart, 2014a). Extra nutrients (especially nitrogen and oxygen) are often needed for the yeast to function optimally. Yeast does not multiply as well under high-gravity conditions so an increase in pitching rate is required. For further details on high-gravity brewing effects on yeast cells, this topic is covered in greater depth in the book *Brewing Intensification* by Stewart (2014a).

OXYGEN DEMAND AND QUALITY CHECKS

Part of ensuring that the yeast culture is healthy is to monitor the oxygen levels during the yeast pitching process. This is the only stage of the wort fermentation process in which the brewer wants oxygen present and the yeast will remove that oxygen from the wort very quickly (in minutes, not hours). An oxygen meter is needed to accurately measure the dissolved oxygen (inline). With higher-Plato wort, it is more difficult to ensure that the dissolved oxygen is at the correct level, and air alone will not suffice to oxygenate the wort. It can be difficult to get enough oxygen into the wort, and when higher levels are required, pure sterile oxygen rather than air must be used. Yeast needs oxygen to synthesize its lipids; lipids are essential for yeast growth. If too little oxygen is present, a poor yeast crop results. The presence of too much oxygen results in an excess of yeast growth with negative effects on beer flavor and stability.

A general guideline on wort oxygenation is 1 ppm of oxygen for each degree Plato. However, it is important to remember that the genotype of the particular strain will affect its specific oxygen requirements, and scientists have grouped brewing yeast into four categories in terms of oxygen demand, from low to very high. Nevertheless, the situation is even more complex and involves the exact timing of wort oxygenation and whether that oxygen is given in stages over time. These factors are strain and environment specific, and once the optimum is determined, monitoring must be put in place to ensure there is consistency in oxygenation from brew to brew to maintain a consistent beer flavor profile.

Not enough oxygenation at pitching can lead to a host of problems such as poor yeast growth and thus poor attenuation, undesirable flavors, aroma compounds, and a yeast population that will perform poorly on repitching. Underoxygenation is a bigger concern than slight overoxygenation. The yeast will very quickly remove all of the oxygen from the wort when pitched. When using air to oxygenate, approximately 8 ppm is the maximum amount of oxygen that will dissolve in a 10° Plato wort at 15°C (Baker and Morton, 1977). To obtain higher values, pure oxygen must be used. In addition, as the temperature and gravity of the wort increases, oxygen solubility decreases.

In summary, it is necessary to determine the best procedure for wort oxygenation for a particular yeast strain, and the particular brewery equipment in use, and then be consistent in when and how the wort is oxygenated. It is also important to ensure that the oxygen level in the wort can be reliably measured with an oxygen meter to maintain the levels at the ideal starting concentration.

WORT NITROGEN AND YEAST

Nitrogen is essential for yeast growth, and the wort must have an adequate amount of useable nitrogen for a successful fermentation. With all malt worts, this is rarely a problem, but with the use of adjuncts and high-gravity brewing, it is important to monitor that the required amounts of nitrogen are present. The genetics of the yeast's nitrogen uptake system are complex (Lekkas et al., 2007). Certain nitrogen sources are preferred (ammonia, urea, amino acids and small peptides), and these will be taken up first and in the order described by Jones and Pierce (1964). Yeast uses the nitrogen in the wort as a nutrient and assimilates it into proteins, RNA, DNA, and other compounds, but first the nitrogenous compounds must traverse the plasma membrane. They do this with the help of specific permeases embedded in the plasma membrane. Yeast can synthesize its amino acids from simple carbon skeletons plus a nitrogen source and assimilate sulfur from the medium to produce the sulfur-containing amino acids (cysteine, methionine, homocysteine).

Free amino nitrogen (FAN) refers to the group of nitrogenous compounds available for consumption by the yeast. It is the sum of the individual wort amino acids, ammonium ions, and small peptides (di-, tripeptides). FAN can be measured using standard chemical methods (Lie, 1973) with a target of 160–190 mg/L wort (based on 1040 wort). The wort FAN has a direct influence on beer quality but optimization of the nitrogen content of wort is a complex issue and knowledge is still inadequate in this area. It is not only a matter of the initial wort FAN content, but equally the amino acid and ammonium ion equilibrium in the medium, and a number of yet undefined fermentation parameters that will influence the final beer flavor and quality.

WORT MINERALS AND YEAST

Zinc is an essential element for yeast. It acts as a co-factor for many yeast enzymes, including the zinc-metalloenzyme alcohol dehydrogenase, the terminal step in yeast alcoholic fermentation. Zinc also helps with stress tolerance. A lack of zinc in the wort can prevent yeast from budding. Typical zinc levels in wort can vary widely depending on the brewing raw materials. How much zinc a strain requires is determined to some extent by the particular strain's genetic makeup, with some strains being very sensitive to low levels of zinc in the wort. Wort zinc levels required for optimal fermentation range from 0.48 to 1.07 ppm (De Nicola and Walker, 2011). Stuck fermentations (incomplete) can often be resolved by the addition of a source of bioavailable zinc (usually zinc chloride or sulfate).

Magnesium is also important for key metabolic processes and is directly involved in ATP synthesis. If magnesium is absent, the yeast cannot grow. Yeast has a high demand for magnesium but not for calcium. A high magnesium-to-calcium ratio in the wort is important, as high calcium levels can antagonize magnesium-dependent functions. Magnesium, similar to zinc, also protects the cell from stress (Walker et al., 2006). Enough manganese in the wort is also important when the wort is deficient in FAN (Bourke and Datsomor, 2013). The control of flavor for beer consistency depends on a number of the metallic ions in the wort, in which either too little or two much can pose problems.

FINAL BEER pH

It is a given that closely monitoring the pH of the mash, the wort, and the pH during fermentation is key to a successful final product. However, what is the influence of the final beer pH on quality parameters? As the yeast ages, it becomes increasingly difficult for it to pump hydrogen ions from the cell into the beer (lowering the beer pH while maintaining constant the cell's intracellular pH) as this is a process that requires energy. Consequently, healthy fresh yeast will lower the wort pH much faster than older cells of poorer viability. The pH of the final beer will vary with the beer type but all malt lager beers are usually in the pH range of 4.25–4.6, with adjunct beers sometimes being as low as pH 4 (lower adjunct buffering capacity). Ales tend to have a wider pH range. A lower-final beer pH favors faster maturation, faster uptake of diacetyl, and better clarity. The beer pH affects flavor stability, taste, and drinkability. Lower-pH beers are often rated more positively than the higher-pH beers with comments of "crisp lively" beer, whereas the higher-pH beers tend to elicit comments of "dull flavored" beer. Again, much will depend on the actual beer. For example, beers made with dark malts tend to have lower, more-acidic pH levels, and slight pH differences will affect the hop components, and the bitterness of the flavor. An unusual decrease in the beer pH is always of concern in case bacterial contaminants are producing the acid. Sour beers such as Lambic beers can have a pH as low as 3, this being a desirable pH for these unique products.

WASHING THE YEAST

Although microbial contamination is covered in "Chapter 14, Microbiology", it is important to discuss the technique of yeast washing in specific detail. It is preferable to be in the situation in which it is not necessary to wash a yeast culture to remove contaminating bacteria, but when yeast washing is needed, careful adherence to the methodology will yield yeast that is cleaner and that will be useable when there are no other options for pitching a particular brew. Washing will not remove all of the bacteria but it will reduce the bacterial load present. Some breweries follow the practice of always washing their yeast to remove some of the bacterial contaminants. Indeed, some breweries maintain that their beer "does not taste right" unless their yeast culture is acid washed every cycle! It is important to remember that washing will not remove wild yeast—only bacteria are removed—as bacteria are more susceptible to a lower pH than yeast. What is key is to follow the procedure carefully with the emphasis being on controlling time, temperature, agitation, and correct addition of the washing solution.

A number of methods can be used including the use of sterile water or acidified ammonium persulfate. A commonly used method is described by Simpson and Hammond (1989) using cold phosphoric acid. A cooled yeast slurry (in water or beer at 2–5°C) is acidified with cold food-grade dilute acid (citric, tartaric, and sulfuric are also options) to a pH of 2.0/2.2 (important to check the pH to make sure it is correct) and agitation should be continuous through the acid addition stage and during the wash. The yeast is exposed to the solution for a maximum period of 2 h and then it is pitched right after the washing stage. Washing an unhealthy yeast is not recommended and if it is yeast from a high-alcohol fermentation, extra care must be taken to reduce the alcohol concentration to 5% (v/v) or less (Cunningham and Stewart, 2000).

There are some alternatives to acid washing. One is the use of chlorine dioxide, which is available for this purpose from various manufacturers. With this method, it is even more critical to take care as it is very easy to kill the yeast in the washing process if there is poor mixing or if the concentration of the chlorine dioxide is not properly adjusted for the amount of yeast present. Instructions must be followed very closely especially in terms of time, temperature, and pH in order not to damage the yeast cells in the process (Johnson, 1998; White and Zainasheff, 2010).

Breweries have different rules for washing their yeast. Some never wash their yeast, some only wash when there are high levels of contamination, and some wash their yeast after every cycle. The choice is very dependent on the particular yeast strain and the environment. Proper washing has been shown to have no negative effects on the yeast but many things can go wrong with improper washing techniques resulting in problems with viability in the subsequent fermentation when using incorrectly washed yeast (Cunningham and Stewart, 2000).

DIFFERENTIATING BETWEEN PLANT ALE AND LAGER YEAST STRAINS (AND WILD YEAST)

Assuming that a brewer is using a single pure culture for a particular beer, it is important to have quality tests in place to ensure that the culture is not contaminated with another yeast, either a nonbrewing strain (wild yeast from the environment) or with a brewing strain from a different beer being produced in the same facility.

To test if a lager yeast strain is contaminated with an ale yeast strain, plate growth on nutrient agar at 37°C can be used. Lager strains do not have the ability to grow above 37°C. Plating the lager strain onto agar plates of peptone yeast extract nutrient media or malt extract, yeast extract, glucose, peptone media and then immediately (not longer than 10 min after inoculation) incubating that plate in an incubator at 37–39°C permits you to determine if there is contamination from an ale strain, as the only growth of colonies on the plate will be from any contaminating brewery ale strains or wild yeast. Contaminating ale/wild yeast will grow on the plates incubated at that temperature (ASBC Method of Analysis Yeast 10B, 2011). If only lager yeast is present in the tested culture, nothing will grow on the 37°C nutrient agar plates.

The reverse—determining if your ale culture is contaminated with a lager strain is more difficult. A number of media can be used to detect wild yeast; some of the more popular ones include lysine media, Lins wild yeast differential media, and copper sulfate media (ASBC Methods of Analysis—Microbiological Control 5, 2011; also see Chapter 14: Microbiology).

To test if an ale yeast is contaminated with lager yeast, the X-α-gal test plate media can be utilized. Lager yeast possess the enzyme melibiase (which ale yeast do not) and by using an X-α-gal medium—any lager yeast colonies present will appear blue-green in coloration on the plate, as lager yeast can break down the indicator compound (ASBC Method of Analysis Yeast 10A, 2011).

Giant-colony morphology can be used when time allows. Colonies of the yeast are grown on very large plates with a gelatin medium for 3–4 weeks. This results in very distinctive colony morphologies (Fig. 4.2). Because the plates have to be grown over a longer period, this is not a rapid test but still a useful one for examining the purity of a culture (ASBC Methods of Analysis, Yeast 9, 2011).

Morphology on Wallerstein Laboratory Nutrient media can also be used to differentiate strains (European Brewery Convention (EBC) Analytica Method 3.3.2.2). Distinguishing wild yeast from the plant yeast often requires a mix of methods and media depending on the contaminating yeasts (see Chapter 14: Microbiology for list of media).

DNA fingerprinting is a rapid technique to differentiate and identify yeast strains and a number of molecular biology methods are available that can be used. One that has been shown to be simple and reproducible is the ASBC Method (ASBC Methods of Analysis – Yeast 13, 2011). As molecular biology methods become simpler to use and are marketed in user friendly kits, and as genome sequencing continues to drop in price, one can expect to see many more new molecular biology methods that were once only feasible in research environments adopted as routine in brewing

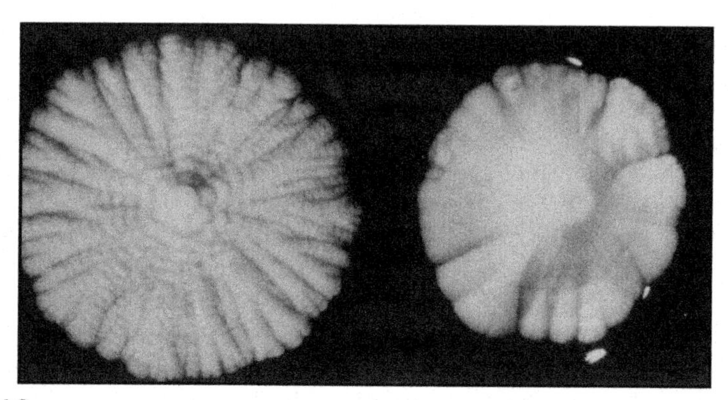

FIGURE 4.2

Example of two ale yeast morphologies after three weeks of growth on a giant-colony plate.

quality laboratories (Driscoll, 2014; Vogeser, 2014). Please see "Chapter 14, Microbiology" for a more detailed discussion on molecular biology methods for brewing yeast and bacteria.

DETECTION OF WILD YEAST

It is very easy to determine if a lager strain is contaminated with an ale strain using a simple plate technique. All media formulations have advantages and disadvantages, which is why more than one media type is usually employed in the quality control laboratory to ensure that no wild yeasts are overlooked. In addition, a number of molecular biology techniques can be used to identify wild yeast contaminants (Pham et al., 2011; see also Chapter 14: Microbiology) and new molecular methods are constantly being published as molecular biology methods become adapted for use with brewing yeast strains.

BRETTANOMYCES AND SPECIALTY BEERS

Specialty beers such as Belgian ale styles (Lambic, Gueuze) utilize yeast and bacteria outside of the normal range of *Saccharomyces* brewing strains. Many craft beers are now produced with cultured strains of *Brettanomyces* and lactic acid bacteria that function in conjunction with *Saccharomyces*. The *Brettanomyces* are slow growers and can utilize carbohydrates, such as starch, that are left behind by *Saccharomyces* yeasts. Contamination of regular beers is usually not an issue because *Brettanomyces* is such a slow grower. *Brettanomyces* can produce lactic and acetic acid, as well as a number of traditional aroma compounds such as ethyl acetate, and some untraditional aroma compounds for mainstream beers. Some of these aromas are pleasant (sour cherry pie) and some are not so pleasant (sweaty horse, mousey). Beer fermented/aged in wooden vessels is often tainted by *Brettanomyces,* because it can survive for extended periods in the wood. For more information on recognizing and testing for *Brettanomyces,* the reader is directed to a recent review by Schifferdecker et al. (2014).

KILLER YEAST STRAINS

Although brewing production strains are not routinely screened to see if they contain the killer factor [encoded by double-stranded RNA (dsRNA) virus-like particles]—from a quality point of view, this would be good information to have about the yeast used on a daily basis in the brewery. It only needs to be carried out once on each of the strains in use and is a simple plate test (ASBC Method 8, 2011). Wild yeast very often are "killer positive" and if the brewing yeast in use is "killer sensitive", the fermentation is at a much higher risk from a wild yeast infection because a "killer positive" strain can secrete a compound (harmless to humans) that can very quickly destroy a "killer sensitive" population and take over the fermentation. A 1% deliberate contamination of fermentation with a "killer positive" yeast has been shown to kill the entire production ale yeast in that fermentation in less than 24 h (Russell and Stewart, 1985).

PHENOLIC OFF-FLAVOR

Many wild yeasts contain a gene for phenolic-off-flavor (POF), which allows them to decarboxylate ferulic acid (naturally present in wort) to a compound that produces a strong phenol note in the beer. In most beers, this is considered a defect and a serious problem (Ryder et al., 1978). However, in certain specialty beers, this can be a desirable characteristic and yeasts are selected with this particular decarboxylase activity. Again, if such a yeast is present in the brewhouse, care must be taken to avoid cross contamination with the strains without this characteristic, as often these POF-positive strains are also "killer positive" strains.

FLOCCULATION AND YEAST CROPPING

Selecting a strain with the correct flocculation characteristics should in theory be relatively easy, but because the yeast is a living evolving organism this characteristic is one that can change easily and result in fermentations that encounter problems. This happens because either it is hard to induce the yeast to fall out of suspension (if you are relying on natural flocculating ability that has mutated to nonflocculent) or the yeast has mutated to an even more flocculent form and falls out of suspension before the fermentation is complete. Not only is there the issue of the genetics of the yeast strain [what flocculation (FLO) genes does it contain—and there are many], but how do these genes interact with the environment? For a recent review on yeast flocculation, see Vidgren and Londesborough (2011). The nutrients present (or lacking) in the fermentation medium (especially minerals such as Ca, Mg, and Zn), the time at which the yeast has been harvested, the storage method, mutations in the strain (eg, respiratory deficient/petites tend to be less flocculent), malt issues—all can change the flocculation behavior of the strain.

Flocculation has many definitions but for brewing this one captures the concept well: "it is the phenomenon wherein yeast cells adhere in clumps and either sediment from the medium in which they are suspended or rise to the medium's surface" (Stewart et al., 2013). The correct flocculation properties are needed to crop a yeast culture at the end of primary fermentation so that it can be reused in subsequent fermentations. Flocculation changes in a yeast culture can happen for many reasons and even the method of harvesting can select specific populations (Stewart, 2014b). Premature yeast flocculation (PYF) can also be malt related (Kaur et al., 2012). How quickly mutations occur in the yeast's flocculation ability is very strain dependent with some strains being more stable than others.

It is important when slight flocculation problems start to be observed to examine the possible causes before they progress into major issues. What has changed in the environment? This is when excellent record keeping can help to identify the cause. Regular use of a fresh yeast culture after a set number of repitchings avoids using yeast that has accumulated too many harmful mutations over time.

YEAST CROPPING FROM A VERTICAL FERMENTOR

Ensuring that the yeast is cropped from the middle of the cone, and not the very top or the very bottom of the cone of a cylindroconical fermentor, is important. If the yeast is harvested from the very bottom, you are selecting yeast cells that have fallen out of suspension early in the fermentation. If you crop from the very top, you will select those that have remained in suspension too long.

Collecting the active healthy cells from the middle of the cone for repitching will yield the highest chance of a good viable pitching yeast culture with the desired flocculation profile. In addition, the very bottom of the cone is where the trub has accumulated. By not using this portion when collecting the yeast crop, there is less trub carryover into the next fermentation.

When harvesting from an ale fermentation vessel, in which the yeast rises to the surface of the fermentation, the middle skim should be saved for repitching.

It is good practice to avoid harvesting yeast from an atypical fermentation. The collected slurry should also be relatively free of trub and exhibit no off aromas when sniffed. Visual inspection of a sample of the slurry in a beaker, as well as microscopic examination, is important.

GLYCOGEN

Glycogen is a polysaccharide consisting of multibranched chains of α-1,4-linked glucose residues. It is an energy-storage form of glucose and an analog form of starch. Over 20% of the yeast dry weight can be glycogen and as the major storage polysaccharide for yeast, it serves as a system of reserve carbohydrate that the cell can call on early in the fermentation, as it can be quickly mobilized when there is a sudden need for glucose for a healthy start to fermentation. Glycogen is a main source of energy during the lag phase when there is a need to build cell components and it is rapidly used up in the first few hours of fermentation and not replenished again until later in the fermentation. During storage, glycogen is used for cellular maintenance, so it is important that after storage the yeast's glycogen reserves are replenished before it is pitched into a subsequent fermentation. Low glycogen levels in pitching yeast usually result from poor yeast-handling practices (high storage temperatures, long time periods, too much agitation during storage) and, as a consequence, the results are lower cell viability levels, longer fermentation times, and at the end of fermentation higher levels of diacetyl, acetaldehyde, and sulfur dioxide.

A simple iodine stain test (Lugol's iodine) using a light microscope has been used in the past to examine the yeast. Glycogen in the cell will stain dark brown and yeast cells low in glycogen will only stain very pale brown, but this is not a very reliable quality test. More sophisticated glycogen tests can be performed in a number of ways but results are still difficult to interpret. Cahill et al. (2000) looked at the pattern of glycogen assimilation within a brewing yeast population using image analysis and an iodine stain and suggested that this particular test could be a useful indicator of yeast quality; however, they noted that just determining the mean cell glycogen content as an indicator of overall yeast health is of limited value.

TREHALOSE

Trehalose, a disaccharide, also acts as a storage carbohydrate for the cell but for a very different purpose from glycogen. A molecule of trehalose consists of two glucose residues linked by an α-1-1 glycosidic bond. It plays a very important role as a stress protectant in the yeast, especially against osmotic stress and ethanol stress. Trehalose has a stabilizing effect on the plasma membrane, providing it with increased tolerance to desiccation, dehydration, temperature changes, and high temperature. When buying dried yeast that will require rehydrating, the levels of trehalose in the yeast are very important to maintain viability during storage. Disappearance of trehalose correlates with a rapid loss in viability. When rehydrating yeast, it is important to pitch it into wort promptly (following the manufacturer's instructions for procedures), as on reactivation the yeast will start to use its store of trehalose and it will quickly weaken if there is not a fresh source of carbohydrate available for it to use.

PITCHING RATE AND CONSISTENCY

To have consistent fermentation, it is important to be able to calculate the correct pitching rate for the particular wort, in the specific brewing environment (micro- or megabrewery). Too low a pitching rate can cause problems due to underattenuation and long lag times and affect the flavor of the final product. There can be an increase in esters and volatile sulfur compounds. There is also a higher risk of infection by bacteria if the yeast is not fermenting rapidly and vigorously as the environment remains friendly to bacteria longer. Overpitching can lead to yeast autolysis flavors, poor head retention, and again flavor differences, such as low esters.

There are many ways to calculate the pitching rate once the desired cell number is known from experience with the different levels of pitching researched as optimal for the particular product. Not only must the number of cells pitched be correct but they must be in a viable state and able to reproduce. Cell counts can be accomplished in a number of ways—simple cell counts under the microscope using a hemocytomer and methylene blue stain (ASBC Methods of Analysis-Yeast 3) or methylene violet (EBC Analytica Method 3.2.1.1)— then a calculation of how many cells are present that are viable. In addition, automated cell counts using advanced staining techniques are available (Hutter et al., 2005).

Pitching by wet weight is carried out by some breweries. A yeast cell weighs $\sim 8 \times 10^{-11}$ g. If yeast slurry is allowed to settle—the general rule is that it will contain (depending on the size of the yeast) 1–3 billion cells per ml. Therefore, a slurry that is 40–60% yeast solids correlates to 1.2 billion cells/ml. When working with dried yeast, the general rule is that it contains ~7–20 billion cells per gram. A dried yeast must first be rehydrated, and viability counts are very important to calculate what is required for the optimal pitching concentration.

For yeast being repitched from a previous fermentation, a general brewer's guideline is to use 1×10^6 cells per degree Plato (and with oxygen a similar guideline is 1 ppm of oxygen for each degree Plato—so a 12°P wort would be pitched to a level of 12×10^6 cells/ml and 12 ppm oxygen/ml and an 18°P wort would require 18×10^6 cells/ml and 18 ppm oxygen, etc.). Ales tend to require a slightly lower pitching rate than lagers (not surprising because the fermentation progresses at a warmer temperature). The pitching yeast from a previous fermentation should always be used as soon as possible and stored cold (1–2°C) in an oxygen-free environment, with care taken not to accidently freeze it. Storage can be under a layer of water or beer—the goal being to avoid the presence of oxygen during the storage phase. A blanket of CO_2 with low positive pressure and minimal trub levels all help with the yeast storage process.

THE MASTER CULTURE—CRYOGENIC STORAGE

Although many brewers buy their yeasts on a regular basis from a yeast manufacturing company in the business of supplying such a service and can choose their yeast strains from the large catalogs available—some prefer to employ the yeast that they have been using historically and to keep their cultures stored locally in their quality control laboratory. It is advised that the yeast should also be kept in an offsite location under cryogenic storage conditions so that it is always possible to go back to the original culture should the need arise. Cryogenic storage (−196°C under liquid nitrogen or in a −80°C freezer) is the storage method of choice for brewer's yeast (EBC Analytica Method 3.4.1) as it holds the yeast in suspended animation with few genetic changes (if any) over time and also minimal cell damage on resuscitation from storage (Russell and Stewart, 1981). The storage technique that should be avoided with brewer's yeast strains is freeze-drying—it results in very low cell viabilities.

After recovering a culture from ultralow temperature storage, for convenient short-term storage (up to 6 months) for a master culture, a large number of nutrient agar slopes can be prepared and stored in a refrigerator at 4°C (±2°C). In this way, one slope can be used each time to start a new propagation. It is important to remember that this slope should not be subcultured repeatedly; it is intended for one-time use (EBC Analytica Method 3.4.2, 2011d).

YEAST PROPAGATION FROM SLOPE (LABORATORY STAGE)

Propagating yeast from a slope to obtain sufficient yeast for pitching a propagation tank in a brewery requires an extra measure of care. Only sterile equipment should be used, the manipulation for scale up is performed under aseptic conditions (preferably in a laminar flow hood if possible), and the media should be autoclaved to ensure no contamination enters in these very early laboratory stages. Orbital shakers are ideal for the first stage, as stirring on a magnetic plate can cause the temperature to rise. Cells are transferred to the next scale while in exponential growth phase. The dilution during scale up is usually kept to 10–20 times, as a large jump in dilution will result in poor growth. Details of how best to perform this propagation are given in EBC Analytica 3.4.4 (2011f).

KNOWING YOUR YEAST STRAIN'S BACKGROUND

DNA fingerprinting of a proprietary strain is a service offered by many yeast storage companies and allows for the identification of proprietary strains. It is always useful to know as much as possible about a strain's background, its sugar fermentation ability, flocculation properties, growth temperature characteristics, ability to produce phenolic off flavors, and whether it contains the "killer factor", which makes it resistant to being killed by other wild yeast. Yeast colony morphology can be useful for identification using giant colony plates. Ale strains vary greatly in their morphology and are very distinctive and recognizable from each other, whereas lager strains tend to be smooth with less variation and are more similar in appearance, which is not surprising based on what is known regarding their evolutionary history.

The rate at which a culture forms respiratory-deficient mutants is important to know especially because after storage these numbers can increase. The mutant cells (RDs) tend to produce smaller sized colonies. These cells often have altered flocculation and sugar uptake characteristics (Stewart, 2014c), so when selecting colonies to use for pitching from agar plates, avoid small colonies. The percentage of these colonies present can be ascertained using the triphenyltetrazolium chloride overlay method (ASBC Methods of Analysis Yeast 3C, 2011).

The previous are all pieces of knowledge that should be known about the master yeast strains that are in use in the brewery.

PURCHASING YEAST

The quality of the yeast purchased for pitching a brew will depend largely on how it is stored before use. If the yeast is in the form of active dried yeast and packaged with a nitrogen flush or under vacuum and stored cold, it will lose viability very slowly. However, if moisture and air enter the package, the viability will drop much more quickly. Proper storage is key to its use in successful fermentations. Documentation of the storage conditions (age of the product and monitoring of storage temperatures) is part of a good quality system.

YEAST SHIPPING FROM CENTRAL FACILITIES

Yeast can be at a high quality when it is first produced in a central brewery or culture plant but lose viability or grow unacceptable levels of bacteria in transit. Common sense precautions should be taken when a shipping process is used. The culture strain that is selected should be free of contaminants. Any residual fermentable material should be removed by washing with chilled sterile water prior to pressing or centrifugation to avoid growth of bacteria enroute. The yeast should be chilled and stored at 0°C prior to shipping. If it is a short distance (time), a slurry can be used, but for longer times a pressed cake (~70%) is better. It is easier to keep smaller containers of yeast cooled for shipping than larger containers.

On receipt, the culture should be examined microscopically and monitored for temperature, viability, and bacterial contamination before the culture is used. For shipping small quantities of yeast, similar rules apply (EBC Analytica Method 3.4.3, 2011e). Disposable time/temperature monitors can be attached to the shipping container to record and alert you regarding any extremes that occurred in temperature during transport. The receiver must also be aware that the shipment is arriving. Shipping over a holiday or to a location where it sits on a receiving dock in the heat or cold for any length of time must be avoided at all costs.

VIABILITY AND VITALITY

It is important to understand what is meant by the terms viability and vitality. Yeast that is viable is an organism that still has the ability to perform certain functions such as fermentation, although it may not be able to reproduce (bud), because there is a maximum to the number of times a yeast can reproduce as discussed earlier. Vitality looks at the health of the yeast—how metabolically active it is. It is about the vigor of the cell, whereas viability is merely about whether it is "dead or alive."

Yeast can be viable but not very vital. An analogy to illustrate the concept is a healthy 20 year old female athlete jogging for one mile versus a 90 year old female athlete walking to cover that same one mile distance. Both can accomplish the task, but there are differences in their vigor. There are also differences in their genetic ability to reproduce due to chronological age!

When yeast is pitched, the goal is not only a high number of viable cells but also cells with high vitality. There is not yet an ideal method that is widely accepted that gives us total cell count, viable cell count, and vitality—all in one easy and inexpensive test. However, there are a number of good methods, some are automated, that allow you to set a baseline for your particular yeast and then any change from that is an alert that there could be a possible problem (Heggart et al., 1999, 2000; White et al., 2008).

There are a number of staining methods used to determine viability, the most common being methylene blue, in which in theory only dead cells stain blue (Fig. 4.3), and methylene violet, which is said to provide less ambiguity in color

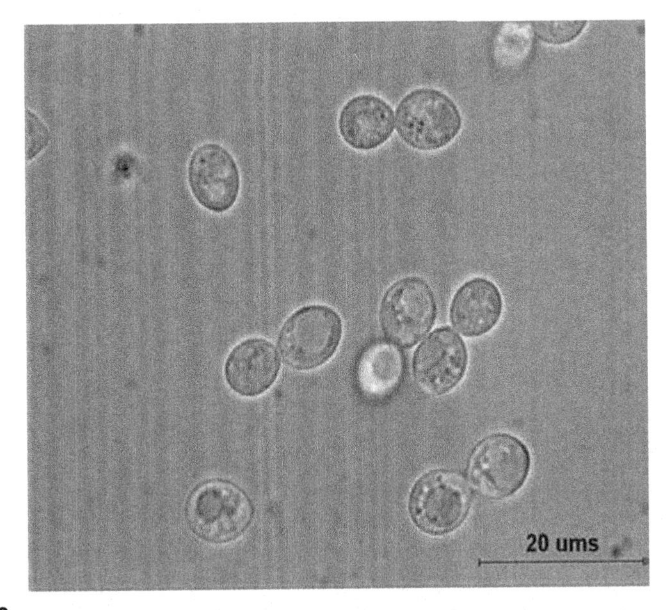

FIGURE 4.3

Methylene blue-stained yeast cells viewed under a light microscope.

variation. In addition some researchers prefer the stains fluorescein diacetate (FDA), acridine orange, or alcian blue. There is no ideal staining method, but for counting cells under the microscope, methylene blue due to its simplicity remains popular. However, there are a number of rules to remember when using it. As the viability of the cells decreases, the stain becomes less and less accurate; at more than 15% dead cells (85% viable count), the true dead cell numbers are much higher, so viability tends to be overestimated as viability decreases. The methylene blue stain can report viabilities as high as 30% to 40% when the true viability is 0%, so it is important to remember how unreliable this test is with cultures that are in very poor shape and it is not to be relied on under those circumstances (White et al., 2008). The stain is also toxic to the cell and counts should be performed within 10 min of mixing the stain with the yeast. The methylene solution should be made up fresh at least every 3 months, as in liquid form it will deteriorate over time (the powder is shelf stable).

There are methods that look at whether the yeast can replicate itself. A dead yeast cell cannot produce a colony-forming unit. The plate culture method is helpful to look at vitality but there is a time delay, so this method's results can only be viewed retrospectively. Also this method does not capture how many cells are indeed still viable and can ferment but just can no longer replicate. However in terms of time, the microscope slide culture technique can be very helpful as one can see within a few hours what the budding % capability is in a yeast population and a high % of cells capable of budding suggests that it is a very vital culture (ASBC Methods of Analysis-Yeast 6, 2011).

There are a large number of metabolic activity tests in the literature testing different aspects of the cell to try to capture the concept of vitality. They include some of the following tests: adenosine triphosphate (ATP) content, vicinal diketone (VDK) reduction, intracellular pH (ICP), enzyme activity (alcohol dehydrogenase (ADH), pyruvate dehydrogenase (PDH), pyruvate decarboxylase (PDC), maltase, protease, etc.), oxygen reduction capacity, oxygen consumption rate (OCR), ethanol, and CO_2 generation rate (respirometer), potentiometric titration, acidification power test (APT), and magnesium release. Flow cytometry to monitor yeast cell physiology is a technique that holds promise for being adapted for use in a quality control laboratory (Sugihara et al., 2006; Chlup et al., 2007). Often, a combination of the above methods is required to obtain a truly accurate picture of the yeast's state of metabolic activity.

YEAST AUTOLYSIS

Yeast autolysis can occur in any fermentation if the yeast is not very healthy when pitched and/or is subjected to high stress during the fermentation—whether that stress is high temperature, high pressure, high gravity, ethanol, old age, or a number of combinations of stress factors acting together. When yeast autolysis occurs, the entire cell contents spill into the beer, including many enzymes. Proteases can degrade the beer foam proteins, fatty acids from the spill can lead to off-flavors, and autolyzed yeast can lead to hazes in the beer.

TRACKING THE HISTORY OF THE YEAST

Tracking is one of the most important aspects of any yeast quality program. It is important to track the yeast from the original starter culture, to recycling it into the next fermentation, and to not only keep accurate notes but to review the notes consistently and to use them to troubleshoot when poor fermentations are encountered. A good yeast tracking system involves documenting the entire yeast's history carefully so that when problems occur it is possible to look at that history for clues as to why fermentation problems might have arisen.

The tracking record (handwritten or electronic) is critical to a good quality system. Tracking for the propagation stage should include the following: strain identification code (and source); starter culture if slope—stored how, where, and how long before use; starter if bulk purchase—stored how, where, and how long before use; details of revitalization for propagation; date of propagation from starter; length of time in propagator; temperature; oxygen; final cell count; and viability.

Tracking for the yeast collection stage should include at minimum the following: brew and fermenter number from which the yeast was harvested; age of the yeast; brew details—style, volume, date, temperature of fermenter at harvest; was the yeast washed? If yes, record time, temperature, and pH; cell count and viability at pitching; volume of yeast pitched; any special visual, taste, or smell observations noted; laboratory test results.

Tracking the fermentation performance of the yeast over time serves as an alert if differences are noted from the normal pattern (Bamforth, 2009). Examples of useful data to track include: wort cooling and run-in time fermenter volume, wort aeration, daily temperature, and gravity readings, time to half gravity or set target, the final gravity, cooling time and date, sterile wort sample and forced ferment sample, trub and yeast removal details.

When the viability of the pitching yeast is not ideal, the number of cells can be increased to compensate, but if increased too much then flavor differences in the final product will start to be seen. Generally, a cycle of five times can be used if the goal is to avoid bacterial problems and 5 to 10 cycles is the maximum when trying to avoid genetic drift issues (a problem with some strains, but not all).

Every brewery has its own parameters for the number of times a yeast crop can be harvested and repitched. The key is to always be on the alert for a reduction in performance, which can be due to a number of factors such as mutations in the strain. However, it is very often just due to mishandling of the yeast at some stage of the process, such as incorrect washing, bacterial contamination, or nonadherence to good harvesting and storage techniques.

SUMMARY

It is always important to remember when you are dealing with yeast and quality issues, stress factors are additive. Although it can appear that everything is close to specifications in terms of temperature, time, pH, osmotic pressure, etc. if too many of the stress factors are present at once—they can have a significant effect on yeast performance. It can be very difficult to trace the problem to only one source in the quality laboratory because it is not just one source! Often, it is a cumulative effect that has built until the yeast is no longer able to function.

REFERENCES

ASBC (American Society of Brewing Chemists) Methods of Analysis, 2011. Yeast 9. Morphology of Giant Yeast Colonies. The Society, St. Paul, MN.

ASBC (American Society of Brewing Chemists) Methods of Analysis, 2011. Microbiological Control 5. Differential Culture Media. The Society, St. Paul, MN.

ASBC (American Society of Brewing Chemists) Methods of Analysis, 2011. Yeast 10. Differentiation of Ale and Lager Yeast. (A) By X-α-gal Medium. (B) By Growth at 37°C. (C) By Growth on Melibiose. The Society, St. Paul, MN.

ASBC (American Society of Brewing Chemists) Methods of Analysis, 2011. Yeast 13. Differentiation of Brewing Yeast Strains by PCR Fingerprinting. The Society, St. Paul, MN.

ASBC (American Society of Brewing Chemists) Methods of Analysis, 2011. Yeast 3. (A) Dead Yeast Cell Stain (B) Yeast Spore Stain (C) Triphenlytetrazolium Chloride for Identification of Respiratory Deficient Cells. The Society, St. Paul, MN.

ASBC (American Society of Brewing Chemists) Methods of Analysis, 2011. Yeast 6. Yeast Viability by Slide Culture. The Society, St. Paul, MN.

ASBC (American Society of Brewing Chemists) Methods of Analysis, 2011. Yeast 8. Killer Yeast Identification. The Society, St. Paul, MN.

Baker, C.D., Morton, S., 1977. Oxygen levels in air-saturated worts. Journal of the Institute of Brewing 83, 348–349.

Bamforth, C.W., 2009. Practicalities of achieving quality. Beer – A Quality Perspective. Elsevier, UK, pp. 255–277.

Bourke, E.J., Datsomor, J., 2013. Effect of manganese and some FAN supplements on fermentations. In: IBD Africa Conference 2013. Paper available from: http://www.ibdafrica.co.za/wp-content/uploads/2013/05/Bourke-Manganese.pdf.

Cahill, G., Walsh, P.K., Donnelly, D., 2000. Determination of yeast glycogen content by individual cell spectroscopy using image analysis. Biotechnology and Bioengineering 69, 312–322.

Chlup, P.H., Wang, T., Lee, E.G., Stewart, G.G., 2007. Assessment of the physiological status of yeast during high- and low- gravity wort fermentations determined by flow cytometry. Technical Quarterly of the Master Brewers Association of the Americas 44, 286–295.

Cunningham, S., Stewart, G., 2000. Acid washing and serial repitching a brewing ale strain of *Saccharomyces cerevisiae* in high gravity wort and the role of wort oxygenation conditions. Journal of the Institute of Brewing 106, 389–402.

De Nicola, R., Walker, G.M., 2011. Zinc interactions with brewing yeast: impact on fermentation performance. Journal of the American Society of Brewing Chemists 69, 214–219.

Driscoll, D., 2014. Leveraging next Gen sequencing to improve brewery quality control. The 2014 Brewing Summit Proceedings Chicago. Presentation A-27.

EBC Analytica, 2011a. Method 3.2.1.1 Methylene Blue/Violet Stain. Fachverlag Hans Carl, Nürnberg. Online from: http://www.analytica-ebc.com/.

EBC Analytica, 2011b. Method 3.3.2.2 Morphology on WLN Agar. Fachverlag Hans Carl, Nürnberg. Online from: http://www.analytica-ebc.com/.

EBC Analytica, 2011c. Method 3.4.1 Yeast Storage at Ultra-Low Temperatures. Fachverlag Hans Carl, Nürnberg. Online from: http://www.analytica-ebc.com/.

EBC Analytica, 2011d. Method 3.4.2 Yeast Subculturing for Short-Term Storage. Fachverlag Hans Carl, Nürnberg. Online from: http://www.analytica-ebc.com/.

EBC Analytica, 2011e. Method 3.4.3 Yeast Strain Transport. Fachverlag Hans Carl, Nürnberg. Online from: http://www.analytica-ebc.com/.

EBC Analytica, 2011f. Method 3.4.4 Yeast Propagation (Laboratory Stages). Fachverlag Hans Carl, Nürnberg. Online from: http://www.analytica-ebc.com/.

Heggart, H.M., Margaritis, A., Pilkington, H., Stewart, R.J., Dowhanick, T.M., Russell, I., 1999. Factors affecting yeast viability and vitality characteristics: a review. Technical Quarterly of the Master Brewers Association of the Americas 36, 383–406.

Heggart, H.M., Margaritis, A., Stewart, R.J., Pilkington, H., Sobczak, J., Russell, I., 2000. Measurement of brewing yeast viability and vitality: a review of methods. Technical Quarterly of the Master Brewers Association of the Americas 37, 409–430.

Hutter, K.-J., Miedl, M., Kuhmann, B., Nitzsche, F., Bryce, J.H., Stewart, G.G., 2005. Detection of proteinases in *Saccharomyces cerevisiae* by flow cytometry. Journal of the Institute of Brewing 111, 26–32.

Johnson, D., 1998. Coming clean: a new method of washing yeast with chlorine dioxide. New Brewer. September/October, Available online from: http://www.birkocorp.com/brewery/white-papers/coming-clean-a-new-method-of-washing-yeast-with-chlorine-dioxide/.

Jones, M., Pierce, J., 1964. Absorption of amino acids from wort by yeasts. Journal of the Institute of Brewing 70, 307–315.

Kaur, M., Bowman, J.P., Stewart, D.C., Sheehy, M., Janusz, A., Speers, R.A., Koutoulis, A., Evans, D.E., 2012. TRFLP analysis reveals that fungi rather than bacteria are associated with premature yeast flocculation in brewing. Journal of Industrial Microbiology and Biotechnology 39, 1821–1832.

Lekkas, C., Stewart, G.G., Hill, A.E., Taidi, B., Hodgson, J., 2007. Elucidation of the role of nitrogenous wort components in yeast fermentation. Journal of the Institute of Brewing 113, 183–191.

Lie, S., 1973. The EBC-ninhydrin method for determination of free alpha amino nitrogen. Journal of the Institute of Brewing 79, 37–41.

Liu, B., Larsson, L., Caballero, A., Hao, X., Öling, D., Grantham, J., Nyström, T., 2010. The polarisome is required for segregation and retrograde transport of protein aggregates. Cell 140, 257–267.

Pham, T., Wimalasena, T., Box, W.G., Koivuranta, K., Storgårds, E., Smart, K.A., Gibson, B.R., 2011. Evaluation of ITS PCR and RFLP for differentiation and identification of brewing yeast and brewery 'wild' yeast contaminants. Journal of the Institute of Brewing 117, 556–568.

Russell, I., Stewart, G.G., 1981. Liquid nitrogen storage of yeast cultures compared to more traditional storage methods. Journal of the American Society of Brewing Chemists 39, 19–24.

Russell, I., Stewart, G.G., 1985. Valuable techniques in the genetic manipulation of industrial yeast strains. Journal of the American Society of Brewing Chemists 43, 84–90.

Ryder, D.S., Murray, J.P., Stewart, M., 1978. Phenolic off-flavour problems caused by *Saccharomyces* wild yeast. Technical Quarterly of the Master Brewers Association of the Americas 15, 79–86.

Schifferdecker, A.J., Dashko, S., Ishchuk, O.P., Piškur, J., 2014. The wine and beer yeast *Dekkera bruxellensis*. Yeast. http://dx.doi.org/10.1002/yea.3023.

Simpson, W.J., Hammond, J.R.M., 1989. The response of brewing yeast to acid washing. Journal of the Institute of Brewing 96, 347–354.

Smart, K., 2000. The death of the yeast cell. In: Smart, K. (Ed.), Brewing Yeast Fermentation Performance. Blackwell Science, London, pp. 105–113.

Stewart, G.G., 2014a. High gravity brewing. In: Brewing Intensification. American Society of Brewing Chemists, St. Paul, MN (Chapter 2).

Stewart, G.G., 2014b. The concept of nature-nurture applied to brewer's yeast and wort fermentation. Technical Quarterly of the Master Brewers Association of the Americas 51, 69–80.

Stewart, G.G., 2014c. Yeast mitochondria – their influence on brewer's yeast fermentation and medical research. Technical Quarterly of the Master Brewers Association of the Americas 51, 3–11.

Stewart, G.G., Hill, A.E., Russell, I., 2013. 125th Anniversary review: developments in brewing and distilling yeast strains. Journal of the Institute of Brewing 119, 202–220.

Sugihara, M., Imai, T., Muro, M., Yasuda, Y., Ogawa, Y., Ohkochi, M., 2006. Application of flow cytometer to analysis of brewing yeast physiology. In: IBD Asia Pacific Conference, Hobart, Tasmania. Paper available from: http://www.ibdasiapac.com.au/asia-pacific-activities/convention-proceedings/2006/program_tues.htm.

Vidgren, V., Londesborough, J., 2011. 125th Anniversary review: yeast flocculation and sedimentation in brewing. Journal of the Institute of Brewing 117, 475–487.

Vogeser, G., 2014. Microbiology in process control made easy – is PCR finally available for everybody? In: The 2014 Brewing Summit Proceedings Chicago, Poster A-49.

Walker, G.M., DeNicola, R., Anthony, S., Learmonth, R., 2006. Yeast metal interactions: impact of brewing and distilling fermentations. Proceedings of the Institute of Brewing and Distilling, Asia Pacific Section 2006 Convention. Tasmania, Hobart, pp. 1–19. Available from: http://www.ibdasiapac.com.au/asia-pacific-activities/convention-proceedings/2006/program_wed.htm.

White, C., Zainasheff, J., 2010. Yeast growth, handling and storage. In: Yeast-the Practical Guide to Beer Fermentation. Brewers Publications, CO, p. 170.

White, L.R., Richardson, K.E., Schiewe, A.J., White, C.E., 2008. Comparison of yeast viability/vitality methods and their relationship to fermentation performance. In: Smart, K. (Ed.), Brewing Yeast Fermentation Performance, second ed. Blackwell Science, UK, pp. 38–147.

CHAPTER

Water

5

M. Eumann, C. Schaeberle
EUWA Water Treatment Plants, Gärtringen, Germany

INTRODUCTION

WATER QUALITIES IN BREWERIES

The main component of beverages is water and beer is no different. Almost every brewery has a water treatment plant, but is the treated water of the right quality for the purposes of the brewery?

Some of the parameters for good brewing water are obvious. The water must comply with the legal limits like the World Health Organization (WHO) guidelines for drinking-water quality. In particular, the water must be clean and free of germs and harmful substances.

Other parameters are linked with its use as brew water in a brewery. Values of interest are calcium content and alkalinity. In former times, the water in different regions determined the individual beer style. So if one wanted to produce a certain beer style, the corresponding water quality was demanded. A popular method for predicting the characteristic of brew water is the empirical formula from Kolbach for residual alkalinity. Residual alkalinity is defined according to the following formula:

$$RA = m - \frac{Ca + 0.5 \cdot Mg}{3.5}$$

RA = residual alkalinity, ppm $CaCO_3$
m = m-alkalinity, ppm $CaCO_3$
CaH = calcium concentration, ppm $CaCO_3$
MgH = magnesium concentration, ppm $CaCO_3$

The value of the residual alkalinity in the water determines whether the pH will rise or fall during mashing.

In this chapter, we will focus on practical issues of water analysis and treatment. The reader seeking a broader understanding of water in brewing should consult the specialist books mentioned in the appendix.

Brewing Materials and Processes. http://dx.doi.org/10.1016/B978-0-12-799954-8.00005-8

WATER ANALYSIS AND MONITORING

The first practical requirement is a water analysis. As mentioned above the water quality required depends on the purpose. Therefore, as a first step, raw water needs to be analyzed completely. At the very least, the analysis according to the drinking-water regulations is needed. For water sources with fluctuating parameters (usually waters impacted by surface waters or that are a blend from different wells), the chemical composition of the water needs to be monitored frequently. It is recommended to analyze the raw water at least once a month.

With the knowledge of raw-water composition one can compare the individual values with the requirements for individual consumers. It is useful to have an approved standard in a company for the different qualities. Particularly for the brew water, different brands need different qualities, and a brewer needs to keep the quality and flavor within the desired range.

As an indication for such a standard, the values in Table 5.1 can be consulted. The mentioned values are proven and based on practical experience.

Based on the comparison between raw-water analysis and quality standards, the target of the required water treatment is obvious. Based on the target, a suitable treatment can be chosen between different technologies.

Comparing the current possibilities of water treatment with the state of the art 10 years ago, it is obvious that there has been a development in technology. Membranes for filtration, ion removal, and degassing/deaeration are reliable alternatives to the traditional systems.

Table 5.1 Other Parameters According to WHO/EU Drinking Water Guideline

Parameter	Service Water	Brew Water	Dilution Water	Boiler Feed Water
TH [ppm $CaCO_3$]	<70			<2
Ca [ppm]		70–90	20–40	
Mg [ppm]		0–10	0–10	
Na [ppm]	<200	<20	<20	
HCO_3 [ppm $CaCO_3$]		10–50 (target 25)	10–50 (target 25)	<50
Cl [ppm]	<50	30–80 (target 50)	<50	
SO_4 [ppm]	<250	30–150 (target 100)	<50	
NO_3 [ppm]	<50	<25	<25	
SiO_2 [ppm]		<25	<25	<20
THM [ppb]	<10	<10	<10	
Fe [ppm]	<0.1	<0.1	<0.02	<0.1
Mn [ppm]	<0.05	<0.05	<0.02	<0.05

TH, *total hardness;* THM, *trihalomethane.*

DIFFERENT WATER TYPES

Of course, not all of the water used by a brewer is brew water. There are in general two types of water: product water becoming finally part of the beer and service water.

In the case of high-gravity brewing, the product water needs to be split into brew water and dilution water because the requirements are different. Brew water usually has higher salinity especially in permanent hardness. Dilution water needs to be deareated and should contain less hardness than the wort water to avoid oxalate precipitation.

Service water will be used for many different purposes in the brewery. Some of the water is used at elevated temperatures. Bottle washers or hot clean in place (CIP) are examples. Hard water can cause precipitation, whereas high chloride contents of over 50 ppm can cause corrosion. Therefore, service water also needs treatment to avoid these issues.

PRACTICAL REQUIREMENTS FOR WATER TREATMENT

GENERAL REQUIREMENTS

Sometimes, water treatment plants (WTP) cause issues due to a bad design. Robust chemical and microbiological quality is essential, especially for the water that gets directly into the product without any possibility of correction.

One general principle is the avoidance of dead ends in the entire installation. In stagnant water, organisms can develop without hindrance. Biofilms can develop in the dead ends and disperse into the entire installation. It is easier to "design in" ideal installations rather than trying to correct a poorly configured plant.

If there is a possibility to sanitize or sterilize a unit, then this solution is preferred; for instance, it is easier to sterilize an ultrafiltration setup as opposed to a packed-bed filter. Sometimes, this depends on the basic material of the equipment. For vessels and piping, the use of stainless steel avoids microcracking in coating materials. This also avoids "hotbeds" in the installation. Compared to installations constructed from plastics, stainless steel allows also fewer edges and a smooth surface. This facilitates easier cleaning.

Finally, the possibility for CIP for the installation is recommended. This should include also the piping.

REQUIREMENTS FOR DISINFECTION

Water is used over a wide network in the brewery, rendering a substantial risk of contamination. Therefore, a disinfection step is recommended upstream of the different draw-off points. In the past, chlorine was usually used. Chlorine gas in water becomes hypochlorite (ClO^-) ions, whereas dosing hypochlorite uses salts from ClO^- ions (sodium hypochlorite or calcium hypochlorite). If the water is disinfected before it is stored, the system downstream is protected.

More recently, chlorine dioxide has replaced chlorine. One reason is that the efficiency of chlorine depends on the pH value, increasing pH leading to decreased efficiency. A pH of 7 renders 75% of the chlorine available for disinfection, whereas a pH of 7.5 reduces this value to 48%. At a pH of 8, only 22% is available. Conversely, chlorine dioxide is a soluble gas, so it is able to penetrate the cell membranes of microorganisms, rendering it more effective than chlorine. Due to the diffusion capabilities of gas, it is also possible to remove biofilms by continuously using chlorine dioxide. Finally, chlorine dioxide does not form hazardous byproducts like trihalomethanes (THMs) with organic matter in the water compared to chlorine as disinfectant.

Chlorine dioxide is usually formed from hydrochloric acid and chlorite, not to be confused with the mineral group of the same name. Because the chlorite can disaggregate and form chlorate, the reaction product needs to be used promptly or stored cool in the dark. The proportion of acid and chlorite needs to be adjusted carefully, and it is recommended to monitor the chlorine dioxide and compare the measured value with the theoretical amount. For water that will enter the beer, whether chlorine dioxide can be used or needs to be removed prior to using that water is debated. Removal is the more frequent approach.

"The more the merrier" seems to be the attitude for disinfection by some operators. Nevertheless, almost all disinfection agents have byproducts. Even from chlorine dioxide, chlorite remains in the treated water. So it is recommended to have clean piping, well-filtered water, and a chlorine dioxide concentration of 0.2 ppm in the pipeline system (design value of 0.4 ppm for the chlorine dioxide plant).

Disinfection using ozone is unusual because it is more expensive. The advantage of ozone is that it reacts to form oxygen, so no byproducts remain in the water. Ozone is also a strong oxidizing agent, so it is not advisable to use ozone in the presence of chlorine (for instance, if a municipal supplier treats the city water with chlorine), because ozone will oxidize the chlorine. The result would be that neither ozone nor chlorine would survive, and the water would not be protected.

A third possibility is irradiation with ultraviolet (UV) light at 254 nm. This kind of disinfection is needs clear filtered water, because particles shield the organisms from UV light. The turbidity should be below 0.3 nephelometric turbidity unit. In addition, metal oxides and humic acid hinder the radiation because they absorb UV light. Therefore, it is recommended to check the light transmission of the water. In well-pretreated water, 98%/cm can be expected.

The intensity of the UV lamps decreases during their life span. After 8000 h of operation, the UV lamps are usually replaced. A UV plant needs to be equipped with overheating protection because the lamps get hot during operation. Additionally, a sensor for the irradiation checks is needed if the life span is reached.

Although this approach has the advantage of leaving no chemical residues, it also means there is no "depot effect." In other words, no disinfectant agent protects the water from recontamination downstream of a UV plant. Therefore, it is best to install a UV setup at the point of use and circulate the water during periods of nonconsumption.

REQUIREMENTS FOR FILTRATION

The first step in water treatment should be filtration. Some suggest that there should be an initial sterilization stage, but oxidizable water consumes a lot of disinfection agents, and the large quantities needed lead to a high content of disinfection byproducts remaining in the treated water. It is necessary to initially oxidize metals such as iron using an oxidation unit and remove them before the water is disinfected. Some types of raw water also need a flocculation stage.

The removal of particles by packed-bed filters is the oldest technology in water treatment but still one of the most widely used. This robust and simple technology uses gravel of different sizes. The smaller the gravel size the better is the removal of particles, but also the pressure drop in the packed bed increases for smaller gravel size.

Apart from sand filters, multilayer filters consisting of different grades of sand, gravel, and anthracite are used. Anthracite is a coal-based material suitable for depth filtration. This means that particles are not only removed at the surface but also within the bed. This makes the removal capacity higher, and this type of filter is capable of dealing with higher concentrations of particles.

At intervals, the removed particles are rinsed out from the packed bed, and a new cycle of filtration can start. This rinsing step is carried out by air and water or a simultaneous air/water backwash.

In recent years, ultrafiltration (Fig. 5.1) has started to replace sand filters. Membranes with a pore size of 0.02 μm enable a high quality of filtered water. The difference from a sand filter is obvious comparing this value with the gravel size of 400–800 μm in a packed bed. Usually, hollow fibers made out of polyethersulfon (PES) are used. They are robust chemically and mechanically.

FIGURE 5.1

Ultrafiltration with stainless steel housing.

At this pore size there is a guarantee of the removal of not only particles and colloidal substances but also bacteria and even viruses. Lifetime of the membranes is high and the automated operation allows a continuously high-quality supply of water. Even if there are fluctuations in the raw water, the same high quality is still generated after the filtration.

Compared with a packed-bed filter, an ultrafiltration setup can be disinfected easily. The chemical resistance of the membranes allows a proper CIP using acid, caustic, and hypochlorite.

REQUIREMENTS FOR DECHLORINATION

As mentioned earlier, the product shall usually be free of disinfection agents, hence the water needs to be dechlorinated. Usually, packed beds with activated carbon are used—known as an Activated Carbon Filter (ACF). Chlorine or chlorine dioxide is removed by a catalytic reaction on the surface of the carbon. Sometimes, cartridges with activated carbon are also used. However, the contact time for this kind of filter is too short to achieve proper dechlorination. The contact time should be at least 2.5–3 min.

The surface area of activated carbon is very large, which enables a complete removal of chlorine. In addition, ozone can be removed with a similar catalytic reaction.

To remove fines, an intensive backwash needs to be carried out, at least after installing or replacing the carbon. A polishing filter downstream of the ACF can avoid fines downstream. Even so, if the filter is not backwashed properly, the filter media will clog quite often. Operating the ACF with half of the design flow is not an issue; however, downtime with the filter can lead to contamination. Accordingly, it is advisable to operate all the filters if there it is more than one ACF on site. This avoids microbiological growth during periods of standby.

The surface of activated carbon is an ideal matrix in which microorganisms can grow. Therefore, it is advisable to sterilize the carbon filter once a week or if microbiological conditions deteriorate. Chemical disinfection agents are inappropriate as the carbon removes them. Elevated temperature is the recommended solution: hot water at 85°C or steam. The vessel needs to be constructed from stainless steel, with a grade depending on the chloride concentration in the water.

A new approach for dechlorination is removing the chlorine with UV light. Different wavelengths are mentioned in the literature; 185 nm can be used, as well as multiwave emitters. An intensity of at least 2500 J/m^2 is required.

REQUIREMENTS FOR ADSORPTION

Some raw waters contain objectionable substances such as THM, odors, or pesticide residues. Such substances are usually removed by adsorption in a packed bed. One of the most common materials for packed-bed adsorbers is, again, activated carbon (Fig. 5.2). Adsorption of an individual substance depends on the concentration,

FIGURE 5.2

Activated Carbon Filter with steam sterilization unit.

the polarity, and the interaction with the surface. However, the presence of other adsorbed substances can also result in competition on the surface of the activated carbon. Usual contact times are 10–20 min.

It is recommended to check the system in advance. A small chromatographic column can be used for establishing the contact time and the life span of the activated carbon. In this way, the gradient of the substance to be removed can be determined and the size of the actual installation designed. The carbon needs to be replaced when the substance to be adsorbed exceeds the limit value, but at least after 2–3 years.

It is also advisable to have sampling valves within the column to detect the gradient, because there is no other possibility to monitor the condition of the adsorbents. An unpredictable breakthrough could cause immediate reaction otherwise.

Issues with not only organic contaminants but also inorganics can be treated by adsorption on active surfaces. Aluminum oxide or iron hydroxide can be used for removal of inorganics like arsenic or antimony. In this case, there is usually no way of avoiding piloting for the determination. Packed beds must not be mixed because contaminated adsorbents would cause values over the limit at the outlet. This is the reason for avoiding intensive backwashing of such columns and the need for a proper prefiltration.

REQUIREMENTS FOR DESALINATION

Frequently, the ionic composition of the water does not match that required for a given purpose. Often, the water has too high an alkalinity or else other ions are above the limits. The traditional treatment is ion exchange in which cations are exchanged with H^+ ions, whereas anions are exchanged with OH^- ions. Using this principle, demineralized water can be produced. Depending on the deviation of the water analysis from the limits, one or both steps are necessary.

Depending on the water composition, different exchange methods can be used. If only a dealkalization is required, weak acid resin can be used. In this case, only Ca^{2+}, Mg^{2+} ions bound to HCO_3^- are exchanged. The bicarbonate is converted to free carbon dioxide which can be removed easily using a degassing system.

If cations other than Ca^{2+} and Mg^{2+} or larger amounts bound to bicarbonate need to be removed, strong acid resin is used. Both kinds of resin are regenerated with acid.

Anions can be removed by a basic resin in a second step. This kind of exchange process is prolonged, because first the cations are exchanged, then the anions, and finally both resins need to be regenerated. The availability of membrane methods has increasingly led to a replacement of ion exchange.

Staying first with ion exchange, though, some of the plants are still regenerated purely on the basis of the treated volume. Accordingly, it is necessary to trigger the regeneration before the resin is totally exhausted, which represents a waste of chemicals and of water. Therefore, if ion exchange is used, it is recommended to use a design in which the load is monitored and the regeneration is triggered as necessary.

The vessels of ion exchange plants are usually rubber lined on carbon steel construction. If the lining is damaged, the vessel will corrode rapidly. If there is no proper maintenance, the lifespan can drop.

Not only can the exchangers corrode, but also the surroundings of the plant. The storage of the hydrochloric acid (most common regeneration agent) demands care. Fumes from hydrochloric acid can even corrode the construction of the building and the equipment. In addition, stainless steel gets attacked. Therefore, fume absorbers need to be filled with water, and personnel filling the storage tanks must take all sensible precautions.

The chemicals used for regeneration can also cause issues. THM problems can be caused by using contaminated hydrochloric acid. The resin is sensitive to oxidizing agents, so life span is shortened if chlorinated water is introduced to ion exchange systems. Each ion exchange resin should be protected from oxidizing disinfection agents. Additionally, fines can increase the pressure drop of the column. Therefore, it is important to perform an efficient backwash from time to time, which will rinse the fines out of the system. The design of the exchanger columns must match this requirement; hence, it is advisable to use a setup with single valves and sufficient freeboard.

The resin can be assessed using tests such as iron content, capacity, and moisture, making it possible to recognize the end of the useful life of the resin.

A very traditional treatment, especially for brew water, has been lime precipitation. Nevertheless, even though the operational costs are low, such systems are rarely installed nowadays due to the high demand for space and the high capital expenditure.

Nowadays, the most popular system for desalination is reverse osmosis (RO) (Fig 5.3). The system uses semipermeable membranes that separate permeant from the brackish side. Without a pressure, the desalinated water would dilute the more concentrated water through osmosis. Using a pump, it is possible to overcome the osmotic pressure in the aqueous solution. Almost all molecules and ions are rejected, with only nonpolar small molecules having a low rejection rate.

An RO module consists of a flat-sheet membrane connected to a permeant tube in the center. The membrane is wound up on the center tube. Raw water flows though small channels between the membranes, whereas permeant inside the membrane is conducted to the center tube. This construction enables a high surface area within the modules. However, the small channels can clog easily. Therefore, it is important to have a proper pretreatment to avoid fouling. Ultrafiltration, as described previously, is the preferred pretreatment because particle removal is most efficient. The quality of the pretreatment can be measured by the so-called silt density index (SDI). This kind of test uses a 0.45-μm filter with constant raw-water pressure. The decrease of flow rate per minute (as a percentage) is averaged over the test period of 15 min.

Using RO means taking desalinated water out of a feed flow. So the remaining water on the feed side becomes more and more brackish. The higher the amount of desalinated water (or, in other words, the higher the recovery), the higher is the salinity in the remaining water. So, there an optimal ratio of permeant and brine needs to be identified for each individual raw water unit. Using low recovery means wasting water. Using high recovery increases the risk of oversaturated brine. In an oversaturated solution, salts will crystallize causing deposition on the membrane.

FIGURE 5.3

Reverse osmosis.

The crystallization can be retarded by dosing a small amount of antiscalant or scale inhibitor. Furthermore, the solubility can be affected by adjusting the pH value. Acidification of the raw water can improve the recovery rate.

If one compares the operational costs of a desalination plant, it is obvious that for reverse osmosis the most important contribution comes from the wastewater. The wastewater increases the need for freshwater, an important fact in regions of water shortages. Additionally, high wastewater flow leads to high costs for disposal. Therefore, a reliable desalination plant is operated at an optimized recovery rate while avoiding trouble with precipitation.

The best approach is to take care of this matter during the design of the plant, calculating the saturation of the raw water and determining the conditions with the best recovery rate. However, during operation the conditions must also be monitored and adjusted. The individual flows at an RO plant must be checked and compared with the designed flow rate. To get the information about the plant conditions, it is necessary to identify the pressures and flow rates of all relevant parts of the plant. At the very least, RO should be equipped with manometers at each stage and flow meters for the individual stages and the total permeant and brine flow. In addition, a conductivity meter is important to get the information about the quality of permeant.

During operation, as a daily routine, flows and pressures should be monitored to recognize issues before they worsen. A decrease of the permeant flow indicates scaling, especially if this happens on the membranes directly at the concentrate outlet—the point with the highest salinity. Usually, this is linked with an increase of the conductivity of the permeant. Fouling on the membranes (caused by particles, metals, or organic matter) can be recognized by an increase of the pressure drop on the feed side. In particular, the first membranes can clog. It is also necessary to keep this background in mind and compare the used raw water with the water used for designing the plant. If the water composition changes, the scaling calculation needs to be reworked with the actual values.

Scaling or fouling on membranes is a relatively quick process; hence, it is important to deal with it rapidly. If one waits too long, the membranes will be destroyed. The deposition needs to be removed regularly. It is advisable to have a CIP system for an RO system. Usually, the cleaning is carried out one stage after the other, starting with a caustic cleaning step. This step can also use ethylenediaminetetraacetic acid (EDTA) for better efficiency. After the caustic step, an acidic cleaning follows. In addition, the acidic cleaning is done stage by stage. The first mud is drained, then the cleaning solution is circulated via a cleaning tank. Different cleaning agents can be used because different salts are soluble in different materials. The most common solutions are NaOH, HCl, or citric acid. Cleaning should not be too frequent because it can reduce the life span of membranes.

Unfortunately, the membranes of RO systems are sensitive to oxidizing agents like chlorine, meaning that disinfection is not easy. Again, the importance of the pretreatment is obvious. The water needs to be free of microorganisms so that the unit can be protected from contamination. One approach for a disinfection of the RO plant is the circulation of a solution of 0.5 to 1% sodium disulfite (using permeate) similar to the aforementioned cleaning procedure.

Disinfection is necessary when the RO system has not been in operation for more than a week or when the membrane elements suffer biofouling. Instead of a chemical disinfection, heat can be used for sanitization. But this approach can only be used if the plant was designed to tolerate such treatment. As well as the construction of the RO unit completely in stainless steel, a special type of membrane with higher temperature resistance is needed.

REQUIREMENTS FOR Ca ADAPTION

Calcium is one of the most important components of brew water as mentioned above. Calcium interacts with phosphate from the malt during mashing to lower the pH, and combines with oxalic acid to precipitate calcium oxalate, thereby obviating problems in the ensuring beer such as gushing and the development of beerstone.

The traditional way to add calcium is by dosing either calcium chloride ($CaCl_2$) or calcium sulfate (gypsum, $CaSO_4 \cdot 2H_2O$) during mashing or into the wort. However, there are some disadvantages to doing so. The solubility of $CaSO_4 \cdot 2H_2O$ is limited, so a lot of powder is removed with the spent grains.

The preferred approach is to add limewater and mineral acids for better control. The limewater and acid need to be dosed according to the flow of brew water:

$$Ca(OH)_2 + 2\,HCl \rightarrow CaCl_2 + 2H_2O$$

$$Ca(OH)_2 + H_2SO_4 \rightarrow CaSO_4 + 2H_2O$$

In this way, the noncarbonate hardness in soft raw water (for instance, RO permeant) can be increased without increasing the alkalinity.

Two approaches are possible: either dosing the solution into brew and dilution water or dosing the solution into the mashing vessel and the lauter tun. This allows a customized water quality for individual beer brands. The approach allows lower operating costs.

REQUIREMENTS FOR DEGASSING

Membranes are nowadays available for degassing (Fig. 5.4). They enable the removal of carbon dioxide to prevent corrosion of mild steel piping and vessels as well as oxygen to prevent a negative impact on the product. In particular, high-gravity brewing demands a low concentration of oxygen (below 10 ppb) in the dilution water.

The use of CO_2 or N_2 as sweep gas and a vacuum pump allows the almost complete removal of O_2. Values below 10 ppb for oxygen can be realized easily.

No energy for heating is required with such membrane technology and the system is completely closed. The membranes are stable so even a hot CIP is possible. Of course, again, this demands a stainless steel installation of the unit.

FIGURE 5.4

EROX for deaeration.

The membranes are sensitive to oxidizing agents, so the feed must be dechlorinated. As a matter of course, one has to be aware about the risk of reinfection of the treated dechlorinated water.

MAINTENANCE

It is not sufficient to check only that the water treatment setup is running and that all parts are functioning correctly. The quality of the treated water needs to be checked. Some of the parameters for quality and performance can be measured inline. Of course, such sensors increase the cost for a water treatment plant, but better control is achieved if you at least measure flows, pressures, levels, and conductivity. Sensors for pH, redox potential, and chlorine concentration may also be used.

Some of these sensors need frequent calibration to ensure that they are reading correctly. It is also necessary to have a spare in stock, because such sensors can break. These sensors use electrochemical signals and are connected to the control system. However, even if information is available in the CPU, a trending is not included automatically. However, trends give an indication of what is happening in the plant.

Apart from the sensors, some analytical parameters including hardness and m and p alkalinity need to be included in a report. The easiest approach is to include all the information in the operations diary. In a single report, you have operational data and quality parameters.

Instead of a simple operations diary, the data can also be transferred to the Supervisory Control and Data Acquisition (SCADA) system of the brew house. A separate SCADA system for the water treatment is possible for recording the mentioned values. Either way, it is necessary to get a fast overview of the trends. Water analysis by the quality laboratory must also be part of the monitoring.

WATER EXAMINATION

A precondition for water tests is that sampling be carried out properly. When taking samples from lines, the water must run off for some time (rinsing the line). Sampling vessels must be rinsed well by using the water to be tested. Water samples that contain undissolved substances must be pretreated clear prior to testing. Therefore, filters designed for analytical proposes must be used as they will not add any foreign matter to the water. The first part of the filtrate—usually 100 mL—must be disposed of. The water temperature when tested should be approx. 20°C (room temperature) to obtain reliable test results.

For the examination of water, some parameters can be used for an overview. This will be described. Nevertheless, this does not mean that other parameters are redundant. Depending on the raw-water analysis and the targets for the treatment, more examinations are necessary. At the very least, temperature, pH, conductivity, microbiological condition, plus hardness, and m and p alkalinity need to be examined.

For microbiological probes, the sampling valve must be sterilizable before taking the samples. Of course, there needs to be a sampling valve upstream and downstream of each treatment step.

Examination of hardness (Table 5.2)

The indicator Eriochrome Black-T (EBT) is used for the determination of hardness. It is carried out as a complexometric titration with EDTA. During the analysis, alkaline earth ions (Ca and Mg) dissolved in water react with EDTA to form a resistant,

Table 5.2 Conversion of Hardness Units

Conversion	Alkaline Earths (mmol/L)	Alkaline Earths (meq/L)	German Degree (°dH)	ppm $CaCO_3$	French Degree (°f)
1°dH	0.18	0.357	1.00	17.8	1.78
1 mmol/L	1.00	2.00	5.6	100	10
1 meq/L alkaline earths	0.50	1.00	2.80	50.0	5.00
1 ppm $CaCO_3$	0.01	0.020	0.056	1.00	0.100

colorless complex (chelate). The end point of titration will be visualized by the indicator. This method is suitable for all kinds of raw and service water. High concentrations of iron and copper will interfere with the accuracy of the test.

For 100 mL of the water sample to be tested, 2 mL of 25% ammonia solution are added. This solution acts as a pH buffer. Further, the indicator Eriochrome black-T is added. Usually, indicator and buffer are available as combined tablets. The indicator (masked with a yellow dye) forms a red complex with Ca^{2+} and Mg^{2+}. These ions are bound by EDTA at the end of the titration. EBT is free from hardness, and the color changes from brown to green. The total hardness is calculated from the consumption of mL of EDTA solution.

Examination of m*-alkalinity*

Because methyl orange is used as indicator, this value is also known as the *m*-value. The method is suitable for all kinds of water with a pH value >4.3. This determination can be carried out as a titration. First, the water sample is filtered if it is turbid. Then 3–5 drops of methyl orange solution are added to the water sample (100 mL). Methyl orange has the property to color alkaline and neutral water yellow. If the water becomes acidic, it turns red immediately. The point of change is at pH 4.3. If the solution is yellow, hydrochloric acid at a concentration of 0.1 mol/L is used for the titration. This must be done slowly while the sample is shaken or mixed until the yellow coloration changes to orange (not to red). If the water has a pH below 4.3, a similar procedure with caustic can be used to determine the base capacity to pH 4.3.

Examination of p*-alkalinity*

The equilibrium of calcite and carbonic acid represents the most important buffer system in water. Although below pH 4.3 there is no HCO_3^- left in the water (see *m*-alkalinity above), above pH 8.2, all CO_2 is converted to HCO_3^- and a further part to CO_3^{2-}.

This determination is also called base capacity to pH 8.2 (*p*-value/KB 8.2). The method is suitable for all kinds of water with a pH value <8.2. Again the water sample will be filtered if it is turbid. Then 3–5 drops of phenolphthalein solution are added to the water (100 mL). Phenolphthalein as indicator appears colorless when the water is acidic and neutral whereas it appears magenta above a pH of 8.2. The indicator gives the method its name.

The titration is carried out with sodium hydroxide solution. Sodium hydroxide is added dropwise with steady shaking or agitation of the colorless water until a magenta color appears for at least 2 min.

CLEANING AND MEASURES DURING MAINTENANCE

To get the operational data, the plant needs regular inspection. This includes cleaning the glass on the pressure gauges and the electronic measuring equipment as well as the labels for the valves and components so that values and descriptions can always be read off immediately.

The entire installation must be free from leaks and dirt. Due consideration must be paid to safety for personnel and equipment. In particular, refilling of the different chemicals must comply with the safety regulations.

From the operational data, the water consumption of the plant can be monitored, allowing the water footprint of the plant to be determined. For instance, if the amount of rinsed water is monitored, it is obvious how much water can be saved with a recycling plant for the individual unit.

Some of the items have a limited life span. For instance, the UV lamps have a lifetime of some 8000 h. As part of preventive maintenance, replacement should be planned. This is also valid for fillings of packed beds like activated carbon, gravel, and anthracite. On the other side, the condition of resin can be checked and restored if necessary to get a long life span.

TRAINING

A through understanding of the treatment plant is the key to properly operated water treatment. Some tasks for the operator are not very frequent. It is advisable to have trained personnel for the operation of the plant but also a training manual to refresh the knowledge. One part of the training should be an overview of the process and the theory underpinning it.

CHAPTER

6 Wort and Wort Quality Parameters

R. Pahl, B. Meyer, R. Biurrun

Research Institute for Beer and Beverage Production (FIBGP), Research and Teaching Institute for Brewing in Berlin (VLB e.V.), Berlin, Germany

For the production of a high-quality beer, a consistent wort quality is essential. Beer quality is impacted by the different raw materials and the brewhouse process, through an influence on fermentation, maturation, and downstream production steps. Varying qualities of the raw materials are always a challenge for the brewer and sometimes hard to compensate in the brewhouse, so for a consistent high wort quality to be achieved the different parameters of wort have to be regulated.

The quality of wort is impacted by the raw materials (malt, adjuncts, water, and hops), milling, mashing, lautering, wort boiling, and wort clarification and treatment.

A wide range of parameters can be measured to indicate the efficiency of the wort production process and to predict the impact that the wort will have on fermentation and other stages downstream and, thence, the finished beer. The various parameters are now described, whereas Table 6.1 illustrates the range of analytical values that would be expected to be displayed in different types of wort.

EXTRACT

The available extract content is mainly influenced by the quality and quantity of the malt and the proportion of adjuncts in the grist. Further influences on the yield during the process have their origin in the nature of the processing of these raw materials, mainly in the steps of milling, mashing, and lautering. Regarding adjunct use, different adjuncts will make for different amounts of extract, depending on their source. Therefore, it is obviously necessary to adjust their proportion in the total grist. This leads to average equivalent values for the adjunct weight compared to the use of 100% malt. Using wheat malt as a substitute the corresponding amount of the wheat is the same as the amount of barley malt. Substituting the malt with, for example, corn (maize) grits, we have to lower the share in the total grist, because of a higher extract content of the corn. This leads to an equivalent load of approximately 98% to reach the same final extract content in wort. Using rice grits, it will be approximately 95–98%. However, unmalted barley has to be used at up to 128% in relation to 100% malt. Extract yield will also depend on the gelatinization temperature of the cereal starch, which also will have an influence on the mashing program to choose and the possible need for exogenous enzymes during the mashing

Brewing Materials and Processes. http://dx.doi.org/10.1016/B978-0-12-799954-8.00006-X

Table 6.1 Typical Specifications for Worts for a Range of Beer Styles

Parameter	Pils	Pale Lager	Export	Dark Beer	Wheat Beer (Pale)
Original extract (%)	11.3–12.0	10.5–12.5	12.5–13.5	11.5–11.8	11.0–12.0
Final degree of fermentation (%)	84.0–85.0	81.0–82.5	81.0–84.0	78.0–81.0	80.0–82.0
pH	5.2–5.4	5.4–5.6	5.4–5.6	5.4–5.7	5.2–5.4
Wort color (EBC)	7–10	7–15	10–15	50–100	10–20
Bitterness (EBC)	25–45	15–30	25–35	30–45	12–18
Total nitrogen (ppm)	800–1000	800–1000	800–1000	800–1000	1000–1200
Coagulable nitrogen (ppm)	15–25	15–25	15–25	20–30	20–40
Magnesium sulfate precipitable nitrogen (ppm)	200–250	200–250	200–250	250–300	300–400
Free amino nitrogen FAN (ppm)	200–250	200–250	200–250	200–300	200–300
Thiobarbiturate index (TBI)	<45 (after boiling <60 (after cooling)	Idem	Idem	No reference values	Idem pils
DMS + SMM (ppb)	<100	Idem	Idem	Idem	Idem
Solids after whirlpool (mg/L)	<70	Idem	Idem	Idem	Idem
Iodine value	<0.30	Idem	Idem	Idem	Idem

process. Here also, the quality and quantity of those enzymes will directly influence the extract yield.

The quality of the raw material is always the base for high-quality wort production and an optimal process. If the malt is not well modified, the enzymes involved in starch degradation will not be able to function optimally during mashing, resulting in a lower extract yield. Based on growth conditions and climatic influences, variations can occur in the conformation of the starch, leading to small but significant changes (1–2°C) in gelatinization temperatures.

The determination of the extract of malt is standardized, for example in the "Congress" mash method. Here, the extract refers to the total amount of substances dissolved from finely ground malt in a standardized mashing process:

- 30 min rest at 45°C (by agitation 80–100 rpm)
- Heating up to 70°C (1°C/min)
- 60 min rest at 70°C
- Cool down within 10–15 min

Finally, the sample is filtered and the density is determined. Additionally, the sample is used for further analyses, like soluble nitrogen, pH, viscosity, color, etc.

The milling process must be checked and controlled, the required particle-size distribution dependent on the wort-separation system in use. Table 6.2 shows the most common distribution related to Lauter Tuns or Mash Filters (Kunze, 2014):

Simultaneously to the control of the milling, just a simple inspection of the spent grains at regular intervals can be very helpful, because by this you very often can assess whether the milling process was properly controlled.

Further parameters that can lead to changes in the extract yield are mashing temperatures, the ratio of water to milled grist, and rest times, because they are, next to the pH, the main influences on the enzymatic potential of the mash and hence a sufficient starch conversion. The mashing-off temperatures are important, because they impact the ensuing wort-separation process.

The wort-separation process itself has a substantial impact on the recovery of extract. Here especially, the amount of sparging water and the relation of extract

Table 6.2 Milled Grist Size Distributions for Different Wort Separation Systems

	Lauter Tun (%)	Mash Filter (Without Membrane) (%)	Mash Filter (With Membrane) (%)
Sieve 1 (husk)	18–25	11	1
Sieve 2 (coarse grits)	<10	4	2
Sieve 3 (fine grits 1)	35	16	15
Sieve 4 (fine grits 2)	21	43	29
Sieve 5 (flour)	7	10	24
Sieve bottom (fine flour)	<12–15	16	24

Adapted from *Kunze, W., 2014. Technology of Brewing & Malting, fifth ed. VLB, S. 220.*

recovery in lautering and water evaporation in the boiling has to be kept in mind. In addition to these considerations, a key test to be implemented is the iodine test (residual amylose yields a blue coloration) to check for the elimination of starch.

The extract content, divided into fermentable and nonfermentable extract, has therefore a direct impact on the final attenuation and the quality of the fermentation.

The determination of the extract content is necessary to calculate the brewhouse yield and to assess the efficiency of the mashing process and the quality of the grist materials. The determination of the extract content is based on the measurement of density, which is most likely achieved these days using a digital density meter.

Density is a measure of mass per unit volume

$$\text{SI unit: kg/m}^3\text{; symbol } \rho \text{ (rho)}$$

$$\rho = m/V$$

$$1000\ \text{kg/m}^3 = 1\ \text{kg/dm}^3 = 1\ \text{kg/L or } 1\ \text{g/cm}^3 = 1\ \text{g/mL}$$

FINAL ATTENUATION

To analyze the extent of final attenuation, the wort is completely fermented in a fermentation tube with the final attenuation degree determined by calculating the difference between starting extract and final extract as a percentage. This provides an insight into the quality of the mashing in terms of the extent to which alpha- and beta-amylases and, to a lesser extent, limit dextrinase have functioned.

Caramelization and Maillard reactions occurring during wort boiling have a small impact by removing fermentable sugars.

The *final degree of fermentation* should be, depending on the beer style, in the range of 78 to 85%. According to the Middle European Brewing Analysis Commission (MEBAK) method, 200 mL of the finished wort is fermented at 20°C. After 24 h, the extract content is measured for the first time and again 3 h later. For the later calculation, the lower value is used. According to the European Brewery Convention (EBC), the wort is fermented at 25°C.

$$\text{Apparent attenuation [\%]} = (E_0 - E_a)/E_0 \times 100$$

E_0 = extract of wort in % Plato; E_a = apparent extract of fermented wort in % Plato.

The real attenuation value is obtained by multiplying the apparent attenuation by 0.81.

NITROGEN

TOTAL NITROGEN

The main nitrogen-containing substances in wort are amino acids, peptides, polypeptides, proteins, derivatives of nucleic acids, ammonium salts, nitrate, and aliphatic and aromatic amines. Amino acids and ammonium salts are the most important nitrogen sources for yeast nutrition, impacting the extent of growth and attenuation and the formation of flavor-active products. The medium and higher molecular weight nitrogen compounds influence the colloidal stability, palate fullness, and head retention. The improvement of the colloidal stability of beer later on by additives should only be a final measure. The main steps involve the optimized selection of raw materials, the composition of the brew water, a controlled mashing process, fermentation, and the cold storage time. Another main aspect is the pH-value with its influence on enzyme activity and impact on the isoelectric points of proteins, which influences their solubility.

The total nitrogen content and its various fractions are mainly influenced by the quality of the raw materials, notably by the extent of proteolytic modification. Deficits are very hard to compensate for in the brewhouse, especially if there is limited flexibility in mashing times, temperatures, procedure, and plant usage.

The most common method to determinate nitrogen is the so-called Kjeldahl method. This is based on the prior digestion of the sample (oxidation of compound substances to H_2O, CO_2, and NH_3), followed by a distillation of the NH_3 over into a boric acid solution, and finally titration with acid of the amount of NH_3 in the distillate (MEBAK Wort, 2013, S.73).

The total nitrogen fraction (950–1150 mg/L) in wort is composed of 22% high molecular nitrogen (2% coagulable N), 18% medium molecular nitrogen, and 60% low molecular nitrogen with 34% of Formol-N and 22% of free amino nitrogen (FAN).

COAGULABLE NITROGEN

The coagulable nitrogen positively influences palate fullness and the foam stability of the beer. On the other hand, excessive amounts can decrease the colloidal stability through proteins precipitating with polyphenols.

The determination of the coagulable nitrogen is based on the principle of precipitation of the high molecular weight nitrogen present in the wort by boiling of the sample for 5 h at 105–108°C. After this procedure, the sample is analyzed using the Kjeldahl method.

$MgSO_4$-PRECIPITABLE NITROGEN

Magnesium sulfate is able to react with nitrogen compounds of a molecular weight of approximately 2600. A close relationship is claimed between magnesium sulfate precipitable nitrogen and the head retention of beer.

The high molecular weight proteins are precipitated by magnesium sulfate, and the nitrogen content in the sediment is determined (MEBAK Wort, 2013, S.81).

FREE AMINO NITROGEN

As the main source in yeast nutrition for multiplication and yeast growth, the FAN value is a very sensitive parameter. Deficits in FAN directly can lead to an insufficient and slow start of the fermentation, insufficient fermentation performance, and stuck fermentations.

Low molecular weight nitrogen compounds, especially the amino acids in wort, influence the metabolism of the yeast. In particular, they impact the production of higher alcohols and vicinal diketones.

The determination is based on the reaction of amino acids with ninhydrin.

During wort production, deficits of the raw materials are relatively hard to compensate for, because of the importance of the original protein content of the barley and malt and the extent of modification achieved in the malting process.

THIOBARBITURIC ACID INDEX

The TBI is considered a summary parameter for the thermal stress on malt and wort. It is a classification number [next to HMF (hydroxymethylfurfural)] and includes Maillard reaction products and other organic compounds.

The determination of TBI is performed according the MEBAK standard method. The sample is reacted with a solution incorporating acetic acid and thiobarbituric acid by placing the sample in a water bath at 70°C for a period of 70 min and measuring the color increase as compared to an unheated control.

According to MEBAK, normal values for the Thiobarbiturate index (TBI) are: pale wort, beginning of boiling <22; pale cast wort <45; pale wort after cooling <60 (MEBAK Wort, 2013, S. 57).

For determination of the thermal stress of dark worts, a dilution of the sample is necessary, leading to a decrease in accuracy.

HIGH MOLECULAR WEIGHT β-GLUCAN

β-Glucan, the main component of barley endosperm cell walls, is primarily degraded during the modification of malt, but enzymolysis may be continued in low-temperature stands in mashing. Undermodified malt that is not handled effectively in a mash will deliver high molecular weight β-glucan ($>10,000$) to the wort, thereby causing lautering and filtration problems.

The Fluorimetric Method, according to EBC (MEBAK Wort, 2013, S. 64), is based on the principle that the fluorochrome Calcofluor forms a complex with high molecular weight β-glucans. The disadvantage of this method is the instability of the resultant fluorescence derivative. Therefore, the use of an automatic analysis system based on injection (flow injection analysis), like the Beta-Glucan Analyzer or the "Carlsberg System," is recommended.

Enzymic methods are also available, in which β-glucanase enzymes degrade the polymer to produce reducing sugars or glucose, the latter being measured using spectrophotometry-based techniques.

Normal values in wort are in the range of about 200–800 mg/L.

TURBIDITY

HOT BREAK

The hot break consists of protein (50–60%), bitter substances (16–20%), and other organic and inorganic substances, such as fatty acids and minerals. The hot-break content of the finished wort is between 400 and 800 mg/L (dry matter, extract-free), eg, 40–80 g dry trub or 200–400 g wet trub per hectoliter.

COLD BREAK

Cold break can be separated at temperatures below 80°C when cooling down the hot break-free wort. Mainly, it comprises protein (48–57%), tanning agents (11–26%), and carbohydrates (20–36%).

The determination of the quantity of solids is performed by filtration of a defined quantity of sample (for hot break, this is done at a temperature of approximately 85°C) using filter paper within a Büchner funnel, and filtered using a water jet or vacuum pump. After the filtration, the filter papers are dried 3 h at 105°C and the weight difference is determined.

Standard values are considered to be <70 mg/L (after Whirlpool) and <100 mg/L Kettle full (<85°C, before hops addition).

DIMETHYL SULFIDE/S-METHYL METHIONINE

DMS is a significant component of flavor, especially in lagers. In the absence of wort contamination, DMS primarily arises through the thermal breakdown of the precursor S–methyl methionine (SMM). This precursor is produced during malting, especially during germination, and its amount is reduced by kilning. The more intensely kilned the malt, the lower the level of SMM in the malt and therefore the lower is the DMS potential. Malt also contains dimethyl sulfoxide (DMSO) which is freely extracted into wort and which can, under certain circumstances, be reduced to DMS by yeast. Spoilage bacteria convert DMSO as a precursor to DMS. The factors impacting SMM and DMS levels have been described by Bamforth (2014).

The DMS in wort is determined using a headspace analysis method with a capillary gas chromatograph equipped with a detector specific for sulfur (FPD or PFPD detector). Because high levels of DMS are generally considered an "off-aroma" in finished beer, values greater than 100 μg/L should be avoided.

For the determination of SMM, the precursor is converted to DMS by heating strongly under alkaline conditions.

pH

The mash pH has a very big influence on the extract content. If mashing is performed at lower pH values than 5.6, the activity of certain enzymes, especially the endopeptidases, increases, and this leads to a higher yield of soluble nitrogen. The total extract content will be increased, but it is not fermentable extract, because more carbohydrates are not dissolved. However, it is claimed that at lower pHs there is a slight stimulation of limit dextrinase activity. Conversely, the activity of lipoxygenase is less at lower pHs and this may be to the advantage of flavor stability and foam stability of the finished beer. Furthermore, the pH value also influences the solubility of hop components and the color of mash and wort. If the pH value during wort boiling is too high, break formation will be insufficient, resulting in a decreased colloidal stability. Normal pH values for cast wort are in the range of 5.3–5.6. Cast-out wort with biological acidification shows values in the ranges of 5.0–5.4.

The pH is usually determined with a pH-measuring device consisting of the measurement electrode and the reference electrode, both combined into a single unit and provided with a signal amplifier and display. The device must be calibrated frequently, using for this purpose standard buffer solutions. It is also essential to determine pH at a defined temperature: pH is lower at higher temperatures.

COLOR

The color is influenced by the grist materials as well as by all the processing steps in wort production with temperature impact (nonenzymic browning in decoction mashing, wort boiling), as well as oxidation of polyphenols (in part enzymic). Turbid samples have to be filtered and dark beers have to be diluted prior to analysis. In the visual determination method, the color of light from a standard source, following transmission through the test wort, is matched to that of one of a series of calibrated colored glasses. A suitable comparator is the Neo Comparator "Model Hellige." Another possibility is the spectrophotometric method in which the color of wort is determined by measuring the absorbance at 430 nm and multiplying by the appropriate factor. The spectrophotometer should have an accuracy of at least ±0.5 nm. The spectrophotometric method has been adopted as the reference method in preference to the more subjective visual method. Normal values for pale beer are in the average range of 7 to 11 EBC units, for dark beer in the range 40–80 EBC. Color is described in more detail elsewhere in this volume.

BITTER UNITS

The hop resins or bitter substances, especially the iso-α-α-acids, are mainly responsible for the bitterness in beer. Alpha-acids from the hops are converted into the corresponding iso-α-acids during wort boiling.

The determination of Bitter Units (BUs) according to the EBC method is by extraction of the bitter substances from a previously acidified sample using iso-octane, and their concentration in the extract is determined using a spectrophotometer (measurement at 275 nm); multiplication by 50 yields BUs (MEBAK Wort, 2013, S. 240).

VISCOSITY

The Viscosity is the resistance of a fluid that is deformed under shear stress. It describes a fluid's internal resistance to flow. The higher the viscosity, the thicker is the fluid. The viscosity in fluids always depends on the temperature.

The unit is based on water having a viscosity of 1.0 cP at 20°C (1 P = 100 cP = 1 g/m*s = 0.1 Pa*s).

High viscosity values cause problems during lautering and filtration.

The viscosity of wort is determined by measuring the time taken for a fixed volume at 20°C to flow through a capillary under reproducible conditions. The viscosity is then calculated, either by comparing the flow time of water or, if the viscometer constants are known, by use of the appropriate factor. Another possibility for measuring the viscosity is the use of a falling-ball viscometer. In this measurement, the viscosity η [kg/m/s] is dependent on the factors of gravitational force, the radius of the ball, the velocity of fall, the density of the ball, and the density of the fluid.

ZINC

In the case of fermentation problems, it might be necessary to check the zinc content of wort. Zinc is an essential trace element for the yeast and should be available in extents of about 0.1–0.2 mg/L. The zinc content can be determined by atomic absorption spectrophotometry.

REFERENCES

Bamforth, C.W., 2014. Dimethyl sulfide – significance, origins and control. Journal of the American Society of Brewing Chemists 72, 165–168.

Kunze, W., 2014. Technology of Brewing & Malting, fifth ed. VLB, S. 220.

MEBAK Wort, 2013. Volume Beer and Beer-based Beverages, 57, 64, 73, 81, 240.

CHAPTER

Alcohol and Its Measurement

7

G. Spedding

Brewing and Distilling Analytical Services, LLC, Lexington, KY, United States

ON ALCOHOL, ALCOHOLIC STRENGTH, AND MEASUREMENTS

Alcoholic strength is the term used to denote the measure of the amount of ethyl alcohol (ethanol) in beer. It may be reported as percent by mass of beer (% weight/weight—w/w or mass/mass—m/m) or as percent by volume (% volume/volume—v/v). These values are expressed at either 15.56°C (60°F) or at 20°C (68°F) depending on country, regulatory authority, or other local requirements.

The analysis of beer for alcohol content is important both for quality assurance programs and for legal reporting purposes. Results, however, are subject to appreciable variation and the analyses are often time-consuming and expensive. Over 180 years of research, laboratory practice, and a number of established verified tables of data and extensive algorithms help the brewing chemist best determine alcohol content in beer. In fact, up until about 40–50 years ago many brewers would have fielded out their samples to one of the major brewing institutes available to serve their research functions and analytical testing needs. These institutes were the ones actively engaged in the research that led to many of the methods officially adopted back then, many of which are still in use today. Such methods have evolved through new technological developments with modern instrumentation now simplifying the measurement of alcohol by brewers in their own facilities. That said, the grounding principles and theories behind alcohol determinations remain largely unmodified since their discovery, as will be demonstrated.

THE PROPERTIES OF ALCOHOL AND THEIR MEASUREMENT

Ethanol is miscible with water in all proportions, with ethanol molecules fitting within the "spaces" in the three-dimensional structure of water. This "space filling" property means that the final volume of any mixture of water and alcohol is less than the sum of their individual volumes, ie, volume contraction occurs upon mixing (Hu et al., 2010; Meija, 2009). This volume contraction has to be accounted for when determining alcohol by volume in aqueous mixtures. Solution volumes are temperature dependent and so temperature compensation has to be allowed for in measurements. Brewers using hydrometers are aware of this with the need to obtain the

Brewing Materials and Processes. http://dx.doi.org/10.1016/B978-0-12-799954-8.00007-1

actual temperature and consult instrument manufacturer's tables to make density or specific gravity measurement adjustments. Increasingly sophisticated instrumentation compensates for temperatures of measurement and can deliver reportable data at specified temperatures. In the United States, alcohol by volume is sometimes still reported at 15.56°C (60°F), though more frequently 20°C (68°F) is the required temperature. The temperature volume change is effectively so small between these two temperatures that the difference for most beers in the range of 3.5–8% alcohol by volume is only 0.02–0.03%. The lower value here is at 15.56°C (60°F). Given this small range, and because the Alcohol and Tobacco Tax and Trade Bureau (TTB) in the United States allows ± 0.3% alcohol by volume tolerance in beer measurements, it is unclear why some authorities are still adamant on the 15.56°C (60°F) recorded value. Early alcohol tables were generated based on determinations made at 15.56°C (60°F) and are still available as needed [Association of Official Analytical Chemists (AOAC, 1995)]. Most modern instruments and methods are now measuring and reporting values at 20°C (68°F), and this will be the value largely discussed here.

Ethanol (CH_3CH_2OH: C_2H_5OH) has, at 1 standard atmosphere pressure, a boiling point of 78.29°C (172.92°F), an ignition or flash point of between 9 and 11°C (48.2–51.8°F), a freezing (or melting) point of −114.14°C (−173.45°F), a density (d_4^{20}—see Table 7.1) of 0.78934 g/ml (or 789.34 g/L at 20°C), a density (d_{20}^{20} — see Table 7.1) of 0.78927 g/ml at 20°C and a refractive index (n20/D: 20°C measured at a wavelength of 589 nm using the sodium D-line) of 1.3611 (Criddle, 2005; Lide, 2005; Tonelli, 2009). These properties may be used to determine the concentration of ethanol as outlined in the following text. Density values may be expressed as relative density or specific gravity values as defined in Table 7.1 and will be important numbers expressed throughout this chapter. Specific gravity values are unitless numbers as they are defined relative to a standard substance, usually water, and so the unit terms of density g/ml (or kg/L) cancel as seen in the definition in Table 7.1. For equations expressed herein it is noted that 0.7907 as a unitless number, or the rounded values of 0.790 or 0.791 are the specific gravity values most commonly used in brewing circles for pure alcohol. The number is derived from the density value of alcohol at 20°C (68°F) and the density of water also at 20°C (68°F): 0.998203 (Schroeder et al., 1982).

These listed properties of ethanol will be employed in methods for its quantitation throughout this chapter. Some definitions pertaining to its properties also appear in Table 7.1.

PRECISION AND ACCURACY

Looking at any analytical technique demands at least a cursory understanding of terms related to the reliability of the information returned or reported. The topic is covered in depth in many analytical textbooks (Skoog and West, 1986; Fritz and Schenk, 1987, for example). The reader following up on any method described will need to determine the applicability of any approach they choose to report alcohol data to regulatory authorities, etc.

Table 7.1 Some Definitions and Terms Pertaining to Measures of Alcohol and Extracts in Beer

Alcoholic Strength: A measure of the amount of alcohol in beer. The alcohol content of a beer typically refers to the amount of ethyl alcohol (as opposed to higher or fusel alcohols). The analysis of beer for alcohol content is an important part of brewing laboratory work for quality assurance programs and legal reporting purposes. Results, however, are subject to appreciable variation and under official methods the analyses are time-consuming and expensive.

Apparent (AE) and Real Extract (RE): Brewers measure changes in density as sugars are consumed and converted into alcohol. Measurements are obscured (the true gravity is "hidden") by alcohol (of lower density than water, sugar solutions, or beer) causing "buoyancy effects" with hydrometers, for example. Thus, false or apparent readings of gravity are made when instruments measure beer (containing water, sugars, and alcohol); hence, "apparent extract" (AE). The RE is a true(r) measure of remaining sugars—and proteins, etc. to be exact about the nature of the extract—as determined in the absence of alcohol (removed via distillation or boiling) or compensated for the actual alcohol content via its determination and/or via algorithms in instrumentation. Such equations are described and discussed in the text. When yeast has finished fermenting the brewer's wort, or the brewer terminates the fermentation, the final gravity reading is sometimes also called the terminal gravity or present gravity (this will also be in "apparent" or "real" terms depending how measured or calculated—see text). RE is real extract in grams per 100 g of beer (Plato).

Apparent Attenuation and Real Degree of Fermentation: The expression of the percentage reductions in the wort's specific gravity caused by the transformation of the sugars (higher specific gravity than for water) into alcohol, carbon dioxide, and yeast biomass. Real and apparent values exist for the reasons discussed under Apparent and Real Extract. It is simply a measure, in percent terms, of the amount of original extract which has been consumed or converted. The real degree of fermentation equation: $RDF = 206.65/(2.0665 + RE/ABW)$ is a fundamental equation also underpinning the theory behind many equations presented here (see also Original Extract or Gravity). ABW = alcohol by weight expressed in %.

Gravimetric and Volumetric Distillations: Prior to sophisticated digital density meters and NIR (Near-infrared) devices to measure ABW beer samples had to be distilled as described in the text. Laboratory-scale distillations were or can be used to obtain the alcohol from the beer—(largely) free of any solids or extract present. Officially approved methods describe the process in detail; however, it should be noted that other alcohols and volatiles do distill over and reactions take place in the heat of distillation that can affect the subsequent reading of alcohol; it is not as simple as expected that the final distillate is simply a binary mixture of ethanol and water. Most alcohol vs. specific gravity tables were generated over 100 years ago by careful distillations of alcohol and water blends. The tables, generated from specific gravity determinations of the pure alcohol/water mixtures, were rapidly adopted by the alcohol beverage industry and have served well since then.
Early on, the density values and specific gravity values were obtained gravimetrically—the latter values using an instrument known as a pycnometer (discussed in the text). Gravimetric distillations (based on weight) giving true alcohol by weight values for a sample of water and ethanol but requiring a correction for alcohol by volume. Volumetric distillations (starting typically with exactly 100 ml) would give a true alcohol by volume content (temperature dependent) but would require a compensating correction for weight. The sample specific gravity includes the extracts present, and alcohol values need to be corrected for the actual sample specific gravity (SG), a fact often overlooked today with some modern methods of alcohol determination.

Continued

Table 7.1 Some Definitions and Terms Pertaining to Measures of Alcohol and Extracts in Beer—cont'd

Original Extract or Gravity: The initial SG of brewer's wort is known as the original extract (OE) or original gravity (OG) depending upon units implied by these terms (OE is in degrees Plato, whereas OG is in numerical "gravity units"—see the following text and under "Plato"). OE is an expression of sugar content—grams of sugar per 100 g of wort [equivalent to % weight/weight (w/w)]. In the brewing industry, this is denoted as "degrees Plato" (°*P*) and, to a first degree of approximation, beer responds in the same way as the original solutions of sucrose and water used to derive the Plato scale. Brewers often determine or record and report this using numerical SG values rather than in degrees Plato and it is this that is more correctly the OG value. OG represents the ratio of the density of wort at 20°C to the density of water at the same temperature (see "Specific Gravity"). Plato and SG values can, however, be interconverted using appropriate formulas as illustrated in the text (example, for our purposes here, 12.00 °Plato is an SG of 1.04838—actually 1.048311 at 20°C/20°C). See Plato in this table. When reading the literature and interpreting equations, it is important for the reader to understand the differences in units applied to OE and OG terms as this can otherwise cause some confusion as to which values to use in the equations. $OE = (RE + 2.0665 \times ABW)/(1 + 0.010665 \times ABW)$. This is a fundamental formula covered in the text underpinning alcohol determinations and which can be used to test derivation of formulas and cross-check values (see Anger et al., 2009; Hackbarth, 2009; Nielsen et al., 2007; Weissler, 1995).

Plato—(see also Original Extract or Gravity). The typical way brewers report extract in wort and beer (as weight percent); Plato values may be obtained directly from hydrometers calibrated in degrees Plato, by means of SG measurements as related to standard Plato (or extract) tables and (to varying degrees of approximation) by calculation—some equations seen here in the text. The Plato scale and tables are themselves approximations for complex physical–chemical reasons but have stood the test of time for most brewing purposes. See references (Anger et al., 2009; Hackbarth, 2009, 2011; Weissler, 1995) for more on this topic.

Specific gravity (SG) or relative density is an intensive property of a substance (be it solid or liquid) and is, historically, the ratio of the density of a substance at the temperature under consideration to the density of water at the temperature of its maximum density (4°C). The actual density in theory—for the purposes of most discussions with respect to beer—is generally 20°C/20°C relative to water at unity, though elsewhere in the alcohol beverage world it is typically expressed as at 20°C/4°C. The complexity of the temperature associations could not be presented in depth here—full details may be found in the literature (see references). An SG value is numerically equal to the density in grams per milliliter (or kg per liter) but is stated as a pure number (because the division by the water's own density value leads to a cancellation of the units), whereas density is stated as mass per unit volume. Water has an SG of 1.0000 at 20°C. Base and derivative density values form mathematical grounding in many of the formulas used in alcohol and extract calculations. The density value 0.998201 will be seen in the literature as pertaining to the formulas and conversions of alcohol density, specific gravity, and alcohol by weight and volume and, although understood within this article, is also not covered in any depth here.

Sample Density or Gravity: The sample density or SG of a beer is taken to be that value measured directly in the beer containing the alcohol, water, any sugars (protein) or additives, and any mineral content, etc. For brewers, the measurements always include the alcohol and extract content of the beer and the latter being related to an expression of sugar content—grams of sugar per 100 g of wort [equivalent to % weight/weight (w/w)]. In the brewing industry, this is denoted as "degrees Plato" (°*P*) and to a first degree of approximation beer responds in the same way as the original solutions of sucrose and

Continued

Table 7.1 Some Definitions and Terms Pertaining to Measures of Alcohol and Extracts in Beer—cont'd

water used to derive the Plato scale. Sample densities and specific gravities can be interconverted using equations described in the text. See prior discussion of "Plato" and related terms in the following text.

Tabarie: A mathematical relationship that goes back 180 years—derived by a Mr. M.E. Tabarie. Tabarie noted the relationship between the SGs of the alcohol and of the real and apparent extracts for beer.

$$SG_{Alc} = SG_{Beer} - SG_{RE} + 1 \quad \text{or } SG_{Beer} = SG_{RE} + SG_{Alc} + 1 \quad \text{or}$$
$$SG_{RE} = SG_{Beer} - SG_{Alc} + 1$$

These rearranged equations work well if any two of the values are known with any degree of accuracy for "normal strength" beers. The details and history are complex and are reviewed elsewhere (see the text and cited references). It is to be noted that this relationship only works within a typical range of 2.5–8% alcohol by weight and real extract in beer in the region of 3–9 Plato or real degree of fermentation between 65% and 85% (Nielsen and Aastrup, 2004, and personal observations and communications; see also Hackbarth, 2009, 2011). This is discussed further in the text (section Density Meter/NIR Analyzer Calculations and the Tabarie Relationship—An Important Interlude). The astute reader will note how they can use the Tabarie relationship if they have obtained the wort OG and the beer SG and then the computed RE [Eq. (7.3)] to solve for the alcohol SG and use that to help confirm the validity of other equations described in the text. The Tabarie formula is crucial to the set of calculations within the Alcolyzer/NIR units to ultimately solve for the OE to confirm Balling's mass relationship as seen in Eq. (7.1). This is best described in MEBAK (2013).

ACCURACY

Denotes the nearness of a measurement to its accepted (or "true") value and is normally expressed in terms of error. Accuracy thus involves a comparison with a true or accepted value. The values determined by an instrument may be relayed dependent upon a calibration of that instrument. If functioning and used correctly, there will be a range over which it will deliver acceptably accurate results. Accuracy then is the nearness of a single result or a mean of a set of results to the true value. The true value may be known through the use of other methods via repeated tests, which may include extended collaborative efforts and statistical evaluation, or through the use of test samples with defined physical parameters (Anger et al., 2009; Bamforth, 2002; Fritz and Schenk, 1987; Sharpe, 1991; Skoog and West, 1986).

PRECISION

A term used to describe the duplicability of results; defined as the agreement between the numerical values for two or more measurements that have been obtained in an identical manner (Skoog and West, 1986). Precision is also measured by statistical means and expressed in terms of the deviation of a set of experimental results from the arithmetic mean of the set (Anger et al., 2009; Bamforth, 2002; Skoog and West, 1986; Fritz and Schenk, 1987). Good precision is often an indication of good accuracy but it is possible to obtain good precision with poor accuracy and vice versa (Fritz and Schenk, 1987).

When measuring samples for alcohol content or for any other parameter, the laboratory staff should understand how the instruments at their disposal are to be used. Competence in calibration of instruments, along with an understanding of the mechanics of their operation, including their degree of accuracy and delivery of precision, is an important skill for all staff.

Control samples are required and very often overlooked by brewers. Modern instruments often represent the "black box" scenario, the inner workings for which are not known by the operator. As such, without understanding controls, accuracy, and precision, the operator will not know when the available results are correct, fully or barely acceptable, or completely wrong.

OUTLINE OF METHODS TO MEASURE ALCOHOL

Over the years many physical and chemical methods have been used to determine the amount of alcohol present in solution. These methods relied on measuring some physical or chemical property of ethanol, and they evolved in sophistication along with microchip technology and microscale developments. In general, alcohol has been or may be determined via:

- Chemical and Catalytic Oxidation Methods
- Chromatography—Gas and Liquid Methods
- Density and/or Specific Gravity Measurement
- Ebulliometry
- Enzymatically
- Flow injection and through the use of Biospecific Detectors or Sensors
- Refractometry
- Spectroscopy

Before detailing the methods just listed, it is to be noted that alcohol was traditionally measured following its separation from beer by distillation. During distillation alcohol is boiled out of the beer and collected by condensation. Along with alcohol and water, some other volatile components of the beer will be distilled over, sometimes impinging upon the specificity of detection. The concentration of alcohol in the distillate is estimated and then converted by calculation into alcohol content of the beer.

The various methods discussed in the following text have all been used to determine alcohol in distillates or were designed to circumvent the need for distillation. It will, however, be seen that most brewers and modern instrumentation will utilize or rely upon tables and algorithms relating distillate or alcohol—water mixture specific gravity values to their corresponding alcohol values as determined by extensive work done over a century ago. Some of the technologies as used in the laboratory have now also been extended to inline instruments and measuring alcohol and beer extracts during production of the beer (Daniels, 1990; Kobayashi et al., 2005, Sharpe, 1991). These inline units rely on the same principles of detection and measurement as the laboratory methods.

THE METHODS USED TO MEASURE ALCOHOL CONTENT AND CURRENT APPLICATIONS

Chemical Oxidation of Alcohol

Ethanol can be readily oxidized to yield ethanal (acetaldehyde) or completely oxidized to acetic acid (ethanoic acid) [$CH_3CH_2OH \rightarrow CH_3CHO \rightarrow CH_3CO_2H$] (Criddle, 2005) and this provides quite rapid methods for measuring ethanol content. Although used quite early on and largely in terms of wet-bench chemistries, chemical oxidation methods do find application today in microscale sensor technologies in more sophisticated methods and instrumentation including modern breath- and blood-alcohol analyzers (Criddle, 2005).

Titration of separated alcohol via distillation followed by its oxidation to acetic acid using powerful oxidizing agents, such as acidified potassium dichromate or alkaline permanganate, represented common assays to determine alcohol content (Joslyn, 1970). Subsequent to separation and oxidation, the determination of the acid formed was made via acid–base titration, or the equivalents of dichromate or permanganate reduced were quantitated either by titration of residual oxidizing agent or its determination by colorimetric or spectrophotometric procedures. For example, a color change as chromium is reduced from the +4 oxidation state (yellow) to the +3 state (green) can be monitored via titration or spectrophotometrically and is stoichiometrically related to the amount of alcohol consumed (Caputi, 1970; Caputi et al., 1968; Caputi and Wright, 1969; Crowell and Ough, 1979; Joslyn, 1970; Zimmermann, 1963). Their modern applications in highly specific sensor technologies have recently been detailed (Criddle, 2005; Tonelli, 2009).

Catalytic Oxidation of Alcohol

An earlier instrument called an SCABA unit (Servo Chem Automatic Beer Analyzer) measured the density of the beer and also alcohol by its catalytic oxidation. The alcohol was transferred to the vapor phase within the unit and measured with a special sensor (Korduner and Westelius, 1981). Despite comments to the contrary (Korduner and Westelius, 1981), it was said to be cumbersome to deal with (personal communications) and other components of beer could be a problem during detection. It was officially adopted as a method by the American Society of Brewing Chemists (ASBC) (Austin, 1989) but is being phased out by most breweries and replaced by digital density meters and Near-infrared (NIR) instruments (see section Coupled Oscillating Density Meters and NIR Alcohol Meters). A newer unit also based on thermal analysis, is current—called the FermentoStar—and relies on measuring the specific heat capacity of beer samples heated to two temperatures. Through such measurements and internal algorithms, alcohol and extract values may be obtained—further details are given elsewhere [Central European Brewing Analysis Commission (MEBAK, 2013)].

CHROMATOGRAPHIC METHODS FOR ALCOHOL DETERMINATION (GAS CHROMATOGRAPHY AND HIGH-PERFORMANCE LIQUID CHROMATOGRAPHY)

Alcohol has been determined utilizing chromatographic methods, which separate and analyze mixtures of chemical components (Spedding, 2012). Requiring expensive instrumentation, this may be reserved for larger brewery operations; the author reserves some judgments on these methods based on the need to also obtain the specific gravity or density of the beer sample to make appropriate corrections to the alcohol values. Furthermore, these methods often rely on densitometry of distillates for calibration purposes and are more frequently used in wine- and distilled spirits-testing facilities rather than breweries. However, such methods could not be overlooked as they may play a wider role in the future.

Gas Chromatography

This relies on selective adsorption and desorption of volatile components on a stationary phase. Components are carried through the column by means of an inert gas (or more frequently today by the use of hydrogen) to a specific type of detector; for alcohol detection, a flame ionization detector is used (Buglass and Caven-Quantrill, 2011; Schwiesow, 1995; Tonelli, 2009). Alcohol and other components are identified and quantitated by weight (wt/vol) (via calibration with known standard compounds) based on retention (residency) time in or on the column. Described as an AOAC (Association of Official Analytical Chemists) approved method accurate to about $\pm$ 0.2% v/v (Jacobson, 2006), gas chromatography (GC) has been described for measuring ethanol at low concentrations in beer distillates (MEBAK, 2013) and in beer by the US agency TTB (Masschelin, 2010) and by Clarkson et al. (1995). GC sees some application in evidential breath and blood analyses (Criddle, 2005) and in other forensic applications (Logan et al., 1999). Regulatory authorities may rely on GC determinations of alcohol based on both the method's sensitivity and rapid turnaround.

High-Performance Liquid Chromatography (HPLC)

This uses a liquid mobile phase to transport compounds of a mixture injected under high pressure onto a packed-column stationary phase and resolved into its constituents via adsorption and release from the column matrix. The eluting molecules are then detected and quantified via a suitable detector; for ethanol, this is usually an ultraviolet (UV) or refractive index detector (García and Alfonso, 2013). Again, the system must be calibrated using known standards. As HPLC is better suited to less-volatile compounds, it is usually used to determine sugars and organic acids with ethanol also coming along for the ride (Klein and Leubolt, 1993). Several recent references and reviews are available for the reader to consult if interested in adopting this method, though it finds more frequent use in the distilling industry than the brewing industry (Cacho and Lopez, 2005; Castellari et al., 2001; García and Alfonso, 2013; Klein and Leubolt, 1993).

DENSITY AND/OR SPECIFIC GRAVITY MEASUREMENTS FOR ALCOHOL DETERMINATION

Historically, alcohol measurements were grounded in physical measurements of mass and volume through density or mass per unit volume intensive properties (see Table 7.1 for definitions). Through density and specific gravity relationships (also defined terms in Table 7.1), instruments and devices such as density bottles, hydrometers, densitometers, refractometers, and pycnometers were used to establish a recognized and officially accepted body of work. This extensive research effort culminated in the derivation of algorithms and tables which define the relationships between density values and specific gravity readings and alcohol by weight and by volume.

Hydrometers

Hydrometers are instruments that work on the principles of buoyancy and liquid displacement (MEBAK, 2013). Hydrometers are hollow-body objects with a constant volume that have been calibrated over a certain range. Buoyancy relates directly to the weight of a solution displaced by the hydrometer, with this being directly proportional to the density of the displaced liquid (beer, wort, or alcohol solutions); readings are subject to errors based on the presence of suspended solids and gases. Liquid surface-tension effects with hydrometer cylinders and the hydrometers themselves also need to be addressed when using such instruments. Furthermore, due to the thermal expansion of liquid volumes and of glass, temperature compensation is also required when using hydrometers and tables of temperature correction are supplied with each calibrated hydrometer. Hydrometers for measuring alcohol or for wort to determine extract are available (the latter based on Plato's experiments using pure sucrose solutions; Plato measurements are covered in detail in sections Fermentation and Alcohol Production: The Mass Balance of Fermentation, Carl Balling, and the Theoretical Basis for Brewing Parametrics and Measuring Alcohol Content in the Modern Brewery and Brewery Laboratory: Tables, Algorithms, and Software. Thus, hydrometers may be marked in specific gravity, density, or Plato units (grams per 100 g of solution). Refer also to Table 7.1 for details defining Plato. As with many modern developments, electronic versions of hydrometers are beginning to appear on the market, so it seems that they will see continued application within the industry, especially in smaller brewery operations.

Pycnometry

In pycnometry, a glass-bottle device engineered to contain a defined volume of a solution (wort, beer, or alcohol) is weighed against the same volume of a reference solution at an exact temperature (usually water at 20°C for brewing work). The ratio of the densities (masses per unit volume) is determined; dividing one density by the other cancels the units. This ratio is expressed as a specific gravity (SG) under the defined conditions of measurement eg, $SG_{20°C/20°C}$ or simply $SG_{20/20}$, with values expressed to five decimal places (Dyer, 1995; MEBAK, 2013).

Standard tables of extract (sucrose solutions which correlate well with beer extract) were made relating the sugar or extract content to SG values. Furthermore, the values pertaining to alcohol concentrations in water and the alcohol obtained from beer samples via distillation were obtained essentially via such gravimetric means using density bottles or the pycnometer. Brewers can obtain SG values for wort or beer samples or distilled beer and refer to tables or use algorithms (discussed in sections Fermentation and Alcohol Production: The Mass Balance of Fermentation, Carl Balling, and the Theoretical Basis for Brewing Parametrics and Measuring Alcohol Content in the Modern Brewery and Brewery Laboratory: Tables, Algorithms, and Software) to obtain the real extract or alcohol by weight or volume of their samples.

Densitometers

Modern densitometers have largely replaced the classic methods based on distillation and either refractometers or hydrometers for measuring extracts and alcohol in samples (MEBAK, 2013). These are "Oscillating U-tube" density meters, and these units are highly sophisticated and expensive instruments for measuring density. They are accurate to five or six decimal places and can be used to measure the density and specific gravities of both wort and finished beer or distillates, to obtain original extract (OE) values, final beer SG (apparent extract) values, or the alcohol content of the beer (LaBerge, 1979; Strunk et al., 1982).

Oscillating U-tube densitometers work on the principle of electronic excitation of a measuring cell—the U-tube filled with a solution to be measured. Based on a fixed volume within the cell, similar to the old pycnometer, and defining the density as mass per unit volume, an increase in mass within the same volume leads to an increase in density. The harmonics or oscillation of the tube (frequency of resonance) is affected by solutions of different density and the density is then calculated from the oscillation period (Criddle, 2005; Garrigues and de la Guardia, 2009; Jiggens, 1987; Kovar, 1981; LaBerge, 1979; MEBAK, 2013; Stewart, 1983; Tonelli, 2009). Such instruments will report density and specific gravities of beer or, if distillates are used, the percent alcohol by weight at 20°C based on official tables generated by the Organization of Legal Metrology (OIML; OIML, 2014; see Table 7.2). The percent alcohol by volume may also be reported based on OIML tables at 20°C (see Table 7.2) and percent alcohol by volume based on tables by the Association of Official Agricultural Chemists (AOAC International). The AOAC tables were later adopted by the American Society of Brewing

Chemists (ASBC), and these tables record the alcohol contents as they would be at 15.56°C (60°F) (AOAC, 1995; ASBC, 1940; OIML, 2014). These units are easy to maintain and need only to be calibrated on dry air and pure water (a fluid for which densities are known to the desired degree of accuracy) (Jiggens, 1987; Schroeder et al., 1982; Strunk et al., 1982).

EBULLIOMETRIC MEASUREMENTS OF ALCOHOL

Used more in the wine industry (Jacobson, 2006), based partly on the fact that wine generally contains a higher amount of extract than beer, this method also allows determination of alcohol without having to first distill the sample. An ebulliometer utilizes the difference in the boiling points of water and dilute alcohol solutions; the elevation of the boiling point of liquids is caused by the presence of solutes. After calibrating with pure water and then applying a test sample, a slide rule is used to read the percentage alcohol by volume or weight based on the boiling point of the beer. The physics are discussed by Weissler (1995) and the method itself by Applegate (1993) and Joslyn (1970). Although early units were quaint in design, some electronically controlled units are available which might provide some impetus to look again at these instruments for routine and less-expensive monitoring of alcohol content.

ENZYMATICALLY MEASURING ALCOHOL VIA SPECTROSCOPIC METHODS AND BIOELECTROCHEMICAL CELLS

Enzymatic assays have proven popular and very useful for alcohol determinations. Essentially, these assays are *in vitro* (in the test tube) biochemical assays relying on the natural enzymes and coenzymes (factors) involved in ethanol metabolism in living organisms. As discussed earlier, under oxidation methods, ethanol (CH_3CH_2OH) is first oxidized to ethanal (acetaldehyde, CH_3CHO) using the biochemical compound nicotinamide adenine dinucleotide (NAD) in the presence of the alcohol dehydrogenase enzyme instead of a chemical oxidizing agent. As the reverse reaction is the thermodynamically favored process, the overall reaction is driven to completion by removing the acetaldehyde. This is done in a second step in the presence of aldehyde dehydrogenase which involves the quantitative oxidation of the acetaldehyde to acetic acid, again with NAD involved. The NAD is reduced to NADH (and a proton: H^+) in each reaction (two molecules of NAD are consumed for every ethanol molecule oxidized to acetic acid) which then affords the quantitation of the alcohol spectrally via NADH's absorbance of energy from wavelengths within the UV/visible spectrum; 334, 340, or 365 nm (Anger et al., 2009).

This particular enzyme assay system is described in kit manufacturer's instructions and in methods of analysis manuals from the American Society of Brewing Chemists (ASBC, 2004 and later editions), the European Brewery Convention (EBC, Analytica-EBC, 1998), and the Institute of Brewing (now the IBD—Institute of Brewing and Distilling) (Institute of Brewing, 1997) and in other works (Anger et al., 2009; Criddle, 2005; Powers, 2000; Upperton, 1985; Tonelli, 2009). As it is a well-established method with kits and instructions readily available, the experimental details are not discussed further here.

Although the traditional assay system was based on the aforementioned NAD-to-NADH coupled assay, a new paired alcohol oxidase and peroxidase enzyme system has also been developed (Gonchar et al., 2001; Prencipe et al., 1987). This system has been described as a less-expensive approach than the former enzyme assay. In this alternate enzyme assay, alcohol oxidase is used to oxidize ethanol to acetaldehyde plus hydrogen peroxide (note chemical oxidation methods and the NAD-enzyme assay described previously both carry the alcohol through acetaldehyde to acetic acid for a complete oxidation). A peroxidase enzyme is then involved to catalyze the reaction between hydrogen peroxide and a colored compound (a chromogen) to produce water and a colored dye. The dye is measured spectrophotometrically; then, the alcohol concentration may be calculated, via a stoichiometrically related equation (Gonchar et al., 2001; Prencipe et al., 1987).

Obviating the need for the NAD and other cofactors mentioned previously, the use of a dye-linked alcohol dehydrogenase, with high affinity for ethanol, coupled with a bioelectrochemical cell as detector, has also been proposed as a means to detect and quantify the ethanol content in beer (Waites and Bamforth, 1984). The method is very specific, providing data that correlate well with the NAD-linked alcohol dehydrogenase method, yet doing so more rapidly and economically according to the authors. As miniaturization of electrical-based probes come to play a role in chemical analysis (briefly noted in the following text), this approach will likely be amenable to such application.

Such assays were originally developed to test for the presence of alcohol in alcohol-free or low-alcohol products (with anything below 0.5% alcohol by volume considered nonalcoholic). However, with suitable and careful dilution, higher alcohol-containing beverages and foods can be tested via these very sensitive enzymatic assays. Again, note that to obtain the true alcohol by volume from such tests, the sample specific gravity must be known. The methods supplied with test kits make no assumption in this regard and do not give true alcohol by volume values (discussed in the following text) unless a separate sample of beer is tested for its current specific gravity.

Finally, it is noted that sophisticated enzyme assay systems relying on sensitive probes or sensors have been developed for rapid assays and monitoring of ethanol within inline systems (Tonelli, 2009). Biosensor-based enzyme chemistries and other miniaturized and modern sophisticated sensor technologies for measuring alcohol are not further discussed in this short review as they apply more to inprocess or breathalyzer situation-type analyses rather than laboratory testing for alcohol (see Criddle, 2005; Tonelli, 2009, for more on this topic).

REFRACTOMETER MEASUREMENTS FOR ASSISTING IN ALCOHOL CONTENT DETERMINATIONS

The principles of refractometry and refractometric analysis may be found in physics textbooks with a useful, though short, discussion presented in the brewing literature by Kunze (2010). Basically, light rays change direction when passing from an optically less dense medium into an optically denser medium. The refractive index of a material is a value that reflects how much light is bent or refracted as it passes from the less dense liquid into an optically denser prism in the refractometer. This value is dependent on the concentrations of alcohol and extract, the temperature of the wort or beer, and the wavelength of light (MEBAK, 2013). Based on these factors, it is possible to obtain the original gravity (OG), percentage alcohol by volume, and the residual gravity of beer by simultaneously measuring the density of the beer and its refractive index and applying a series of polynomial equations to the data. The density or apparent gravity is often obtained by use of the hydrometer (Kunze, 2010; MEBAK, 2013).

Refractometers found early use by brewers in measuring beer parameters by making use of the refractive index (RI) of solutions (Brown, 1977; Buckee, 1982; Essery and Hall, 1956; Koestler and Hagen, 1999). The accuracy and precision of the equations developed were described as "adequate to the task" for "normal quality control operations" (Brown, 1977). Craft brewers are showing a growing interest in the use of refractometers, and whereas many references are quite old, they will be of use should brewers choose to use some of the newer units on the market (Brown, 1977; Buckee, 1982; Criddle, 2005; Essery, 1954; Essery and Hall, 1956; Hudson, 1975; Joslyn, 1970; MEBAK, 2013). Particular attention must be made to calibrating the refractometer and often to determining factors for each beer to be tested. A nomograph for estimating the OG, percent alcohol by volume, and residual gravity from refractometer scale readings, and the present gravity of beer, was provided by Brown (1977) and a detailed discussion was presented by Joslyn (1970).

One issue being highlighted for research today is that of obtaining refractometric readings based on truer wort sugar compositions rather than on Plato's sucrose determinations. Although beer wort and sucrose are stated, and have long been held, as close approximations of each other with respect to extract sugar concentrations, this is not entirely true (Hackbarth, 2009, 2011). This issue is being addressed by some manufacturers of newer refractometers which are being developed specifically for brewing work.

ALCOHOL MEASUREMENTS USING SPECTROSCOPY

Infrared (IR) spectroscopy is a method which utilizes the energies of the infrared light spectrum to promote transitions within the various functional groups of molecules. Within certain regions of the electromagnetic spectrum, chemical compounds may absorb the infrared radiation and specific vibrations may be measured (Buglass and Caven-Quantrill, 2011; Garrigues and de la Guardia, 2009; Mehrotra, 2000; Osborne, 2000). By measuring the vibrations of atoms and bond stretching, those functional groups can be determined. The frequencies and intensities of the infrared bands exhibited by a chemical compound uniquely characterize the material—generating a fingerprint—and thus, infrared spectra can be used to not only identify a particular substance in an unknown sample but also can be used to quantify that substance (Garrigues and de la Guardia, 2009; Mehrotra, 2000). As such, the use of the energies associated with the mid-range and near-range of the infrared spectrum may be used to determine the content of alcohol in beverages. A short, yet nice discussion of the method is provided elsewhere (Tonelli, 2009). Moreover, it is noted that IR spectroscopy in both the mid-infrared (MIR) and near-infrared (NIR) regions is gaining popularity both qualitatively and quantitatively as an analytical technique with potential in other areas of alcoholic beverage and beverage raw materials testing; near-infrared spectroscopy techniques have been implemented for malting and brewing since the early 1990s (Proudlove, 1992). Instruments require calibration, but, once set up, several components in samples can be measured simultaneously with little to no sample preparation needed. However, a major limitation of NIR spectroscopy in alcohol beverage and food analysis is its dependence on less-precise reference methods (Osborne, 2000). Also, once again, we note the need to obtain independently the density values of samples to perform subsequent calculations. This measurement of density, alongside the alcohol by volume determination, is covered in section Coupled Oscillating Density Meters and NIR Alcohol Meters.

Infrared devices are also finding application within inline measurements in the brewery, and recent developments in the area apply to flow injection methods (Tipparat et al., 2001). Most significant to the present discussion is that highly accurate NIR spectrometers are now on the market that can measure the alcohol content of beer in the range of 0–12% v/v. These units are often used today along with coupled density meters.

COUPLED OSCILLATING DENSITY METERS AND NIR ALCOHOL METERS

As seen earlier, some coupled instruments today rely on two fundamental properties of alcohol; namely, its density and its absorption intensity in the near-infrared region of the spectrum. In principle, the alcohol by volume at 20°C is determined by the NIR instruments based on a specific function of the absorption intensity of the NIR line of alcohol (section Alcohol Measurements Using Spectroscopy). The coupled instrument's software programs reference the OIML tables for solving for the percentage of alcohol by weight (Boulton, 2013; OIML, 2014, and see section Fermentation and Alcohol Production: The Mass Balance of Fermentation, Carl Balling, and the Theoretical Basis for Brewing Parametrics and Table 7.2).

The combined instruments can then determine the ethanol concentration (weight and volume), the specific gravity of the sample, and then, via calculation, the original extract for the beer. The calculation for weight of alcohol relies on the density value for pure alcohol at 20°C (taken as 0.78924 g/cm^3) and the density of the sample measured in the density meter.

FERMENTATION AND ALCOHOL PRODUCTION: THE MASS BALANCE OF FERMENTATION, CARL BALLING, AND THE THEORETICAL BASIS FOR BREWING PARAMETRICS

For brewers and sophisticated measuring instruments to be able to determine the alcohol by weight and by volume for beer, the following values must be known or obtained: the present or apparent gravity of the beer, the real extract of the beer, the specific gravity of the alcohol contained within the beer, and the original extract of the wort (terms defined in Table 7.1). To help solve for these various terms, it is noted that an arithmetic relationship exists between them through which Carl Balling first laid down the foundation for brewing calculations over 160 years ago (MEBAK, 2013; Nielsen et al., 2007). The relationship known as Balling's formula is:

$$\mathrm{OE} = \frac{(A\%\ \text{mass} \times 2.0665 + \mathrm{RE}) \times 100}{(A\%\ \text{mass} \times 1.0665 + 100)} \qquad (7.1)$$

in which: OE is original extract in degrees Plato (Plato; g/100 g or mass/mass), A% mass is alcohol by weight, RE is the real extract, and the numbers in the numerator and denominator are as described in the following text.

The above formula (Eq. (7.1)) is based on an understanding of the mass balance relationship in brewing—simply, the chemical relationship dealing with the conversion of fermentable sugars to alcohol, carbon dioxide, and yeast biomass as noted by Balling and Plato (MEBAK, 2013; Nielsen et al., 2007; Weissler, 1995). Theoretically, 1 g fermentable sugar will yield 0.51 g of ethanol and 0.49 g of carbon dioxide. In fact, some sugar is needed for cell growth and so, more realistically, the ethanol yield is more likely 0.46 g, and carbon dioxide 0.44 g from 1 g sugar. In summary: 2.0665 g sugar yields 1 g ethanol, 0.9565 g CO_2, and 0.11 g yeast and—for the equations following: 0.9565 and 0.11 g sum to 1.0665 g extract not converted to alcohol (Anger et al., 2009; MEBAK, 2013; Nielsen et al., 2007; Spedding, 2013; Weissler, 1995). These relationships have been reviewed and slight revisions suggested or postulated, and the interested reader should consult Nielsen, et al. (2007), Cutaia, et al. (2009), and Hackbarth (2009, 2011) for more current and much more exact formulations. However, as these changes have not yet affected the routine calculations or methods being used in practice, we stay with the currently accepted and utilized theory and math.

Table 7.2 Tables and Polynomials Used by Brewers (Referenced in the Text)

Table and Organization	Table Title	Polynomial or Derivation Details	Notes
ASBC Table 1 (ASBC, 1940) (Also known as the Goldiner/Klemann, Kampf sugar table, see MEBAK, 2013)	"Specific Gravity and Degrees Plato of Sugar Solutions or Percent Extract By Weight"	AE [°P, % w/w] = $-202.414 \times SG^2 + 662.649 \times SG - 460.234$ (Eq. (7.9) in the text) SG = Beer SG (20°C/20°C) The SG is apparent SG (see text and notes)	The table and equations report "Plato" extract: grams extract in 100 g of solution AE: Apparent Extract or "Present Gravity". See Table 7.1
ASBC Table 2 (ASBC, 1940) [Also from a modified OIML Table expressing SG vs. Alcohol content (see following text)]	"Relation Between Specific Gravity at 20°C/20°C and Alcohol Content of Mixtures of Ethyl Alcohol and Water"	$ABW = 517.4 \times (1 - SG) + 5084 \times (1 - SG)^2 + 33503 \times (1 - SG)^3$ (Eq. (7.12) in the text)* SG = Alcohol SG (20°C/20°C) (MEBAK, 2013)	The OIML Tables are described below. The ASBC table may be based on a different equation
AOAC Table 913.02 AOAC, 1995)	"Percentages by volume at 15.56°C (60°F) of ethyl alcohol corresponding to apparent specific gravity at various temperatures"	None officially available though some quick rule of thumb equations exist for converting ABV 20°C (68°F) to ABV 15.56°C (60°F) (These ROT's are not described here; consult author for details)	The table provides a good ABV value based on SG to 4 decimals and requires some interpolation. It is little used today
OIML Table VI (or Va) (OIML, 2014)	Alcohol strength by mass of a mixture as a function of its density at 20°C	See ASBC Table 2 previously $ABW = 517.4 \times (1 - SG) + 5084 \times (1 - SG)^2 + 33503 \times (1 - SG)^3$ (Eq. (7.12) in the text) SG = Alcohol SG (20°C/20°C) (MEBAK, 2013)	Note the tables themselves require the Density not the SG values for the alcohol. SG and density interconversions are discussed in the Text
OIML Table VII (or Vb) (OIML, 2014)	Alcohol strength by volume of a mixture as a function of its density at 20°C	Obtained from Table V1 (or Va) and a formula relating the alcoholic strength by volume and the alcoholic strength by mass (involves the density of pure ethanol at 20°C)	Details and formulas are included in instructions to the tables from the OIML

A SUMMARY OF BEER CALCULATIONS

An application of Balling's formula and a minimal number of analytical measurements allows the brewer, without the full range of sophisticated instruments, to obtain some quite accurate values for alcohol and extract through the use of a series of approximate beer calculations (Weissler, 1995). Furthermore, an understanding of these calculations allows us to see how more sophisticated instruments may arrive at accurate alcohol by weight and volume values without the requirement of a prior distillation.

In simplifying the previous discussion, it is known that a quite accurate estimate of alcohol content can be derived by subtracting the final or present gravity (PG) as "real extract" (RE—not the apparent extract, AE) from the original extract (OE). This assumes no process dilution, and the application of conversion factors which infer the alcohol content from the drop in wort gravity occurring during fermentation. Using an equation, derived from Balling's formula noted earlier, the alcohol by weight can be calculated:

$$\text{ABWt} = \frac{(\text{OE} - \text{RE})}{(2.0665) - (1.0665 \times \text{OE}/100)} \tag{7.2}$$

in which ABWt is the percentage alcohol by weight (as w/w, grams of alcohol per 100 g of beer), OE, original extract, and RE, real or present extract, in degrees Plato (see Table 7.1 for definitions) and the numerical values are derived as discussed previously.

The original extract of wort is the amount of material extracted from the mash and is measured in grams of extract per 100 g of wort. The brewer will determine the original extract by measuring the specific gravity of the wort and will relate this to the established sucrose tables to report the original extract content in degrees Plato. The original sugar tables are attributable to Goldiner/Klemann and Kampf (MEBAK, 2013; Rosendal and Schmidt, 1987) and were presented in a form as published by the ASBC (ASBC, 1940; see Table 2 for details). The terms Plato and extracts are covered in Table 7.1.

For the brewer limited to the use of a hydrometer, RE can be approximated using the OE (of the original wort) and AE (apparent extract of the final and degassed beer) and by using another empirical equation from Balling: RE, % w/w = (0.1808 × OE) + (0.8192 × AE). However, it is believed that this last equation is not adequate for current purposes and the following modification is proposed instead:

$$\text{RE} = (0.1948 \times \text{OE}) + (0.8052 \times \text{AE}) \tag{7.3}$$

AE = apparent extract—that extract value determined when alcohol is present and as defined in Table 7.1.

This RE assessment will always be an estimate because minerals from water, malt, and any nonethanol volatiles produced during fermentation, will alter the apparent extract (AE). This alternative expression to those seen elsewhere in the literature, and which is based on the relatively constant ratio: (OE−RE)/(OE−AE) = 0.8052 ± 0.0012 (95% confidence interval) for the region OE 9−18 (Plato) and RDF 58−70 (%), is derived from Hackbarth (2009, 2011, and personal communications). A general reassessment of Balling's formula has been presented (Nielsen et al., 2007) and Hackbarth provided a rigorous treatment of SG in ternary solutions of ethanol, sucrose, and water (Hackbarth, 2009, 2011) which led to the updated factor for the "Balling relationship" seen earlier. Using this equation, the brewer will find satisfactory answers to RE values compared to those obtained from official instrumentation over the typical ranges of OE, real degree of fermentation (RDF), and alcohol for most beers (see Spedding, 2013 for a fuller discussion on limitations to the above approaches).

By substituting Plato values in the above formula [Eq. (7.3)], it is possible to solve for the alcohol by weight without directly inputting a calculated or determined RE value:

$$\text{ABWt} = \frac{0.8052 \times (\text{OE} - \text{AE})}{(2.0665) - (1.0665 \times \text{OE}/100)} \tag{7.4}$$

And then, if the final SG value for the beer has also been determined or is known, the alcohol by volume can be deduced with a correction to account for the specific gravity of the beer:

$$\text{ABV} = \frac{\text{ABWt} \times \text{SG beer}}{\text{SG ethyl alcohol at } 20^\circ\text{C}/20^\circ\text{C}} \tag{7.5}$$

in which SG: specific gravity, eg, for the beer or pure ethanol.

This simplifies to:

$$\text{ABV} = \frac{\text{ABWt} \times \text{SG beer}}{0.7907} \tag{7.6}$$

Variant equations are available but alcohol should be reported both % by weight (wt) and by volume (vol) to two decimal places. For reporting purposes, brewers are allowed a tolerance of ±0.3% alcohol by volume.

To use the above set of equations, the brewer needs to have obtained both the OE of the original beer wort and the AE in Plato of the beer. If the brewer uses a specific gravity-reading hydrometer rather than one calibrated in degrees Plato, the Plato value may be determined from sugar tables (see Table 7.2) or via calculation. For Eqs. (7.5) and (7.6), the brewer will need the final SG of the beer.

Two equations exist for interconverting Plato and SG values. These have become known generally as the Lincoln equations (or derivations thereof) with more detail described elsewhere (Lincoln, 1987; Weissler, 1995). One useful equation to convert any Plato value to SG for typical strength beers is:

$$\text{SG} = \frac{^\circ P}{258.6 - \left[\frac{^\circ P}{258.2} \times 227.1\right]} + 1 \tag{7.7}$$

And to convert an SG value to its corresponding Plato value:

$$\text{Plato Extract} = \left[(\text{SG})^2(-205.347) + (668.72)(\text{SG}) - 463.37\right] \quad (7.8)$$

The alternate equation (the Goldiner equation as mentioned earlier) is often found in the European literature and is provided for reference (Kobayashi et al., 2005, 2006; MEBAK, 2013; Rosendal and Schmidt, 1987). Both Eqs. (7.8) and (7.9) yield closely similar values:

$$\text{AE} = -202.414 \times \text{Beer SG}^2 + 662.649 \times \text{Beer SG} - 460.234 \quad (7.9)$$

The ASBC also provided a table of extract in degrees Plato versus SG readings (see Table 2 in ASBC, 1940). It should be noted that these tables are limited to 0–20 g of extract in 100 g of solution (or degrees Plato). See section Measuring Alcohol Content in the Modern Brewery and Brewery Laboratory: Tables, Algorithms, and Software for more on the use of the tables and equations available for the brewer's use.

If instruments are used to determine density values of wort and beer, rather than specific gravity numbers (as SG), an interconversion from density to SG will be required. In these cases, a simplified relationship of Density (ρ) to SG for water, extract, or alcohol solutions can be applied (true density of water is given as 0.998203 g/cm^3, rather than 0.998201 at 20°C, but the latter value is most often used and acceptable):

$$\text{SG} \times 0.998201 = \rho \text{ and,} \quad \text{therefore, } \rho/0.998201 = \text{SG} \quad (7.10)$$

Using the approximate beer calculations noted previously (fully described by Weissler, 1995), the brewer can obtain reliable data for routine work. For more exacting work, the standardized methods in the industry are outlined in the following text. Originally, the officially approved methods called for distillation to obtain the specific gravity of the distillates, then the brewer would look up information about corresponding alcohol by weight or by volume or calculate those values via the related equations (see Table 7.2). Today, for officially accepted values of alcohol content, modern instruments as discussed in sections Coupled Oscillating Density Meters and NIR Alcohol Meters and Density Meter/NIR Analyzer Calculations and the Tabarie Relationship—An Important Interlude are used without the need for prior sample preparation and distillation. Such instruments can calculate the alcohol contents through the use of the various algorithms (see Table 7.2) and through the use of the Tabarie relationship (Table 7.1) which is also detailed in the following.

Without recourse to distillation or the most expensive instrumentation, brewers can make some quite accurate determinations of alcohol content using some basic theory as outlined previously. Furthermore, such theory accounts for the underlying principles behind even the most sophisticated methods used today and provides an understanding as to how these methods work in practice.

DENSITY METER/NIR ANALYZER CALCULATIONS AND THE TABARIE RELATIONSHIP—AN IMPORTANT INTERLUDE

Over 180 years ago, a relationship was determined by E. Tabarie and, although crucial to the determination of alcohol in modern electronic instruments, it has largely been forgotten by brewers. The density and specific gravity values for degassed beer are determined by the density meter. Avoiding the need to analyze distillates, the coupled NIR instruments measure alcohol by volume (see section Coupled Oscillating Density Meters and NIR Alcohol Meters) and then, via the OIML tables (OIML.org, 2014, and see Table 7.2), the corresponding alcohol density is obtained which is equivalent to the density of a volumetric distillate. The concentration of ethanol as percent by weight is computed from the alcohol by volume and the beer density (ABW = ABV*0.78924/Beer Density). As outlined in Table 7.1, the Tabarie equation relates the beer specific gravity (current or present gravity) and the alcohol specific gravity to derive the real extract specific gravity. The SG for the real extract is converted to its Plato value using the sugar tables as outlined in Table 7.2 so other calculations can be performed (see Eq. (7.1)). Next, the alcohol by weight is determined using the alcohol by volume, the beer specific gravity, and the OIML tables. As both density meter and NIR units communicate with each other, a complex series of calculations is performed to solve for the correct values of all parameters (Roman Benes of Anton Paar, personal communication, and see MEBAK, 2013) and the simple outline earlier will suffice for this discussion. The calculators within these instruments can then compute several other parameters of interest to the brewer.

However, the Tabarie relationship does have its limitations. It is to be noted that this relationship only works within a typical range of 2.5% to 8% alcohol by weight, real extract in beer in the region of 3 to 9 Plato, or real degree of fermentation between 65 and 85% (Nielsen and Aastrup, 2004, and personal observations and communications. See also Hackbarth, 2009, 2011). This is a very important issue, often overlooked today, and is addressed elsewhere (Spedding, 2013, and references contained therein). Craft brewers who are making high-alcohol beers much above 10–15% alcohol by volume are thus cautioned about the limitations of modern density meters/NIR-coupled units and should understand the mathematical relationships involved in alcohol calculations including the Tabarie formula (Alvarez-Nazario, 1982; Blunt, 1891; Hackbarth, 2009, 2011; Leonard and Smith, 1897; Nielsen and Aastrup, 2004).

MEASURING ALCOHOL CONTENT IN THE MODERN BREWERY AND BREWERY LABORATORY: TABLES, ALGORITHMS, AND SOFTWARE

Using many of the methods outlined previously, especially pycnometry and hydrometry, extensive investigations were undertaken using defined alcohol–water solutions, distillates, sucrose solutions (Plato), and beer samples to determine their densities and specific gravity values at defined temperatures (historically at 60°F or 15.56°C and later at 20°C or 68°F). Brewing institutes, brewing laboratories, and regulatory agencies collected data sets relating alcohol content and sugar concentrations in solution to their respective gravities. As noted in sections Fermentation and Alcohol Production: The Mass Balance of Fermentation, Carl Balling, and the Theoretical Basis for Brewing Parametrics and Density Meter/NIR Analyzer Calculations and the Tabarie Relationship—An Important Interlude, the tables generated from such data sets, and respective algorithms and equations derived therefrom, have been used in the brewing community for over a century and are also now built into the software of modern instruments to help simplify the work of the brewing chemist (see Table 7.2).

The protocols used by brewers, following officially and internationally adopted recommendations, are now outlined. As these methods have been universally accepted, the American Society of Brewing Chemists' published details will serve to represent the approaches generally taken by all brewing chemists. The beer specific gravity ($SG_{20°C/20°C}$) is obtained as outlined in ASBC method BEER-2, originally by pycnometer and now by the use of a digital density meter (ASBC, 2004 and other editions). The alcohol specific gravity is also determined; originally, this was achieved via distillation, though is now typically done through the use of coupled density meter/NIR alcolyzer devices as discussed in section Coupled Oscillating Density Meters and NIR Alcohol Meters. Method Beer-4A (ASBC, 2004) details volumetric distillation. The grams of alcohol per 100 mL (percent by volume) corresponding to the specific gravity for the distillate is obtained via the table of specific gravity at 20°C/20°C and alcohol content of mixtures of ethyl alcohol and water (ASBC, 1940, and see Table 2). The alcohol as percent by weight is then obtained as follows:

$$\text{ABWt} = \frac{g\ \text{alcohol per 100 mL distillate}}{\text{SG beer}} \tag{7.11}$$

in which, SG beer is the beer specific gravity reading. Alcohol would be reported as % by weight and by volume, to two decimal places.

An alternative to looking up the aforementioned table (Table 2 in ASBC, 1940) would be to apply the following polynomial derived from the OIML tables (OIML, 2014) to determine the percent alcohol by weight from the determined alcohol specific gravity (Rosendal and Schmidt, 1987; MEBAK, 2013).

$$\text{ABWt} = 517.4 \times (1 - \text{SG}) + 5084 \times (1 - \text{SG})^2 + 33{,}503 \times (1 - \text{SG})^3 \tag{7.12}$$

The SG value of alcohol is expressed at 20°C/20°C.

In the above case, the alcohol by volume is then determined (as we saw also in section Fermentation and Alcohol Production: The Mass Balance of Fermentation, Carl Balling, and the Theoretical Basis for Brewing Parametrics) using:

$$\text{ABV} = \frac{\text{ABWt} \times \text{SG beer}}{0.7907} \qquad \text{(Variant Eq. (7.5))}$$

in which 0.7907 is the SG (specific gravity) of pure alcohol at 20°C/20°C (Note: the ASBC Method Beer-4 uses 0.791 for this value).

If the brewer followed the ASBC method Beer-4B to determine the specific gravity from a gravimetric distillation instead of volumetrically, then the ASBC tables would still be consulted (Table 2 in ASBC, 1940), though this time seeking the alcohol by weight corresponding to the specific gravity value. The ASBC then outlines the conversion of the alcohol by weight to alcohol by volume using the equation:

$$\text{ABV} = \frac{\text{ABWt} \times \text{SG beer}}{\text{SG ethyl alcohol at } 20^{\circ}\text{C}/20^{\circ}\text{C}} \qquad \text{(Eq. (7.6) from earlier)}$$

This simplifies to:

$$\text{ABV} = \frac{\text{ABWt} \times \text{SG beer}}{0.791} \qquad \text{(Variant of Eq. (7.5))}$$

which, with the use of 0.791 instead of 0.7907 for the SG of pure ethyl alcohol at 20°C (68°F), is the same approach as using the OIML algorithm to obtain alcohol by weight and then converting to alcohol by volume (MEBAK, 2013).

Today, the coupled density meter/NIR instruments perform all the "steps" described previously and calculate all the parameters to be printed out or sent to a computer for recording of the data. Inline units allow for constant monitoring of the brewing process helping the brewer achieve real-time quality control monitoring of beer production and maintaining brewhouse efficiencies. Laboratory testing will be performed on the final product.

INTRODUCTION TO AN ONLINE CALCULATOR AND BREWERS' CALCULATIONS SOFTWARE

Computer technology and mathematical spreadsheet programs make it easier today for brewers to check their data. In conclusion, we note the availability of an online and robust calculator, the Scandinavian Beer Calculator (Beercalc.com, 2014). The beer calculator, complete with detailed instructions, allows brewers to perform complex equations quickly and allowing them to obtain alcohol and extract values and more. In addition to the Scandinavian beer calculator, it is also noted that James Hackbarth developed the *Master Brewers Toolbox*, a relational database software program for beer specifications and formulations, which is available from the Master Brewers Association of the Americas (MBAA). These are powerful tools to be added to the brewer's arsenal.

SUMMARY

Understanding fermentation principles and the physical, chemical, and biochemical properties of alcohol has allowed brewers to determine beer alcohol content or to determine its "alcoholic strength" for over 180 years. The methods outlined here and the detectors described for alcohol determination (both in the laboratory and with inline process applications) may take on ever more sophisticated forms, including the incorporation of miniaturized versions of sensors and more involved spectroscopic or chromatographic techniques. These may be developed, but the basic principles outlined here—along with a careful review of the references—should provide the reader with enough information to keep ahead in understanding this important topic. Additional robust algorithms may be developed, minor adjustments may be made to the background theory, and ever more sensitive detection levels for alcohol will come to bear here, but the theories developed up to now for monitoring this important analyte will remain sound.

ACKNOWLEDGMENTS

I should like to thank James Hackbarth and Joe Power for providing their insights over many years into the measurement of beer attributes, its constituents, and especially alcohol. Any errors here remain mine though their insights are embedded throughout this work. I also thank Amber Weygandt, Matt Linske, and Tony Aiken for carefully reading through the manuscript and ironing out areas of confusion, certain grammatical errors, or issues. Again, any faults or errors of fact that may have been overlooked remain my responsibility.

REFERENCES

Alvarez-Nazario, F., 1982. Evaluation of Tabarie's formula. Journal of the Association of Official Analytical Chemists 65 (3), 765–767.

Analytica-EBC, 1998. European Brewery Convention. Verlag Hans Carl.

Anger, H.-M., Schildbach, S., Harms, D., Pankoke, K., 2009. Analysis and quality control. In: Eßlinger, H.M. (Ed.), Handbook of Brewing: Processes, Technology, Markets. Wiley-VCH, pp. 437–476.

Applegate, H.E., 1993. Wine analysis. In: The Complete Handbook of Winemaking. The American Wine Society and G.W. Kent, Inc., pp. 72–78.

ASBC, 1940. Specific gravity and degrees Plato of sugar solutions or per cent extract by weight – table 1 and Relationship between specific gravity at 20°C/20°C and alcohol content of mixtures of ethyl alcohol and water – table 2. In: Tables Related to Determinations on Wort, Beer, and Brewing Sugars and Syrups, third ed. The American Society of Brewing Chemists.

ASBC, 2004. Methods of Analysis Manual. American Society of Brewing Chemists.

Austin, E., 1989. Scaba method for alcohol and original gravity content in beer. ASBC Journal 47 (4), 130–132.

AOAC, 1995. Percentages by volume at 15.56°C (60°F) of ethyl alcohol corresponding to apparent specific gravity at various temperatures – table 913.02. In: International Official Methods of Analysis, sixteenth ed. Association Official Agricultural Chemists.

Bamforth, C.W., 2002. Standards of Brewing: A Practical Approach to Consistency and Excellence. Brewers Publications.

Beercalc.com, 2014. The Scandinavian Brewers Calculator. http://www.beercalc.com/ (last accessed June, 2014).

Blunt, T.P., 1891. Note on Tabarie's process for the indirect determination of alcohol. The Analyst 221–224.

Boulton, C., 2013. Encyclopaedia of Brewing. Wiley Blackwell.

Brown, D.G.W., 1977. Estimation of original gravity by refractometer. Journal of the Institute of Brewing 83, 41–45.

Buckee, G.K., 1982. Refractometric measurement of original gravity. Journal of the Institute of Brewing 88, 39–42.

Buglass, A.J., Caven-Quantrill, D.J., 2011. Analytical methods. In: Buglass, A.J. (Ed.), Handbook of Alcoholic Beverages: Technical, Analytical and Nutritional Aspects, vol. 2. John Wiley and Sons.

Cacho, J.F., Lopez, R., 2005. Alcoholic beverages. In: Worsford, P., Townshend, A., Poole, C. (Eds.), Encyclopedia of Analytical Science. Elsevier Ltd, pp. 285–291.

Caputi Jr., A., 1970. Determination of Ethanol in wine by chemical oxidation; 1969 studies. Journal of the Association of Official Analytical Chemists 53 (1), 11–12.

Caputi Jr., A., Wright, D., 1969. Collaborative study of the determination of ethanol in wine by chemical oxidation. Journal of the Association of Official Analytical Chemists 52 (1), 85–88.

Caputi Jr., A., Ueda, M., Brown, T., 1968. Spectrophotometric determination of ethanol in wine. American Journal of Enology and Viticulture 19 (3), 160–165.

Castellari, M., Sartini, E., Spinabelli, U., Riponi, C., Galassi, S., 2001. Determination of carboxylic acids, carbohydrates, glycerol, EtOH, and 5-HMF in beer by high-performance liquid chromatography and UV–refractive index double detection. Journal of Chromatographic Science 39, 235–238.

Clarkson, S.P., Ormrod, I.H.L., Sharpe, F.R., 1995. Determination of ethanol in beer by direct injection gas chromatography: a comparison of six identical systems. Journal of the Institute of Brewing 101, 191–193.

Criddle, W.J., 2005. Ethanol. In: Worsford, P., Townshend, A., Poole, C. (Eds.), Encyclopedia of Analytical Science. Elsevier Ltd, pp. 562–569.

Crowell, E.A., Ough, C.S., 1979. Research note: a modified procedure for alcohol determination by dichromate oxidation. American Journal of Enology and Viticulture 30 (1), 61–63.

Cutaia, A.J., Reid, A.-J., Speers, R.A., 2009. Examination of the relationships between original, real and apparent extracts, and alcohol in pilot plant and commercially produced beers. Journal of the Institute of Brewing 115 (4), 318–327.

Daniels, D., 1990. On-line alcohol and real extract measurement. Brewers Digest, September 14–15.

Dyer, R.H. (Ed.), 1995. Distilled liquors (Chapter 26). In: AOAC Official Methods of Analysis, sixteenth ed. AOAC, International, MD.

Essery, R.E., 1954. Determination of extract and cold water extract in malt by means of the refractometer. Journal of the Institute of Brewing 60, 303–306.

Essery, R.E., Hall, R.D., 1956. Determination of Original gravity of beer by means of the refractometer. Journal of the Institute of Brewing 62, 153–155.

Fritz, J.A., Schenk, G.H., 1987. Quantitative Analytical Chemistry, fifth ed. Allyn and Bacon, Inc.

García, M.J.L., Alfonso, E.F.S., 2013. HPLC analysis of alcohols in foods and beverages. In: Nollet, L.M.L., Toldrá, F. (Eds.), Food Analysis by HPLC, third ed. CRC Press Taylor & Francis Group, Boca Raton, FL, pp. 253–270.

Garrigues, S., de la Guardia, M., 2009. Methods for the vibrational spectroscopy analysis of beers. In: Preedy, V.R. (Ed.), Beer in Health and Disease Prevention. Elsevier/Academic Press, pp. 943–961.

Gonchar, M.V., Maidan, M.M., Pavlishko, H.M., Sibirny, A.A., 2001. A new oxidase-peroxidase kit for ethanol assays in alcoholic beverages. Food Technology and Biotechnology 39 (1), 37–42.

Hackbarth, J.J., 2009. The effect of ethanol–sucrose interactions on specific gravity. Journal of the American Society of Brewing Chemists 67 (3), 146–151.

Hackbarth, J.J., 2011. The effect of ethanol–sucrose interactions on specific gravity. Part 2: a new algorithm for estimating specific gravity. Journal of the American Society of Brewing Chemists 69 (1), 39–43.

Hu, N., Wu, D., Cross, K., Burikov, S., Dolenko, T., Patsaeva, S., Schaefer, D.W., 2010. Structurability: a collective measure of the structural differences in vodkas. Journal of Agricultural and Food Chemistry 58, 7394–7401.

Hudson, J.R., 1975. The Institute of Brewing Analysis committee estimation of the original gravity of beer. Journal of the Institute of Brewing 81, 318–321.

Institute of Brewing, 1997. Recommended Methods of Analysis, Vol. 1, Analytical. The Institute of Brewing.

Jacobson, J.L., 2006. Introduction to Wine Laboratory Practices and Procedures. Springer Science + Business Media, LLC.

Jiggens, P., 1987. Gravity measurement by the Oscillating U-tube method. The Brewer, September, 403–405.

Joslyn, M.A., 1970. Alcoholometry. In: Joslyn, M.A. (Ed.), Methods in Food Analysis: Physical, Chemical, and Instrumental Methods of Analysis, second ed. Academic Press, pp. 447–474.

Klein, H., Leubolt, R., 1993. Ion-exchange high-performance liquid chromatography in the brewing industry. Journal of Chromatography 640, 259–270.

Kobayashi, M., Hiroshima, T., Nagahisa, K., Shimizu, H., Shioya, S., 2005. On-line estimation and control of apparent extract concentration in low-malt beer fermentation. Journal of the Institute of Brewing 111 (2), 128–136.

Kobayashi, M., Nagahisa, K., Shimizu, H., Shioya, S., 2006. Simultaneous control of apparent extract and volatile compound concentrations in low-malt beer fermentation. Applied Microbiology and Biotechnology 73 (3), 549–558.

Korduner, H., Westelius, R., 1981. Rapid Automatic Method for Alcohol and Original Gravity Determination. EBC Congress, Copenhagen, pp. 615–622.

Koestler, P., Hagen, W., 1999. Determination of original gravity and alcohol content of Dark Starkbiers with the refractometer. MBAA-TQ 36 (2), 231–233.

Kovar, J., 1981. Oscillating U-tube density meter determination of alcoholic strength: analysis of parametric errors. Journal of the Association of Official Analytical Chemists 64 (6), 1424–1430.

Kunze, W., 2010. Technology Brewing & Malting, fourth ed. VLB Berlin.

LaBerge, D.E., 1979. Determination of Specific gravity of malt extracts, worts, and beer. Journal of the American Society of Brewing Chemists 37 (2), 105–106.

Leonard, N., Smith, H.M., 1897. Note on the indirect (Tabarie's) method for the estimation of alcohol. The Analyst 225–228.

Lide, D.R. (Ed.), 2005. CRC Handbook of Chemistry and Physics Eighty, fifth ed. CRC Press.

Lincoln, R.H., 1987. Computer compatible parametric equations for basic brewing computations. Technical Quarterly – the Master Brewers Association of Americas 24, 129–132.

Logan, B.K., Case, G.A., Distefano, S., 1999. Alcohol content of beer and malt beverages: forensic considerations. Journal of Forensic Sciences 44 (6), 1292–1295.

Masschelin, J., 2010. GC: validation report – alcohol content by gas chromatography. Alcohol and Tobacco Tax and Trade Bureau.

MEBAK, 2013. Brewing Analysis Methods: Wort, Beer & Beer-Based Beverages (Mitteleuropaische Brautechnische Analysenkommission).

Meija, J., 2009. Solution to Mendeleyev vodka challenge. Analytical and Bioanalytical Chemistry 395, 7–8.

Mehrotra, R., 2000. Infrared spectroscopy, gas chromatography/infrared in food analysis. In: Meyers, R.A. (Ed.), Encyclopedia of Analytical Chemistry: Applications, Theory and Instrumentation. John Wiley and Sons.

Nielsen, H., Aastrup, S., 2004. Improving Tabarie's formula. Scandinavian Brewers' Review 61 (4), 30–34.

Nielsen, H., Kristiansen, A.G., Lassen, K.M.K., Erikstrom, C., 2007. Balling's formula – scrutiny of a brewing dogma. BRAUWELT International 25 (2), 90–93.

OIML.org, 2014. The Alcoholometric Tables. http://www.oiml.org/en/files/pdf_r/r022-e75.pdf/view (last accessed June, 2014).

Osborne, B.G., 2000. Near-infrared spectroscopy in food analysis. In: Meyers, R.A. (Ed.), Encyclopedia of Analytical Chemistry: Applications, Theory and Instrumentation. John Wiley and Sons.

Powers, J.R., 2000. Enzyme analysis and bioassays in food analysis. In: Meyers, R.A. (Ed.), Encyclopedia of Analytical Chemistry: Applications, Theory and Instrumentation. John Wiley and Sons.

Prencipe, L., Iaccherl, E., Manzati, C., 1987. Enzymic ethanol assay: a new colorimetric method based on measurement of hydrogen peroxide. Clinical Chemistry 33 (4), 486–489.

Proudlove, M.O., 1992. The use of near Infra-red spectroscopy in malting and brewing. Ferment 5 (4), 287–292.

Rosendal, I., Schmidt, F., 1987. The alcohol table for beer analysis and polynomials for alcohol and extract. Journal of the Institute of Brewing 93, 373–377.

Schroeder, M.R., Poling, B.E., Manley, D.B., 1982. Ethanol densities between −50 and 20°C. Journal of Chemical & Engineering Data 27, 256–258.

Sharpe, F.R., 1991. In/on-line sensor specifications for brewing industry analytes. Ferment 4 (4), 217–218.

Schwiesow, M.H. (Ed.), 1995. Malt beverages and brewing materials (Chapter 27). In: Official Methods of Analysis, sixteenth ed. AOAC International, Gaithersburg, MD.

Skoog, D.A., West, D.M., 1986. Analytical Chemistry: An Introduction, fourth ed. CBS College Publishing.

Stewart, L.E., 1983. Alcohol proof determination from absolute specific gravity (20°C/20°C) using oscillating U-tube digital density meter with a programmable calculator. Journal of the Association of Official Analytical Chemists 66 (6), 1400–1404.

Strunk, D.H., Aicken, J.C., Hamman, J.W., Andreasen, A.A., 1982. Density meter determination of proof of ethanol-water solutions: collaborative study. Journal of the Association of Official Analytical Chemists 65 (2), 218–223.

Spedding, G., 2012. Chromatography. In: Oliver, G. (Ed.), The Oxford Companion to Beer. Oxford University Press, pp. 248–249.

Spedding, G., 2013. Empirically measuring and calculating alcohol and extract content in wort and beer with a reasonable degree of accuracy and confidence – using a series of inter-related and conversion equations, algorithms, tables and an on-line calculator. Brewers Digest September–October, 39–50.

Tipparat, P., Lapanantnoppakhun, S., Jakmunee, J., Grudpan, K., 2001. Determination of ethanol in liquor by near-infrared spectrophotometry with flow injection. Talanta 53, 1199–1204.

Tonelli, D., 2009. Methods for determining ethanol in beer. In: Preedy, V.R. (Ed.), Beer in Health and Disease Prevention. Elsevier/Academic Press, pp. 1055–1065.

Upperton, A.M., 1985. Determination of ethanol in beverages low in alcohol. Journal of the Institute of Brewing 91, 151–153.

Waites, M.J., Bamforth, C.W., 1984. The determination of ethanol in beer using a bio-electrochemical cell. Journal of the Institute of Brewing 90, 33–36.

Weissler, H.E., 1995. Brewing calculations. In: Hardwick, W.A. (Ed.), Handbook of Brewing. Marcel Dekker, Inc., pp. 643–705

Zimmermann, H.W., 1963. Studies on the dichromate method of alcohol determination. American Journal of Enology and Viticulture 14, 205–213.

CHAPTER

Flavorsome Components of Beer

8

C.W. Bamforth

University of California, Davis, California, CA, United States

The only true and meaningful way in which to assess the flavor of beer is through organoleptic analysis, a topic embraced by Bill Simpson in Chapter 13, Sensory Analysis in the Brewery. Notwithstanding this, there is, of course, always a demand for some form of "back up" in terms of instrumental analysis, generating values that, hopefully, substantiate what a taste and aroma analysis has revealed. Such measurements certainly facilitate control strategies, are useful in terms of achieving consensus (eg, between a brand holder and franchisee), and allow for the long-term monitoring of performance.

Practical issues pertaining to the flavor of beer have been addressed by this author elsewhere (Bamforth, 2014). The focus in the present chapter is primarily on analytical techniques (other than organoleptic) which can be used to assess the myriad of substances in beer that may impact flavor. In addition, it includes the changes that may take place in some of these substances over the shelf life of beer that lead to a perceived change in flavor, whether advantageous or, more usually, disadvantageous.

APPROACHES TO MEASURING SUBSTANCES IMPACTING BEER FLAVOR

The contributors to the flavor of beer can be classified into those that primarily influence taste, entities that are for the most part nonvolatile and those that impact the aroma, which tend to have greater volatility. It is recognized that the greater part of the flavor of beer is detected through the nose, either directly or retronasally. The principal techniques for measuring nonvolatile and volatile components are high-performance liquid chromatography (HPLC) and gas chromatography (GC), respectively.

In considering the measurement of the wide diversity of molecules that can contribute to the flavor of beer, it is perhaps most useful to divide them according to the frequency with which they might be measured in the brewery.

Some of these substances are typically measured on every batch, indeed at the stage in the process in which a deviation from the laid-down specification can be corrected. In terms of specific types of molecule in this category, we include the vicinal diketones, the iso-α-acids (perhaps more usually cumulatively as bitterness units), carbon dioxide (its direct contribution to flavor, of course, being directly though

Brewing Materials and Processes. http://dx.doi.org/10.1016/B978-0-12-799954-8.00008-3

the trigeminal sense—pain), ethanol, and pH (viz. sourness). We might also include sulfur dioxide (SO_2), which certainly can makes a contribution to the aroma of beers (struck matches) but which would primarily be quantified in markets such as the United States owing to the legal expedient of obligatory labeling if the total SO_2 level exceeds 10 mg/L.

It is sufficient to quantify most other substances on a periodic basis, perhaps monthly. In some cases, we can specifically identify some of these materials, eg, esters such as isoamyl acetate, ethyl acetate, and phenylethyl acetate; certain higher alcohols; sulfur-containing substances, notably dimethyl sulfide (there are several brands for which DMS is measured in every batch); sugars; and inorganic ions. However, some of the flavor characteristics of beer are less readily defined and therefore a challenge to quantify by instrumental analysis. Examples here would be hop aroma, whether late or dry, and the spectrum of maltiness as delivered by the range of malts that brewers use from the very pale to the very black. Although it is certainly possible to dissect these malt and hop contributions into various molecules, it is certainly not possible at this stage to nominate one or even a very few compounds to represent marker substances to quantify these flavor attributes, drawing attention to the key role of sensory panels (see also Chapters 1, Malts and 3, Hops).

Perhaps it is at this point that one might mention the attempts that have been made to develop artificial systems that mimic the nose (Ghasemi-Varnamkhasti et al., 2011) and the palate (Blanco et al., 2015). As yet, such devices are very much at a development stage and certainly not ready for routine consideration in the day-to-day quality operations of a brewery.

BITTERNESS

Still, the easiest way to quantify the bitterness level of beer is via the Brenner method [American Society of Brewing Chemists (ASBC, 2015) Beer-23A], in which beer is extracted with iso-octane and the absorbance at 275 nm of the resultant solution multiplied by 50 yields Bitterness Units. Recognizing that the actual bitterness of the six different iso-α-acids differs and that the proportions of the more bitter *cis* isomers tend to be higher in postfermentation bittering materials, beers that are bittered using extracts to the finished product may have a factor of 70 applied in the BU calculation.

Quantification of the specific iso-α-acids can be obtained by high-performance liquid chromatography (HPLC; ASBC Beer-23C, E). It is the present author's contention that this is overkill and that, relative to the precision with which even expert tasters can quantify bitterness organoleptically, the Brenner method is adequate (Bamforth, 2014).

SWEETNESS

There is no current ASBC method for specific sugars in beer, though there is for the assessment of these materials in wort (ASBC Wort-14, ASBC Wort-19, and ASBC Wort-22). In any event, this is an analysis which is perhaps only worthwhile on a troubleshooting basis, or for beers in which sugar is added downstream to afford sweetness. The other example would be those beers (eg, milk stouts) to which lactose is added, although there are no recommended methods for this sugar in the current portfolio. Otherwise, for most beers, the brewer will likely be seeking full conversion of fermentable carbohydrates in wort by yeast, assessing the same by measurement of specific gravity (ASBC Beer-2) and alcohol (ASBC Beer-4).

SOURNESS

This is directly measured by pH (ASBC Beer-9), although may also be quantified by total acidity (ASBC Beer-8).

SALTINESS

The salt character of beer is due to the presence of sodium and potassium, which can be measured by atomic absorption spectroscopy (ASBC Beer-36 and Beer-37, respectively).

OTHER INORGANIC IONS

The taste of beer is also impacted by other ions, notably chloride and sulfate, which can be estimated by ion chromatography (ASBC Beer-43).

VICINAL DIKETONES (VDKs)

Although there are colorimetric procedures, the most reliable approach to measuring the two principal VDKs involves GC (ASBC Beer-25F). It is important to highlight that this method incorporates a heating stage, to convert the precursors, acetolactate and acetohydroxybutyrate, to the free VDKs. This ensures that no precursor is still present in beer that would be converted to VDK during the lifetime of the beer.

DIMETHYL SULFIDE

Although Flame Photometric detection following GC was formerly the method of choice for dimethyl sulfide (DMS), more recently chemiluminescence detection has been used (ASBC Beer-44).

METHODS FOR MEASURING OTHER VOLATILE MATERIALS IN BEER

Aldehydes, alcohols, and esters are also measured by GC; see, for example, ASBC Beer-29, ASBC Beer-48. Of course, ethanol per se has a direct and indirect impact on beer flavor, through its measurement is performed routinely from a control of beer strength and the taxation aspect (ASBC Beer-4).

FLAVOR STABILITY ASSESSMENT

All too often, researchers have focused on the development of a single molecule, E-2-nonenal, as the determinant of staling in beer. Although this substance does indeed possess a cardboard-like note and can be found in aged beer, the chemistry of beer staling is substantially more complicated than this (Vanderhaegen et al., 2006; Bamforth, 1999, 2011). Many substances, including a diversity of carbonyl-containing species, can contribute to the aging of beer. As such, it is hard to recommend any one analytical approach, but rather focus on organoleptic procedures, but even here approaches have been generally inadequate (Meilgaard, 2001). A detailed critique is offered elsewhere (Bamforth, 2016). If there is one instrumental technique to have found favor, it is the application of electron spin resonance spectroscopy: see, for example, ASBC Beer-46, which explores the oxidative resistance of beer.

REFERENCES

American Society of Brewing Chemists, 2015. Methods of Analysis 14th Edition. http://methods.asbcnet.org.

Bamforth, C.W., 1999. The science and understanding of the flavour stability of beer: a critical assessment. Brauwelt International 17, 98–110.

Bamforth, C.W., 2011. 125th anniversary review: the non-biological instability of beer. Journal of the Institute of Brewing 117, 488–497.

Bamforth, C.W., 2014. Practical Guides for Beer Quality: Flavor. American Society of brewing Chemists, St Paul MN.

Bamforth, C.W., 2016. Practical Guides for beer quality: Freshness. American Society of brewing Chemists, St Paul MN forthcoming.

Blanco, C.A., de la Fuente, R., Caballero, I., Rodríguez-Méndez, M.L., 2015. Beer discrimination using a portable electronic tongue based on screen-printed electrodes. Journal of Food Engineering 157, 57–62.

Ghasemi-Varnamkhasti, M., Saeid Mohtasebi, S., Siadat, M., Lozano, J., Ahmadi, H., Hadi Razavi, S., Dicko, A., 2011. Aging fingerprint characterization of beer using electronic nose. Sensors and Actuators B 159, 51–59.

Meilgaard, M., 2001. Effects on flavour of innovations in brewery equipment and processing: a review. Journal of the Institute of Brewing 107, 271–286.

Vanderhaegen, B., Neven, H., Verachtert, H., Derdelinckx, G., 2006. The chemistry of beer aging – a critical review. Food Chemistry 95, 357–381.

CHAPTER

Dissolved Gases

9

C.S. Benedict[†]

Dissolved gases play an important role in the quality of finished beer. Fermentation requires adding oxygen at optimal levels, but oxygen (O_2) exposure should be minimized during the rest of the brewing process and in the final product. The carbon dioxide (CO_2) produced during fermentation adds desirable qualities to the finished product and must be measured to maintain sensory consistency and package integrity. Nitrogen (N_2) is sometimes added to certain types of finished beer for its impact on foam and on the palate. This chapter will focus on the practical aspects of dissolved gases in beer and wort, primarily concentrating on dissolved oxygen, but also discussing dissolved carbon dioxide and dissolved nitrogen.

THEORY: LAWS GOVERNING GASES DISSOLVED IN LIQUIDS

Dalton's Law, Henry's Law, and the Ideal Gas Law will give a brewer most of the tools needed to understand how gases dissolve and behave in liquids.

DALTON'S LAW

$$P_{Total} = P_{Carbon\ dioxide} + P_{Oxygen} + P_{Nitrogen}$$

Dalton's law states that the total pressure of all gases in a mixture of gases is equal to the partial pressure of all the gases added together. On a practical level, it can be helpful to think of the partial pressure as the percentage of a particular gas in beer (Silberberg, 2009).

[†]*Sadly, our good friend Chaz Benedict died on March 17th 2015.* As stated in a press release from the American Society of Brewing Chemists: "Chaz was an application development manager for the Hach Company with a decades long passion for understanding and refining the measurement of oxygen in beer. Chaz worked with and befriended many brewing scientists across the industry and was greatly respected for his contributions to the field of package oxygen measurement." Final edits to this chapter were undertaken by the editor of this volume.

Brewing Materials and Processes. http://dx.doi.org/10.1016/B978-0-12-799954-8.00009-5

HENRY'S LAW

Gas concentration = gas solubility$_{\text{(constant temperature)}}$ × gas partial pressure

Henry's Law builds on Dalton's Law. Henry's Law states that the solubility of a gas—how well a gas dissolves into a liquid—is relative to the temperature of the liquid and the concentration of solute in that liquid. With regard to beer and wort, gases dissolve to a greater extent in lower Plato liquids. Dissolved-gas analyzers all use a fixed solubility and assume that the user will develop standards around that solubility, taking into account the known Plato value of the liquid they are measuring.

THE IDEAL GAS LAW

$$PV = nRT$$

The Ideal Gas Law is also related to Dalton's Law. It simply states that the pressure (*P*) and volume (*V*) of a gas multiplied together is equal to the number of moles (*n*) of gas times the temperature ($T\,^{\circ}K$) and a gas constant (R). Many units can define the gas constant, but $R = 0.08205\ L\ atm\ deg^{-1}\ mol^{-1}$ is common and easy to use. The Ideal Gas Law is useful when determining how much gas is being introduced into a liquid, whether it is the injection of carbon dioxide or nitrogen into beer, or oxygen or air into wort.

UNITS: DISSOLVED GAS NOMENCLATURE

The amounts of most dissolved gases are expressed in units of the weight of the gas being measured per volume of liquid. Depending upon the particular solubility of the gas, the units might be mg/L, μg/L, or g/L (which is sometimes noted as g/kg.) These are the typical units for each type of gas:

OXYGEN

Dissolved oxygen units may be expressed as mg/L or μg/L; these are also used interchangeably with "parts-per-million" (ppm) and parts per billion (ppb). Gas-phase units are usually percent O_2 concentration or volume/volume units of gas phase ppm. Units of ppm can be confusing when switching between gas and liquid phases, as 10 ppm gas phase is equal to 0.001% O_2, but would account for less than 1.0 ppb liquid phase.

CARBON DIOXIDE

CO_2 units differ, depending upon where in the world one is making measurements. The United States has adopted the unit volume/volume (V/V), whereas most of the rest of the world uses grams-per-kilogram (g/kg), which is equivalent to slightly less than 2 g/kg (per V/V). The unit V/V is defined as 1 L of CO_2 for every liter of liquid.

NITROGEN

Nitrogen (dissolved) is usually defined as units of ppm.

Fig. 9.1 shows the relative solubility of O_2 and N_2 in water.

WHY MEASURE DISSOLVED GASES?

Carbon dioxide is by far the most soluble gas found in beer. It is a natural byproduct of fermentation and will be retained by beer in the fermenter based on the beer temperature and fermenter overpressure. Consistent CO_2 concentration is a cornerstone of the many predictable sensory attributes of beer. The concentration of CO_2 found in most North American beers is between 2.4 and 2.9 V/V (4.76 and 5.75 g/kg).

N_2 is dissolved into beer to both change the appearance and size of the bubbles (ie, the foam) as well as the sensory attributes of the beer. Nitrogenated beer tends to have a smoother palate than nonnitrogenated beer. Nitrogen concentrations in non-nitrogenated beer run between 0.5 and 2 ppm, whereas nitrogenated beer typically

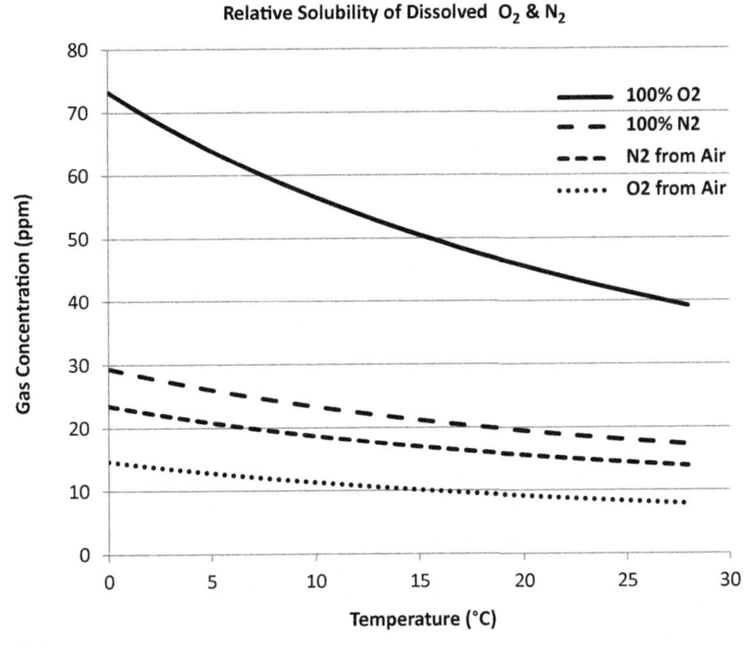

FIGURE 9.1

The relative solubility of pure oxygen, pure nitrogen, nitrogen from air, and oxygen from air, dissolved in water at 1000 mbar of total pressure.

ranges between 30 and 45 ppm. The CO_2 concentration of nitrogenated beer is usually 1.2–1.4 V/V, which is about half the concentration generally found in nonnitrogenated products. The gas mixtures used to nitrogenate beer are typically about 60–70% N_2 and 30–40% CO_2.

Unlike carbon dioxide and nitrogen, oxygen nearly always has a negative effect on beer, including flavor degradation and haze formation. In particular, trace metals in beer catalyze oxidative reactions, thus altering key flavor compounds (Bamforth et al., 1993). Although the damage from oxygen is irreversible, many of the flavors associated with oxidation are slow to reach threshold values. Exposure to high temperatures and other negative storage conditions may accelerate off-flavor development (Saison et al., 2010).

However, this is not to imply that oxygen must be eliminated from every phase of beer making. Oxygen is a critical component of fermentation, and best practices include the injection of concentrated oxygen (8–20 ppm) into wort. Oxygenation of wort is dictated by the specific gravity of the wort and the type of yeast used for fermentation. Postfermentation is when the quest for low oxygen begins, as measuring focuses on eliminating as many unwanted sources of oxygen as possible.

All oxygen measurements must be made in real time, because the half-life of an oxygen reaction will vary depending upon beer type and temperature. Chemical compounds and live yeast will interact with oxygen in beer at a faster rate at warmer temperatures. This is often referred to as oxygen "consumption."

Fig. 9.2 shows oxygen "consumption" in beer, with consumption meaning the rate at which live yeast and trace metals in beer will quickly react with oxygen, thereby making any measurements not taken in real-time invalid. The x-axis of this graph has no units, because the half-life of these reactions can range from minutes to a day. Factors that affect the half-life of oxygen reactions are beer temperature, bottle conditioning, specific gravity, and beer type. Because half-lives of oxidative reactions are so unpredictable, quality checks looking for oxygen concentration are always made with portable and inline instrumentation as the beer is being transferred. Once beer has been sitting in a package or vessel for a day or two, it is impossible to reconstruct its original dO_2 level.

MEASUREMENT TECHNOLOGIES—O_2, CO_2, AND N_2

OXYGEN

Measurement of dissolved oxygen in beer first began in the 1960s using a reaction of oxygen with reduced indigo carmine [American Society of Brewing Chemists (1976)], a process that involved adding disodium indigo disulfonate to a beer sample and taking a colorimetric measurement. In the 1980s, brewers dramatically improved their measuring capabilities with the use of electrochemical sensors (Mitchell et al., 1983). Then around 2008, another advance was made with the widespread adoption of optical sensors, popular among brewers for their robust nature

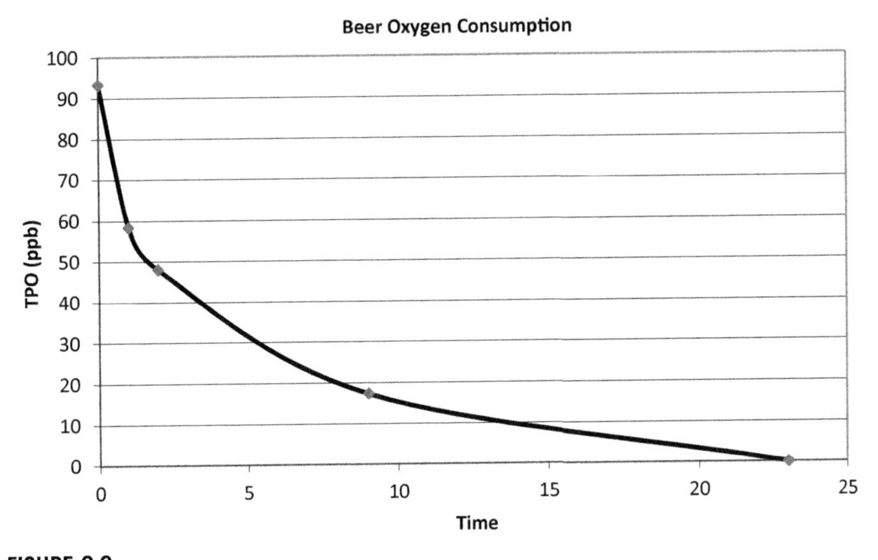

FIGURE 9.2

The oxygen consumption of beer over time.

and ease of maintenance. They are as accurate as electrochemical sensors and now account for over 95% of the devices sold to the beverage industry.

All oxygen-sensing technologies measure the partial pressure of oxygen in liquids. Sensor manufacturers apply Henry's Law using standard beer solubility factors and give the user a measurement in ppm or ppb dissolved oxygen (dO_2).

Optical sensors use a luminescent polymer in direct contact with the gas or liquid being measured (Fig. 9.3). Oxygen permeates into the polymer, and the polymer's florescent properties are quenched when excited by a pulse of light, with the result that the amount of quenching is proportional to the dO_2 concentration of the liquid.

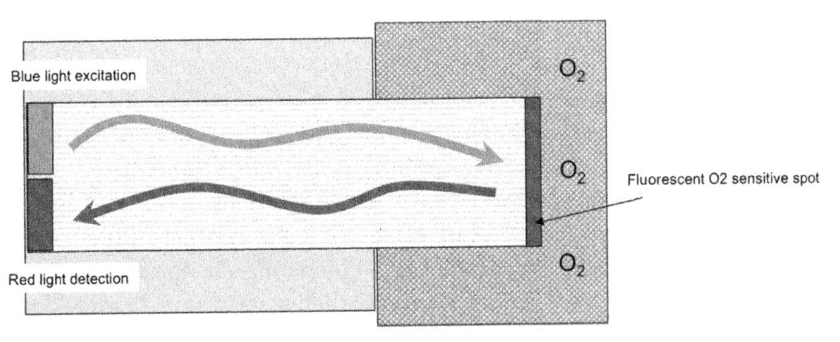

FIGURE 9.3

Optical oxygen sensor.

Optical measurement has an advantage in that the polymer matrix does not consume oxygen. Also, because the polymer is rigid, pressure changes in the liquid do not affect the stability of the readings.

The only real limiting factor in optical sensors is that they have a narrow dynamic range when compared to electrochemical sensors. A good electrochemical sensor can measure from 0.001 to 20 ppm with excellent linearity, whereas optical sensors measure most accurately in a specifically limited range. For example, a sensor that measures well at 0.001 ppm will probably not have great linearity above 2.0 ppm. Likewise, an optical sensor that measures well at high levels may have a low-end sensitivity of only 0.02 or 0.05 ppm. Brewers must take care to pair the right optical instrument with specific brewery applications.

CARBON DIOXIDE

There are four ways to measure carbon dioxide.

Pressure–Temperature

Worldwide today, most carbon dioxide measurements are calculated using a pressure–temperature (PT) formula standardized by the American Society of Brewing Chemists (1947). The pressure and temperature of beer in a vessel are measured with the assumption that 100% of the pressure in the container is CO_2. Some PT analyzers rely upon mechanical shaking, whereas others start a hydrolysis reaction in the beer to create small bubbles that trigger a cascade, thus elevating the CO_2 pressure to an equilibrium that is the same as shaking. Whether using shaking or hydrolysis, if O_2 as air is present in the liquid, it can lead to an overestimation of CO_2 concentration by 0.1 V/V CO_2 for each 1.0 ppm of dO_2.

Multivolume Expansion

The 1990s saw the introduction of a novel and popular method of measuring CO_2 by expanding the volume of a liquid sample multiple times. This method takes into account that the first gases out of solution will be oxygen and nitrogen, leaving behind the CO_2. Using a series of stirred volume expansions, the method accurately determines true CO_2 with minimal interference from O_2 or N_2.

Thermal Conductivity

CO_2 concentration can also be determined using a thermal conductivity (TC) sensor placed under a gas-permeable membrane. First, a reference gas is allowed to flow under the membrane to establish a background. After establishing the background, the reference flow stops, and all the gases in the liquid permeate into the space around the TC sensor. The CO_2 in the mixture is then measured by maintaining the TC sensor at a constant temperature and comparing the thermal load of the measured gas to the background gas. Like multivolume expansion, thermal conductivity gives a true CO_2 reading without interference from O_2 or N_2.

Infrared

Infrared spectroscopy evaluates the amount of energy absorbed by CO_2 when infrared light is projected into a liquid sample. Chemical bonds have specific frequencies that vibrate with distinct characteristics. The Beer–Lambert Law is critical to molecular spectroscopy. Using the formula associated with this law, infrared light is directed into a sample of beer and the amount of energy absorbed by CO_2 is measured.

NITROGEN

Thermal conductivity (TC) is the only direct method for measuring nitrogen in liquid. CO_2 is used as the reference gas. The only drawback to TC measurement of nitrogen in beer is that dissolved oxygen can interfere with the N_2 reading, so it is important to measure nitrogen when dO_2 levels are low.

MEASURING DISSOLVED GASES: CHALLENGES

GENERAL PRINCIPLES

The accurate measurement of gases during the brewing process hinges on one important factor: whatever the gas, it must be measured at a point at which it is fully dissolved in liquid, be it wort or beer. For example, air (or pure oxygen) is injected into wort to start fermentation, but the oxygen may not dissolve completely due to the solubility of air in wort or improper mixing of the gas and liquid.

Likewise, measuring dissolved oxygen in beer can be a challenge because the CO_2 produced during fermentation wants to establish equilibrium with atmospheric pressure, and CO_2 that is off-gassing will carry oxygen with it. Because there is so much CO_2 in beer, all oxygen measurements must be made above the pressure at which the CO_2 will come out of solution. If there are CO_2 bubbles in the beer during measurement, all the gas measurements (oxygen, nitrogen, carbon dioxide) will be underestimated: the greater the amount of degassing, the greater the error. The following guidelines should be considered whenever making portable dissolved gas measurements or placing gas-measuring instrumentation inline in a process stream. They apply to both wort and beer:

- Always make measurements as wort and beer are being transferred. It is impossible to reconstruct the dO_2 content after the liquid has been sitting in a vessel for a day or even just a few hours.
- Take measurements as far as possible from any air/O_2, CO_2, or N_2 injection points.
- Always measure in horizontal or ascending pipes and never in descending pipes, as degassing can occur in descending pipes if the liquid separates and gaps occur in the liquid flow.

- Sample as far from pumps or centrifuges as possible, because these are possible sources of oxygen ingress and putting distance between the ingress source and the measuring point will allow as much time as possible for gases to dissolve in the liquid.
- Inline sensors should be placed where there is laminar flow, preferably toward the end of straight runs and never less than five pipe diameters from any bend or valve.
- Never place any type of sensor in the drip-line of a filler, because when you remove the sensor for maintenance it makes it difficult to keep the sensor cable dry.

PROBLEMS SPECIFIC TO WORT

It can be challenging to measure oxygen in wort. Regardless of whether a brewer is injecting air or pure oxygen, to get accurate measurements readings must be taken only after injected gases have dissolved completely.

Air Injection

Overpressure is the key to dissolving gases in wort. The volume of oxygen in air that will dissolve into wort will always be relative to the overpressure on the fermentation vessel. Even a liquid that is supersaturated with O_2 and N_2 will quickly equilibrate to the overpressure in the tank. Nitrogen has low solubility and can never fully dissolve in wort. If the overpressure on a tank is too low, the nitrogen will off-gas and carry oxygen with it, causing the wort to foam excessively, especially if too much air has been injected into the wort in the first place.

The amount of oxygen that dissolves in wort is directly related to the partial pressure of oxygen in the gas used for oxygenation, regardless of whether it is pure O_2 or air. At 10° Plato and 15°C, oxygen in air will dissolve into wort at about 7.1 ppm dO_2 (Bamforth, 2006). Choosing whether to use air or pure oxygen depends upon the specific gravity of the wort and the requirements of the yeast.

Oxygen Injection

For higher levels of wort oxygenation, brewers inject pure O_2. Oxygen solubility is almost five times the solubility of air (see Fig. 9.1) and should not degas in the fermentation vessel. Pure oxygen will dissolve into 15°C wort with a Plato of 10° at about 33.8 ppm.

WORT OXYGEN MEASUREMENTS

The dissolved oxygen content of wort is most accurately measured at the base of the fermentation vessel. The procedure is simple (Fig. 9.4): immerse a high-level oxygen probe into an Erlenmeyer flask while overflowing wort into the flask from a strategically placed sample valve. This method fairly mimics the dynamics inside the fermenter as long as the overpressure on the vessel is not too much above atmospheric pressure.

FIGURE 9.4

Simply run the wort sample into a flask containing an oxygen probe capable of measuring at the appropriate dissolved oxygen level.

Many brewers do not measure the oxygen concentration of their wort, instead relying on seat-of-the-pants methods such as sparging for a certain amount of time, or simply looking for a visual cue like plenty of bubbles flowing through the fermenter. Although it may not be necessary to measure wort O_2 content, oxygen levels that are key to the type of yeast and specific gravity of the wort will optimize fermentation and can even prevent the formation off-flavors down the line in the finished product. At a minimum, a brewer can determine a theoretical oxygenation rate using the Ideal Gas Law. If one knows the wort aeration or oxygenation flow rate and the flow rate of wort during knockout, then the oxygen concentration of the wort can be calculated, assuming that all of the injected gas has dissolved in the wort.

Flow calculations are simple. To get the weight of the O_2 being injected, use Eq. (9.1):

$$\mathrm{MgO_2} = \frac{PV}{RT} \times 32{,}000\ \mathrm{g\,L^{-1}\,mol^{-1}}$$

Equation 9.1. Determination of the weight of gas injected into a wort stream.

MgO_2 = The weight of gas being injected per minute in mg or ppm

P = Purity of gas being injected in atmosphere's (100% O_2 = 1 atm and air = 0.02095 atm)

V = Volume of gas in liters/min (1 ft^3 = 28.317 L)

R = Gas constant (0.0821)

T = Gas temperature °K (273.3°K = 0°C)

You can convert the weight of the gas added to a concentration by using Eq. (9.2).

$$O_2 \text{ concentration} = MgO_2 / W_{flow}$$

Equation 9.2. Conversion of the weight of gas added to a concentration of mg/L of wort.

W_{flow} = Wort flow in L/min

The flow meter used to measure gases will be calibrated to three specific things. These are: the type of gas being measured, the pressure of the gas entering the flow meter, and the temperature of the environment of the meter. Flow meters used for air and oxygen are calibrated differently, because the molecular weights of the two gases are different. It is also important to compensate for any difference between the calibration temperature of the flow meter and the ambient temperature at the time the wort is being aerated. This can be done using Eq. (9.3).

$$\text{Flow}_{(\text{true})} = \text{Flow}_{(\text{observed})} \sqrt{\frac{P_{(\text{observed})}}{P_{(\text{calibrated})}} \Big/ \frac{T_{(\text{observed})}}{T_{(\text{calibrated})}}}$$

Equation 9.3. Calculation to accurately determine the concentration of gas being delivered by a flow meter when operating at a different temperature than at which it was calibrated.

If calculations show that wort is being over- or underoxygenated, it is best to make changes gradually. Ideally, a brewer will always confirm oxygenation calculations by taking dissolved oxygen measurements at the base of the fermenter. Drastic alterations to oxygenating processes may seem like the right thing to do, but the dynamics of the brewing process must be validated, because other factors can prevent oxygen from going completely into solution.

Now we may ask: if it is possible to calculate the theoretical oxygenation of the wort in the fermenter by simply looking at the oxygen injection rate and flow of the liquid, then why is it important to actually take oxygen measurements? Accurate oxygen measurements can lead to savings in several ways: those who sparge with pure O_2 may be able to cut back on O_2 purchases, whereas those who inject air and have problems with foaming may be able to decrease the energy used during injection. Another compelling reason is that a program of regular monitoring will allow a brewer to build data points that could help with future issues involving flow meters, pressure gauges, and mass flow controllers.

Inline analyzers measuring oxygen in wort should be positioned with special care. A sample valve should be located near an inline sensor so that a portable probe can be used to occasionally validate the inline instrument. Once an inline placement point has been selected, validate the readings at that location against readings taken with a portable probe at the base of the fermentation vessel. Some breweries choose to place their inline wort oxygen sensor where they know it will read lower than the oxygen concentration at the base of the fermenter because they cannot get the sensor far enough from the injection point to ensure that all of the injected gases are dissolved by the time they reach the probe. Although this is not the best scenario, it still allows for consistent measurement over time and is a viable solution in situations in which an ideal placement point simply is not available.

CHALLENGES IN BEER

Measuring Dissolved Oxygen in the Brewing Process

The sensory effects of dissolved oxygen in beer are cumulative. The more a beer is exposed to oxygen, the greater the likelihood that the beer will suffer unwanted oxidative flavor changes at some point before its consumption. Oxygen is metabolized by yeast and is a necessary part of fermentation. By the end of fermentation, the dissolved oxygen concentration of beer has usually dropped to below 5 ppb. It is after fermentation that exposure to oxygen must be minimized. Air contact is the main culprit in the exposure of beer to oxygen and can come from many sources: leaky pump seals, valves, or centrifuges; residual air in piping; filter-aids such as diatomaceous earth (DE) or DE-dosing systems; plate and frame filters; air in finished-beer tanks; and air in packages.

Determining Points of Oxygen Ingress

Most oxygen pickup in beer can be tied to specific locations or actions. By testing for oxygen throughout the brewing process, a brewer can pinpoint and mitigate problem sources of dO_2 before they have a cumulative impact on the product. When looking for a source of oxygen ingress, it is best to measure while beer is pumping through the process and work backward from the point at which high values were discovered. Here is an example:

The first beer to fill a finished-beer tank can pick up air in any number of places. In breweries that are not careful to purge piping, filters, and filter-aid dosing systems with deaerated, carbonated water, the oxygen pickup in the first few barrels of beer can exceed 1–2 ppm dO_2. In a finished-beer tank, it may take a while for the beer to reach the sample valve, and even this short delay can affect steady-state oxygen measurements. Because of this, the best place to start measuring is immediately before the finished-beer tank. If oxygen levels are greater than desired, then work backward while the beer is in process and measure O_2 at each possible source of introduction. First check might be at a seal pump between the fermenter and the filter or centrifuge, next check might be to look for a leaky valve in the process or residual air in the piping. If those things are okay, then a brewer may test the DE-dosing system, filter, and centrifuge. If everything looks tight, then final check would be for residual oxygen in the fermentation vessel.

The best technique for measuring dissolved gas in any type of liquid is to make sure that the beer or liquid being measured is at a pressure higher than the total gas pressure in the liquid. This is typically done by attaching portable instrumentation (an oxygen sensor) to a sample valve, fully opening the sample valve, and controlling flow through the instrument with a valve on the exit side of the oxygen sensor. As the beer exits the instrument, it should foam as the CO_2 equilibrates to ambient pressure, but there should be no visible bubbles in the tubing leading to the instrument.

DE filtration poses one of the greatest challenges to brewers who want to minimize oxygen pickup in their beer. Some breweries dump the initial beer out of the DE system and others coat the filter with deaerated, carbonated water. Small breweries that want to maximize yield will sometimes skip dumping the first beer out of the DE system, but this can lead to serious off-flavor problems down the line because the dissolved oxygen in the initial few barrels of freshly filtered beer can be as high as 1.0 ppm. Because of this, small brewers who want to maximize yield may wish to switch from filtration to centrifugation.

Ambient air in finished (bright) beer tanks is another source of concern for brewers who wish to minimize O_2. Some breweries flush their tanks with very pure CO_2 for a set time, whereas others pressurize the tanks to a particular set point and release the gas in the tank, sometimes for a few cycles. No matter the method, the best results seem to come from performing this routine just before filling the tank with liquid, so the tank is layered with the heavier CO_2 at the bottom and the lighter air at the top. If a tank is flushed and then allowed to sit too long, the gases will eventually mix and oxygen will be introduced into the finished beer.

Many dissolved-gas analyzers can be configured to measure the partial pressure of oxygen in a tank that is being purged with CO_2. This is usually a percent, bar, or ppm gas-phase oxygen reading. The best way to obtain an accurate reading is to keep the analyzer at ambient pressure by controlling the flow to the instrument before the sensor. Keep the gas flow rate through the analyzer low. The correct rate of flow will be just enough so that, if the gas is allowed to exit through a tube in into a beaker of water, only a few bubbles per second will be seen at the exit of the flow path. Using this method, the readings should reflect the percentage of oxygen in the area of the tank in which the sample is being measured. By measuring the partial pressure of oxygen in the tank, a brewery can standardize its air flushing procedure and minimize its CO_2 usage.

Oxygen Pickup from Impure Oxygen during Carbonation

What may appear to be insignificant levels of O_2 contamination in CO_2 may actually lead to significant oxygen contamination of beer, depending on the method used for carbonation.

There are three ways to carbonate beer. The first and "original" method is through natural carbonation during fermentation, and in some cases secondary fermentation in package. The second method is when CO_2 is "injected" into the process pipe postfiltration and before beer is pumped to the finished-beer tank. The third method is the "sparging" of CO_2 into the bright beer tank.

The purity of the CO_2 that is injected into beer—or more precisely, the amount of O_2 contaminating the CO_2—has a profound effect on oxygen pickup. It is easy to calculate the impact using the Ideal Gas Law to get exact concentrations, but the main point to understand is that injected CO_2 must be as pure as possible. At most breweries, the purity specification is a minimum of 99.98%, which has an impurity of 0.02% "air" and a maximum oxygen concentration of around 0.004%, or 40 ppm gas-phase O_2.

Sparging is a much more forgiving process when it comes to CO_2 purity and keeping oxygen out of beer. The concentration of any oxygen imparted to beer during sparging with CO_2 will be limited to the solubility of oxygen as allowed by Henry's Law. It is still wise to use the purest CO_2 possible, but purity as low as 99.9% used in sparging would impart no more oxygen than purity of 99.98% used in injection.

A brewery that carbonates by passing sparged CO_2 though a sintered carbonation stone must be alert to factors that can mimic the O_2-pickup problems of injection. If a tank that is being sparged is pressurized with CO_2, and then held at that pressure until the majority of the gas dissolves in the beer, then O_2 pickup may mimic that of injection. If, on the other hand, the tank is held to low overpressure (not much above ambient pressure) as CO_2 is bubbled into beer, then O_2 pickup will be minimal, though this may pose other issues like loss of volatile flavor compounds due to excess gas loss during sparging (Huige et al., 1985).

Auditing Brewery Oxygen

There are several ways to audit oxygen in the brewery. Checking the wort process can be fairly simple and does not require complicated instrumentation. The easiest way to check dissolved oxygen in the fermentation vessel is to use an oxygen probe submerged in wort that is slowly overflowing an Erlenmeyer flask, as shown in Fig. 9.4. If this is done at the base of the fermenter, it will best mimic the oxygen concentration of the wort in the vessel at the start of fermentation. Next, measure in the fermentation vessel a day or two after the yeast is pitched using a dO_2 sensor optimized for beer, to ensure that fermentation is robust and that most of the oxygen in the wort was consumed before the yeast went anaerobic.

The best place to start measuring postfermentation is in a full fermentation vessel. This will assure you that your fermentation was complete. Then measure the empty finished-beer tank after it has been blown down with CO_2, to make sure the tank will not contribute oxygen to the product. The initial beer going into the finished-beer tank should be measured postfiltration, but before entering the vessel during pumping. If high levels of oxygen are identified, work backward through the process and pinpoint the source of oxygen ingress. These are some typical dissolved oxygen levels found in best-in-class breweries:

- Fermentation Vessel: 3–10 ppb
- Pump pickup: 0–3 ppb
- Filtration: 5–20 ppb
- Finished beer tank: 5–50 ppb (includes all pickup from fermentation, pumps, and filtration)

These are ideal levels and the need for such strict oxygen control depends, to a certain extent, on a brewery's quality goals, including flavor stability and shelf life. A brewery may base its goals in part on its method of delivery to the consumer. A brewery with complete control of the distribution and storage temperature of its beer may have fewer concerns about oxygen contamination than a brewery with little power over how its beer is treated once it enters the distribution system. Accomplishing a high level of control is less important if a brewery knows its finished product will be treated well and quickly consumed. Also, knowledge of a beer's sensory profile and how well it ages in different conditions will allow a brewery to gauge the extent to which current process parameters need altering to minimize dO_2 pickup.

Package Oxygen Measurements

Brewers have long known that minimizing the "air" or "foreign gas" trapped in packaged beer was important. In about 1948, the American Society of Brewing Chemists established a method for measuring the headspace "air" in a package of beer by releasing the carbonation pressure from the headspace into a caustic solution and measuring the volume of the resulting bubble. This method was used at most North American breweries until the late 1990s, when total package oxygen (TPO) measurements became the new standard. The caustic "air" method can provide good feedback on the volume of headspace air, but there is little-to-no correlation between "air" and TPO.

The typical oxygen content of 0.15 mL of air at STP (25°C) is the dissolved equivalent of about 109 ppb dO_2. Fig. 9.5 shows the theoretical oxygen content of "air" if it were completely isolated to either the headspace or the liquid. The air in the headspace will follow Dalton's Law, in which about 20% will be oxygen and 79% will be nitrogen and other trace gases. The dissolved "air" in the liquid

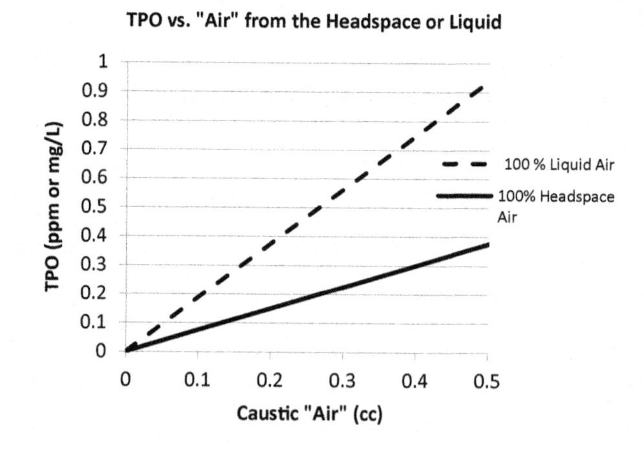

FIGURE 9.5

The theoretical oxygen concentration of equal amounts of "air" from the liquid or the headspace of packaged beer when using a caustic test for gas other than CO_2.

will follow Henry's Law—about 38% oxygen and 61% the remaining inert gases that make up air.

The value of measuring TPO came to the forefront in 1985, when two brewing chemists at the Polar brewery in Venezuela published a paper on the measurement of dissolved oxygen in packaged beer (Vilachá and Uhlig, 1985). In 1997, the measuring of TPO became widespread with the introduction of a Total Package Oxygen analyzer manufactured by Orbisphere Laboratories. Early instruments relied on four parameters to calculate TPO: the dO_2 of a package of beer shaken to gas equilibrium, liquid volume, headspace volume, and beer temperature.

Today there are many TPO "package analyzers" available, but it is still possible to calculate TPO using simple dissolved oxygen instrumentation and a three-step process: equilibrate the dissolved gases in the package, measure the dissolved oxygen, and calculate the TPO using the formula derived by the researchers at Polar.

For brewers who use variously sized packages, one advantage of measuring TPO is that instead of giving a result based on the quantity of gas in the package, the TPO measurement "normalizes" to the values of a 1-L container. This allows a brewer to calculate the relative oxygen concentration in variously sized packages without having to use a different metric for each size.

The widespread use of TPO measuring has correlated with a dramatic decrease in package oxygen concentrations. "Best in class" measurements on many fillers have target values as low as 0.050 ppm in bottles and 0.070 ppm in cans. Although not all fillers will achieve these low values, it is generally regarded that any TPO value under 0.10 ppm is considered excellent.

TPO can be determined using a number of different techniques. One is to measure the dissolved oxygen in packages that have been well shaken and use the dO_2 result to calculate back to the TPO. Another is to measure the headspace gases and the dissolved gasses separately and then add the results.

The advantage of the first method, in which dissolved oxygen is measured in packages that have been well shaken, is that it can be done without sophisticated instrumentation. All that is needed is a dissolved oxygen analyzer that is suited for measuring in packaged beer (Fig 9.6).

The disadvantage of this method is that there are extra steps needed to shake the packages and accurately measure the headspace volume, unless the instrument has the ability to make a direct measurement. Instruments that measure the headspace gas and the dissolved oxygen in the same package do not need shaking to properly determine the TPO, but some require the collapse of the foam to get accurate results.

Assuming that one is using the simplest instrument, which measures dissolved oxygen in liquid, then package preparation is very important. Packages with temperatures above 15°C should be shaken for a minimum of 3 min. Packages with temperatures below 15°C need a minimum of 5 min of shaking. It is best to shake using a rotary platform shaker capable of over 180 revolutions per minute. It is very important to prevent packages changing temperature between when they were shaken and when they are measured. In other words, measurements should be taken without delay.

FIGURE 9.6

A simple dissolved oxygen instrument for measuring package oxygen, coupled with a packaging-piercing device

Image courtesy of Hach Company.

As to how quickly a beer should be measured after packaging, this is entirely dependent on the type of beer being measured and can only be determined by knowing how oxygen-reactive a beer may be. Some beer is very quick to "consume" oxygen and should be measured immediately, ie, rushed from the filler to the shaking platform. (This is especially true of bottle-conditioned beer.) But even if the type of beer allows for measuring without critical urgency, it should still be done in a timely manner, usually no more than half-an-hour after filling.

Measuring Oxygen in Bottle-conditioned Beer

Little research has been done on the effects of high oxygen in bottle-conditioned beer, but it is generally accepted that oxygen levels should be kept as low as with filtered beer (Dekonick et al., 2013). Measuring the oxygen content in bottle-conditioned beer poses specific issues not found when measuring filtered beer, because the added yeast will quickly metabolize the oxygen in the package.

Because the yeast in bottle-conditioned beer metabolizes so quickly, measurements with dissolved oxygen analyzers should be carried out on shaken cold packages as soon as possible after packaging. TPO instrumentation that measures headspace and dissolved gasses separately (no need to shake the package) will also work just fine, but it is important to let any foam in the headspace completely collapse, so that all air in the headspace gets measured.

Validating Filler O_2 Pickup

TPO measurements are perhaps most valuable for their ability to show how well bottle or can fillers supply product to a package without adding air during the process. Dissolved gases may be introduced into the beer in many ways during the filling process, including some or all of the following:

- Initial dO_2 content of the beer before entering the filler
- Filler bowl air pickup
- Residual air in cans or bottles (due to incomplete CO_2 purging or poor vacuum seal performance during bottle evacuation)
- Over- or underjetting during headspace fobbing (overjetting can force air into the liquid; underjetting will fail to push all the air from the headspace)

If a brewer sees high O_2 levels in a batch of packages that correlate to a specific filler, it may be helpful, if at all possible, to further target testing to individual valves on the filler. In general, the percent deviation of a specific valve can be much lower than the percent deviation of the valves collectively. The percent deviation, or percentage difference of all the different filler's valves, can vary from as little as 25 to as much as 100%, so it is important to understand each filler and have realistic TPO criteria based on its capability.

When troubleshooting filler issues using dissolved measurement techniques, these points may help:

- Oxygen pickup in liquid during package filling can be determined by measuring unshaken packages. The unshaken package result will indicate the oxygen content of the beer entering the filler plus the oxygen picked up in the filling process. If a brewer measures beer before it enters the filler, the filler pickup can be calculated by subtracting the base-of-filler dO_2 from the unshaken package dO_2.
- Headspace oxygen concentration will be the TPO of the package minus the unshaken dO_2 of a different package. Use the average of multiple packages when quantifying headspace O_2 using dO_2 (liquid) measurements.

Validating Dissolved Gas Instrumentation

Today's brewer has many useful instrumentation options for measuring dissolved gases in beer. Nevertheless, to use that instrumentation to best advantage, it helps to know how to validate—and thus trust—that instrumentation.

Instrumentation validation is fairly straightforward. Inline instrumentation is best validated against a certified portable instrument that can access a measurement point as close as possible to the inline sensor. This is works fine if a brewery also has a trustworthy portable instrument. But if the reliability of a portable instrument is unknown, then it is important to validate the portable analyzer.

A portable instrument may be validated against containers of zero beer and of known gas concentrations. Validate low values first, so you know the instrument is not biased on the high side. Then validate high samples to ensure that higher readings are in line with measurement expectations.

When validating a portable instrument for its accuracy in dO_2 analysis, the low end may be checked with old beer in packages or fermented beer still in a fermentation vessel. It is highly unlikely that 30-day-old packaged beer in cans or 30-day-old bottled beer stored upside down will contain much more than a few ppb of dO_2. Likewise, fermentation vessels should be in the range of 3–10 ppb. The high range of a portable dO_2 measuring instrument may be validated using a gas with an oxygen residual of 0.1 to 0.2% O_2, or with an old can of beer injected with a known concentration of air (Schmidt, 2013).

Carbon dioxide analyzers can be validated with packages of known CO_2 concentration or by checking finished beer vessels that were measured in the recent past and are known to contain the appropriate concentration of CO_2.

The best brewers understand every aspect of their process and their measuring instrumentation. Frequent testing at many points in the process over months and years combined with accurate record keeping will enhance opportunities for creativity in the brewing process. With full grasp of the nuances of dissolved gases in beer, a brewer who sees high O_2 in packages will have the tools needed to quickly troubleshoot any problem, from the purity of CO_2 in a blanket tank to a faulty filler valve. The reward is the sort of flavor stability and quality excellence that builds a loyal following among those who love beer.

REFERENCES

American Society of Brewing Chemists, 1947. Report of subcommittee on carbon dioxide in beer. Proceedings of the American Society of Brewing Chemists 124.

American Society of Brewing Chemists, 1976. Report of subcommittee on analysis of dissolved oxygen in water, wort, and beer. Journal of the American Society of Brewing Chemists 34, 115.

Bamforth, C.W., 2006. Scientific Principles of Malting and Brewing. American Society of Brewing Chemists, St Paul MN.

Bamforth, C.W., Muller, R.E., Walker, M.D., 1993. Oxygen radicals in malting and brewing: a review. Journal American Society of Brewing Chemists 51, 79–88.

Dekonick, T.M.L., Martens, T., Delvaux, F., Delvaux, F.R., 2013. Influence of beer characteristics on yeast refermentation performance during bottle conditioning of Belgian beers. Journal of the American Society of Brewing Chemists 71, 23–34.

Huige, N.J., Charter, W.M., Wendt, K.W., 1985. Measurement and control of oxygen in carbon dioxide. Technical Quarterly of the Master Brewers Association of the Americas 22, 92–98.

Mitchell, R., Hobson, J., Turner, N., Hale, J., 1983. A new dissolved-oxygen analyzer for in-line analysis of beer. Journal of the American Society of Brewing Chemists 41, 68–72.

Saison, D., Vanbeneden, N., De Schutter, D.P., Daenen, L., Mertens, T., Delvaux, F., Delvaux, F.R., 2010. Characterization of the flavor and the chemical composition of Lager beer after aging in varying conditions. Brewing Science 63, 41–53.

Schmidt, G., 2013. Brewing and Beverage Industry International, 2, pp. 54–57.

Silberberg, M.S., 2009. Chemistry: The Molecular Nature of Matter and Change, fifth ed. McGraw-Hill, p. 206.

Vilachá, C., Uhlig, K., 1985. The measurement of low levels of oxygen in bottled beer. Brauwelt International 1, 70–77.

CHAPTER

Controlling Beer Foam and Gushing

10

L.T. Lusk

Formerly Foam and Flavor Development Scientist, Miller Coors LLC

INTRODUCTION

Foam is part of beer's visual, flavor, and tactile (mouthfeel and CO_2 tingle) properties (Chandrashekar et al., 2009; Evans and Bamforth, 2009; Langstaff and Lewis, 1993; Smythe et al., 2002). The pleasant bitterness from hops concentrates in the foam as well as the aromas of hop oils, sweet corn (dimethyl sulfide), cardboard (oxidation), and spice aromas. Absence of foam accentuates malt, caramel, butterscotch (diacetyl), and fruity aromas (Delvaux et al., 1995; Kosin et al., 2012). Consumers differ in the amount of foam they prefer, but they regard beer without foam or with rapidly collapsing foam as "flat."

Desired foam properties result from the culmination of the ingredients and the brewing process, from malting to package release to the beer pour. Gushing, however, is the explosive release of carbon dioxide, foam, and beer after the gentle opening of a bottle or can. Gushing is a serious quality defect. This chapter will emphasize quality assurance methods and good manufacturing practices (GMP) needed to obtain a good foam head—without gushing—and in the process improve a beer's visual, flavor, and tactile properties. This chapter will not cover beer foam or gushing scientific research. For this, the reader is referred to Bamforth, 1985; Evans and Bamforth 2009; Garbe et al., 2009; Shokribousjein et al., 2011.

QUALITY ASSURANCE METHODS AND TROUBLESHOOTING

Brewers have a vision for the appearance of a brand's poured beer foam head, stability, and cling/lacing. This vision should be realistic and be built into the brand's profile during product development. Quality assurance (QA) measures are necessary to ensure consistency. Despite the results frequently not representing the consumer poured beer experience, brewers prefer rapid instrumental foam analysis techniques for monitoring quality. Many of these methods measure foam stability without regard for typical foam formation (carbonation dependent), creaminess (bubble size dependent), and lacing.

Only a few methods have met stringent criteria for approval by the American Society of Brewing Chemists (ASBC, 2011b), Brewery Convention of Japan

Brewing Materials and Processes. http://dx.doi.org/10.1016/B978-0-12-799954-8.00010-1

Table 10.1 Accepted Foam Analysis Methods of Scientific Organizations

Organization and Method	Brief Description	Manual or Instrumental
ASBC Beer-22	Foam collapse rate (Sigma value) and Foam flashing methods	Manual
MEBAK 2.18.1	Ross and Clark (1939)	Manual
BCOJ Beer 8.29	NIBEM-T	Instrumental
EBC Beer 9.42	NIBEM-T	Instrumental
MEBAK 2.18.2	NIBEM	Instrumental
MEBAK 2.18.3	Lg-Foamtester	Instrumental
MEBAK 2.19.4	Steinfurth Foamtester	Instrumental
MEBAK 2.19.5	NIBEM Cling Meter	Instrumental

(BCOJ, 2004), European Brewery Convention (EBC, 2013), or Mitteleuropäische Brautechnische Analysen kommission (MEBAK, 2012) brewing organizations (Table 10.1). The advantages of instrumental methods are sample throughput and operator-to-operator consistency. Because of instrument cost, smaller brewers may prefer older manual methods. The approved methods were extensively compared by Evans and Bamforth (2009) except for the Steinfurth Foam Stability Tester (FST). With few exceptions, FST stability values correlate with the National Institute for Malting Barley, Malt, and Beer (NIBEM) and with the manual pour method of Constant (Constant, 1992; Wallin et al., 2010).

Lacing or cling is a pleasing aspect of beer foam, though possibly more so for men than women. Lacing can be measured manually (Jackson and Bamforth, 1982) or instrumentally with the NIBEM-CLM (cling meter) (Roza et al., 2006).

QA should include testing for foam protein (Evans and Bamforth, 2009), International Bitterness Units (IBU), and CO_2. These parameters impact foam head, stability, and lacing. A database of normal QA values proves useful when troubleshooting a foam issue.

The hydrophobic protein (Bamforth, 1995; Slack and Bamforth, 1983) and high molecular weight protein (HMWP) (Hii and Herwig, 1982) assays both estimate foam protein levels utilizing the Bradford Coomassie blue assay. Assay kits are available from several vendors. Beer contains some dye-reactive background material. This background material can be removed either by hydrophobic interaction or size exclusion chromatography. This author's opinion is that the assays provide similar results. Once the background level is known, the assay can be performed directly on beer and the known background subtracted for that beer style. The analyst also needs to compare dye lots because of lot-to-lot variability.

The Bradford assay may not be useful if the specific foam issue is a loss of barley lipid transfer protein (nsLTP1). Then the pyrogallol red-molybdate (PRM) dye assay may be more useful because the PRM response to nsLTP1 is greater than the Bradford dye response (Evans and Sheehan, 2002).

The Constant manual pouring method gives foam parameters that are valuable when troubleshooting foam problems, especially when taken together with CO_2, IBU, and HMWP values (Constant, 1992). The parameters are Foam Height, Half Life, Normalized Half Life, Life, and Density (foam creaminess/bubble size), as well as an overall Quality Number. An advantage to the method is that the pour is similar to the consumer experience. However, the method is labor intensive.

Beading measurements are useful for troubleshooting, as is a rapid test to determine beer foam negative materials such as lipids or detergents (Dickie et al., 2001; Lynch and Bamforth, 2002). Beading is the desirable trait of the continuous release of a bubble stream, sweeping aromas to the beer headspace. Beer contamination with foam-destroying surfactants (tensides), lipids, and line or lid lubricants is caused by poor GMP.

GOOD MANUFACTURING PRACTICES AND BREWING PROCESS OPTIMIZATION

To obtain consistently good foam, build it into the process. Start with good ingredients and brewing practices. These steps not only ensure good foam characteristics, but the flavor stability also frequently improves. Automated inline monitoring is useful to ensure consistency in the brewhouse, fermentation, cold cellars, and packaging.

INGREDIENTS

PALE MALTED BARLEY

The economics of brewing is that pale malted barley is chosen on the basis of brewing performance, especially extract yield. Malted barley specifications do not include parameters relevant for foam characteristics because realistic methods that can be employed in a manufacturing setting are not available. Several sophisticated research laboratory methods show modest correlations between various malted barley, wort, or beer components and beer-foam stability (Evans and Bamforth, 2009). However, on a manufacturing scale, malt-batch variability of total and hydrophobic polypeptides or β-glucan content did not correlate with the foam stability (Kordialik-Bogack and Antczak, 2011).

Most brewers use malted two-row barley, but especially in North America malted six-row barley is also used (Briggs et al., 1981). The most apparent differences between two-and six-row malts relate to their level of grain protein and diastatic power. The higher protein and enzyme levels of six-row cultivars allow for the widespread use of cereal adjuncts in major North American breweries and the double-mash system for precooking them (Schwarz and Li, 2011).

LIPOXYGENASE-FREE PALE MALTED BARLEY

Pale malt prepared from lipoxygenase-free barley improves foam, flavor, and flavor stability because the barley does not contain lipoxygenase enzymes (Hirota et al., 2006). Barley varieties in commercial use include CDC PolarStar and Null-LOX (Lipoxygenase) (Canadian Malting Barley Technical Center, 2013; Carlsberg, 2011). Lipoxygenases start a cascade of reactions that lead to beer-soluble foam-destroying di- and trihydroxyoctadecenoic acids (Esterbauer and Schauenstein, 1977; Garbe et al., 2005; Kobayashi et al., 1994, 2000, 2002). These acids are bitter and astringent, decreasing beer smoothness (Kaneda et al., 2002). An additional lipoxygenase cascading-reaction pathway generates carbonyl compounds that give stale aromas, including the *trans*-2-nonenal cardboard aroma (Garbe et al., 2005; Kuroda et al., 2003; Vanderhaegen et al., 2006). Specialty malts that are not prepared from lipoxygenase-free barley retain lipoxygenase activity but not as much as standard pale malt (Coghe et al., 2003).

SPECIALTY MALTS

Specialty malts generally increase beer foam and foam retention but are primarily chosen for their contribution to flavor, color, and increased palate fullness or body (Briggs et al., 1981; Gruber, 2001; Hertrich, 2013; Kieninger and Wiest, 1982). Although used in a small proportion of the malt grist, specialty malt's effect on beer characteristics is appreciable (Kieninger and Wiest, 1982). Contrary to the usual foam-positive impact, inferior foam stability was found for brews prepared from 80% pale: 20% crystal 75 L, or 20% crystal 120 L compared to the foam in the 100% pale malt control (Combe et al., 2013).

Melanoidins are a major contributor to specialty malted barley's foam enhancement properties. Direct and indirect evidence suggest that beer melanoidins are surface active, provide bulk viscosity to beer, and have excellent independent foam stability (Lusk et al., 1987, 1995, 2003). In contrast, the foam-protein level decreases logarithmically with increasing malt color (Fig. 10.1).

Unlike other specialty malt, pale crystal dextrin-style malt improves foam retention while having less influence on color or aroma. This malt also adds body and mouthfeel (Kieninger and Wiest, 1982). Dextrin-style malt is used at a rate of 1–5% of the malt bill. Trade names for dextrin-style malts include Cara-Pils, Carapils, and Carafoam. Dextrin-style malt has low fermentability (Briggs et al., 1981). The malt can either be added to the mash or steeped in hot water. If steeped, it will contribute unconverted starch that may cause starch haze.

UNMALTED BARLEY

Brewing beer with up to 100% unmalted barley can be accomplished with an enzyme mixture containing pullulanase and alpha-amylase for attenuation, protease for free amino nitrogen, β-glucanase and xylanase for filtration, and lipase for wort clarity. A similar enzyme mixture is commercially available. Barley brews have

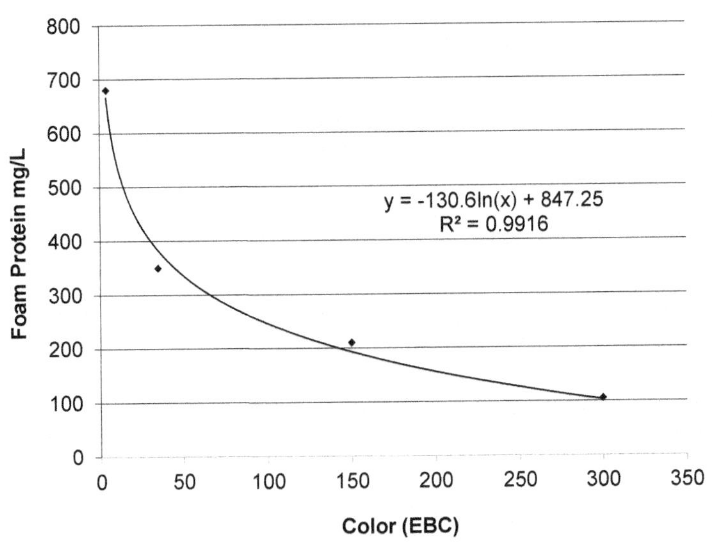

FIGURE 10.1

Relationship Between Malt Color and Foam-Active Protein.

Data from Table 1, Ishibashi, Y., Kakui, T., Terano, Y., Hon-no, E., Kogin, A., Nakatani, K., 1997. Application of ELISA to quantitative evaluation of foam-active protein in the malting and brewing processes. Journal of Amercian Society of Brewing Chemists 55, 20–23; foam protein determined by Suntory Ltd. ELISA method.

good foam stability and possibly better flavor stability than malted barley brews (Glatthar et al., 2003; Kunz et al., 2012; Kunzmann, 2011).

WHEAT

Malted wheat as an adjunct improves foam stability (Evans and Bamforth, 2009), but also has flavor impact (Faltermaier et al., 2014). Blenkinsop (1991) suggested 3–5% medium-modified malted wheat for lagers and ales and 40–100% well-modified malted wheat for wheat/weiss beers. Germany's beer purity law and the World Beer Cup Style Guidelines specify that "white beer" must be made from at least 50% malted wheat and be top-fermented (Faltermaier et al., 2014). For unmalted wheat, increasing the level of protein degradation decreases the level of protein precipitation and a more stable haze forms. Unmalted wheat had a positive impact in beer brewed using overmodified barley malt, but had a negative influence in beer brewed using barley malt with high foaming potential (Delvaux et al., 2004; Depraetere et al., 2004).

OTHER ADJUNCTS

Sorghum, oats, rice, and maize decrease foam stability, as do liquid adjuncts. Triticale (a hybrid of wheat and rye) increases foam stability (Glatthar et al., 2003; Schnitzenbaumer et al., 2012, 2013).

Hops

Iso-α-acids or advanced hop products are necessary for any significant foam stability and cling. Within certain ranges, the level of these bittering acids determines the appearance of the foam and cling and the perceived bitterness (Fritsch and Shellhammer, 2008; Kunimune and Shellhammer, 2008). The range is narrower for tetrahydro-iso-α-acids (Tetra), and hexahydro-iso-α-acids (Hexa) because of solubility limits. A reasonable Tetra limit is 5–6 ppm. Exceed this limit and an unappealing appearance develops (eg, Fig. 1.13 in Evans and Bamforth, 2009). Brewers successfully add low concentrations of Tetra or Hexa postfermentation to iso-α-acid bittered beers to improve foam stability and cling. Low-bitterness α-acids can be added postfermentation to improve foam stability and cling. These improvements are stable through the beer shelf life despite the fact that α-acids are not stable in beer (Wilson et al., 2011). Hop oils from dry hopping have surprisingly little negative impact on foam stability (Kaltner et al., 2013).

Cations

Divalent metal cations can promote foam stability and cling, but many decrease flavor and physical stability (Evans and Bamforth, 2009; Roberts, 1976). Some are toxic. Metal ions are naturally present in beer, and further addition is not recommended, with the possible exception of zinc (Roza et al., 2006; Zufall and Tyrell, 2008). Cations enter the brewing process from vessels, piping, and process aids. Stripping from bimetallic junctions such as a brass fitting on a stainless steel pipe should concern brewers (Beardmore, 2013). Passivation helps control copper leaching from brewing vessels. Diatomaceous earth (kieselguhr) is a well-known source of iron. Bentonite, which is seldom used in modern brewing practice, leaches numerous cations (Catarino et al., 2008).

Foam Aids

Propylene glycol alginate (PGA) is a well-known foam stabilizer (Evans and Bamforth, 2009). Long-term physical stability testing should be performed before putting PGA-treated beer into the trade. The Joseph Schlitz Brewing Company added PGA to papain-treated beer to compensate for papain-destroyed foam proteins, a common practice (Lusk et al., 2003). Schlitz continued adding PGA after changing to silica hydrogel chillproofing (personal communication). White flakes developed several months after packaging. Consumers called the flakes "Schlitz Bits." The flakes were one of several factors in Schlitz's rapid demise (Hintz, 2011). PGA also reacts with isinglass creating a flock in alcohol-free beer (Lewis and Bamforth, 2006). Isinglass, a beer-fining agent, may improve foam stability but not when combined with PGA (Leather, 1998; Leiper et al., 2002; Lewis and Bamforth, 2006).

Saponin-rich extracts derived from *Quillaja saponaria* Molina or *Yucca schidigera* are extensively used as natural foaming agents in food and beverages (Heymann et al., 2010; San Martın and Briones, 2000). Beer and alcoholic beverage formulations with minimal haze-forming tannins are available.

Nitrogen gas and "widgets" are used to enhance foam stability (Evans and Bamforth, 2009). With nitrogen, bittering acids are swept into extremely stable foam creating very bitter foam and sweet beer lacking CO_2 tingle, ie, not your typical lager or ale.

Foam Suppressors

In general, higher ethanol-content beers have poorer foam (Evans and Bamforth, 2009). Poor GMP can lead to contamination with foam-destroying surfactants (tensides), lipids, or line or lid lubricants. In general, package release beer-lipid content from ingredients is negligible (Anness and Reed, 1985) except for the previously discussed di- and trihydroxyoctadecenoic acids.

PROCESS VARIABLES AND CHALLENGES

Both two-row and six-row malted barley varieties provide abundant foam-promoting proteins, but significant foam protein losses occur during brewing (Evans and Bamforth, 2009; Ishibashi et al., 1997). For a six-row malt brew, 400 mg/L of nsLTP1 was found in the mash, 100 mg/L at kettle full (denatured plus native nsLTP1), and less than 20 mg/L at package release (Lusk et al., 2001). A two-row malt brew (personal communication) had 100 mg/L at kettle full and 40 mg/L after the boil (Van Nierop et al., 2004).

MASHING AND LAUTERING

Beer with lower pH has better head retention (Melm et al., 1995). The key pH-control point is mashing, which is influenced by both grist and liquor composition. Additional beer pH-control points are fermentation dissolved oxygen and zinc content, and the high-gravity brewing adjustment liquor (Taylor, 1990).

Acidifying mash to pH 5.0 with lactic, minimizing the protein rest, or mashing-in at saccharification temperature limits the lipoxygenase activity that creates the foam, flavor, and flavor stability destroying di- and trihydroxyoctadecenoic acids (Kobayashi et al., 2000). Minimizing the protein rest also lessens the destruction of foam proteins.

The lautering system or mash filter needs an inline wort turbidity meter to ensure wort of predetermined clarity enters the kettle. Trub entering the fermenter from the whirlpool needs a minimum level of lipids necessary for good fermentation without unduly adding di- and trihydroxyoctadecenoic acids (Kühbeck et al., 2006a,b).

LOW THERMAL HEAT LOAD WORT BOILING

Increased head retention (NIBEM) was obtained when the kettle thermal heat load was reduced while keeping hop utilization, color formation, and volatile removal within normal ranges (Robinson and David, 2008). Similarly, the higher wort boiling temperature (102°C) at a sea-level brewery reduced the final beer foam-promoting nsLTP1 10-fold compared to that found at a high-altitude brewery (wort boiling temperature 96°C) (Van Nierop et al., 2004). These results support claims for dynamic low-pressure boiling systems improving foam stability, as well as flavor stability (Bühler et al., 2003; Buttrick, 2006; Michel and Vollhals, 2002, 2003). The obvious further advantage of decreased thermal load is energy consumption reduction.

FERMENTATION

Normal-gravity brewing minimizes foam losses compared to high-gravity brewing (Brey et al., 2003; Stewart et al., 1997, 2006). For large brewers, the trade-off is loss of brewing capacity. Stewart et al. (2006) showed that there was over 250 mg/L hydrophobic protein at kettle full for both normal (10°Plato) and high-gravity (20°Plato) wort, but only about 60 mg/L in finished beer from normal-gravity wort and approximately 25 mg/L from high-gravity wort. High-gravity brewing also stresses yeast that then releases foam-destroying proteases and lipids.

Fermenter foam-over is extremely detrimental to foam stability. Both foam proteins and hop-bittering acids concentrate in the foamed-over liquid. Contrarily, slow fermentations can result in such high concentration of foam proteins that it becomes difficult to control the foam on beer (fob) during bottle or can filling. Low wort aeration and premature yeast flocculation are examples of conditions that lead to slow fermentations.

BEER TRANSFER

Turbulence during beer transfer is foam destructive. Sources of turbulence include improper pressure control in pumps, lines, or filler bowls. In one unpublished example, a decrease in foam stability was traced to nsLTP1 but not to CO_2, IBU, HMWP, or protein Z. Beer turbulence was detected at a sight glass window. Balancing the line pressure eliminated the turbulence and foam problem.

CHILLPROOFING

Silica hydrogel, polyvinylpyrrolidone (PVP), and tannic acid do not decrease foam stability at typical dosage rates (Rehmanji et al., 2005). A slight foam-protein loss is anticipated with proline-specific endoproteinase (Evans and Bamforth, 2009) but should not be of concern unless used in low-protein, low-IBU products such as American light-style beer. Papain chillproofing destroys both foam and haze proteins.

PACKAGED BEER

Foam stability and IBU levels decrease in parallel if packaged beer is stored at room temperature because of the degradation of iso-α-acids. Aseptically packaged beer will lose foam proteins in addition to IBU because of the action of yeast proteinase A. In both cases, foam stability loss can be minimized by bittering with tetrahydro-iso-α-acids (Lusk et al., 2003).

GUSHING AND OVERFOAMING

Gushing is the explosive release of carbon dioxide, foam, and beer after gently opening a package. In severe cases, the spurt is 2–10 cm high with a large volume loss. Beer does not spontaneously erupt into foam without a nucleation source. The nucleation source is frequently the interaction of complex factors originating with the raw materials, the production process, or even transportation and storage. Timely quality control and preventive solutions are still in development (Amaha and Kitabatake, 1981; Bamforth, 2011; Casey, 1996; Christian et al., 2011; Garbe et al., 2009; Sarlin et al., 2007; Sarlin, 2012; Shokribousjein et al., 2011).

Gushing is classified as either primary or secondary. Primary gushing results from the use of mold-infected grain (head blight, scab), whereas secondary gushing is caused by the presence of colloidal or solid particles. Primary gushing will likely impact the entire production from a given lot of infected barley or malt. Secondary gushing might be more sporadic within a lot, as in the case of particulates resulting from filter breakthrough (Garbe et al., 2009). Causes of secondary gushing that might be minor enough to not cause gushing can exacerbate primary gushing. This makes troubleshooting gushing problems frustratingly difficult.

PRIMARY GUSHING

Primary gushing occurs when using barley or malt contaminated by filamentous fungi, ie, *Fusarium* sp., *Trichoderma* sp., *Nigrospora* sp., *Aspergillus* sp., *Penicillium* sp., *or Stemphylium* sp. (Sarlin et al., 2005). One gushing study showed three distinct scenarios: (a) barley that gushed and yielded malt that gushed, (b) barley that did not gush, but yielded malt that gushed, and (c) barley that did not gush, yielding malt that did not gush. In no instance did barley that gushed correspond to malt that did not (Munar and Sebree, 1997).

Primary gushing is attributed to the presence of excreted hydrophobins, or other less well-characterized peptides (Sarlin, 2012; Shokribousjein et al., 2011). Purported additional factors in primary gushing include beer foam-stabilizing protein barley lipid transfer protein (nsLTP1), glycosylated peptides derived from nsLTP1, and the fungispumin, AfpA. Both nsLTP1 and AfpA are plant-defense proteins (Hippeli and Elstner, 2002; Hippeli and Hecht, 2009; Zapf et al., 2006, 2007). Interestingly, fungal infection can cause premature yeast flocculation which would lead to slow fermentations (MacIntosh et al., 2014). As previously discussed, slow fermentations lead to excess nsLTP1 and foam, possibly resulting in perfect conditions for a gushing trigger such as CO_2-hydrophobin nanobubbles (Shokribousjein et al., 2011).

Numerous methods have been suggested to control, but unfortunately not prevent, primary gushing. Farmers utilize an integrated approach to head blight (scab) management, including spraying with fungicides and appropriate crop rotation (McMullen et al., 2012). A very low level of chlorination and increased overflow can be utilized during steeping to minimize fungal regrowth during germination. Lactic acid bacteria or the yeast-like fungi, *Geotrichum candidum*, can be seeded into malt houses to overcome the growth of *Fusarium* during storage (Boivin and Malanda, 1997, 1999; Laitila et al., 2006; Lowe and Arendt, 2004). *Geotrichum candidum* is naturally prominent in green malted barley (Douglas and Flannigan, 1988).

The best defense against primary gushing is use of sound barley. Finding affordable sound grain during severe scab infections is challenging. Maltsters and brewers typically have specifications for weather damaged, blighted, or moldy grain, as well as for mycotoxins like deoxynivalenol (DON) (Garbe et al., 2009). Routine DON analysis provides some protection (ASBC, 2011b; Barley-11; Neogen, 2012), but DON accounts for less than 60% of observed variability in gushing potential (Casey, 1996). If used, wheat and maize should be inspected and analyzed because they are also subject to mold infections (Engelmann et al., 2012; Schaafsma et al., 1998). An enzyme-linked immunosorbent assay (ELISA) that shows good correlation of hydrophobin level and gushing potential was developed at the Technical Research Centre of Finland (VTT) (Sarlin et al., 2005). The assay has potential for scaling to commercial use, but it is not yet available as a simple kit. The assay detects hydrophobins in both barley and malt but is best performed on malt because of substantial production of hydrophobins during malting (Sarlin et al., 2007). An emerging technique to predict gushing is dynamic light scattering detection of CO_2-hydrophobin nanobubbles (Deckers et al., 2012).

Sound barley or malt can be blended with higher DON-level malt to meet a DON specification (Johnson and Nganje, 2000). This practice correlates with the hypothesis that gushing-inducing substances have to exceed a minimum concentration required to cause gushing (Sarlin et al., 2005; Christian et al., 2011). Brewers may anticipate some minor overfoaming, albeit less than from unblended infected malt.

MEBAK (2006) provides two methods to predict the gushing propensity of barley, malt, wort, and beer (Table 10.2), but neither EBC nor ASBC has approved methods. The MEBAK methods are too slow for high-throughput screening in a large malting company or brewery.

REMEDIAL MEASURES FOR PRIMARY GUSHING

Chillproofing removes secondary factors that can exacerbate primary gushing. PVPP, nylon (PVPP predecessor), and silica xerogel show some gushing suppression (Amaha and Kitabatake, 1981; Garbe et al., 2009). Xerogel used with isinglass might be particularly effective (Leiper et al., 2002). Commercial brewing enzymes also reduce the impact of primary gushing. In-bottle additions of Ceremix (β-glucanase, xylanase, α-amylase, protease mixture), Neutrase (*Bacillus subtilis*

Table 10.2 Accepted Gushing Propensity Procedures of MEBAK

Test Name	Sample Preparation	Gushing Test	Duration of the Test
MEBAK 3.1.4.21.1	Mashing of malt sample to produce wort followed by wort clarification, carbonation, bottling, and 48 h storage at 4°C.	After storage for 48 h at 20°C, bottles are turned four times through 360° in the vertical plane. Bottles are then allowed to stand for 5 min before opening. Volume of collapsed foam is recorded.	About 6 days
MEBAK 3.1.4.21.2	Cold-water extract of malt or unmalted grain coarse grist is centrifuged, the supernatant boiled, filtered, and 50 mL of filtrate is added to carbonated mineral water. Bottles are shaken in horizontal position for 3 days.	After shaking, bottles are allowed to stand for 10 min before 3× longitudinal axis rotation, 30 s rest, and opening. Volume of collapsed foam is recorded.	About 4 days

Table significantly modified from Table 4, Sarlin, T., 2012. Detection and Characterisation of Fusarium Hydrophobins Inducing Gushing in Beer, Dissertation for Doctor of Science in Technology, Aalto University School of Chemical Technology, VTT Science 13. VTT, Espoo, Finland. http://www.vtt.fi/inf/pdf/science/2012/S13.pdf (accessed 13.09.13.).

metalloendopeptidase), and fermentation additions of β-glucanase reduced primary gushing (Aastrup et al., 1996; Burger and Bach, 2001). Interestingly, wort from *Fusarium*-infected malt had a higher β-glucan level (and premature flocculation), but the beer itself had similar β-glucan level as the control (MacIntosh et al., 2014; Oliveira et al., 2012). Neither Ceremix or Neutrase decreased foam stability.

Other processing aids have also been suggested. A calcium-enriched silicate mixture is in commercial use to reduce both primary and secondary (oxalate) gushing (Besier et al., 2013). There must be a taste impact because this mixture replaces the calcium sulfate or chloride normally used in the mash liquor. The higher mash pH would favor lipoxygenase activity, which could be detrimental to foam and flavor stability. Beer brewed with gallotannin addition during wort boil is reported to decrease hydrophobins, have less gushing, and have less iron (Schneidereit et al., 2013). Beer-soluble iron can enhance primary gushing and is also a factor in flavor instability. Charcoal; the clays Fuller's Earth, kaolin, and Tansul (a proprietary bentonite the trademark status for which was canceled in 2002); activated alumina; and nylon are often presented in the literature as successful gushing suppressors despite the authors' own conclusion, "No reliable method of suppressing gushing has been found" (Curtis and Martindale, 1961).

HOPS, ADVANCED HOP PRODUCTS, AND GUSHING

Hops and hop components can act as promoters or suppressors of both primary and secondary gushing. Higher hop additions or hop oil additions suppress primary gushing (Buffin and Campbell, 2013; Gardner et al., 1973; Hanke et al., 2010, 2011). Of course, increasing the hop oil level changes beer flavor. β-acids are gushing suppressors; α-acids are strong gushing suppressors (Laws and McGuinness, 1972). Beer solubility of α-acids, though limited, is greater than that of β-acids. Larger overfoaming volumes are reported in pellet-hopped primary gushing beer with a high ratio of polyphenols to iso-α-acids (Müller et al., 2010). PVPP would remove these polyphenols. Commercial hop antifoam, which contains hop fats and waxes (Ford et al., 2012), did not suppress primary beer gushing (Shokribousjein et al., 2014).

Advanced hop products such as preisomerized iso-α-acids, reduced dihydro-α-alpha-acids (Rho), tetrahydro-α-alpha-acids (Tetra), and hexahydro-α-alpha-acids (Hexa) can promote beer gushing (Carrington et al., 1972). The source of the gushing is abeo-iso-α-acid (oxidation products derived from iso-α-acids). The tendency of advanced hop products to promote gushing can be eliminated by protecting the iso-α-acids from oxygen prior to reduction. Alternately, in the cases of Tetra and Hexa, the iso-α-acids' active olefinic groups are reduced prior to isomerization. For dihydro-α-alpha-acids (Rho), the carbonyl is reduced concurrent with its formation (Buffin and Campbell, 2013). Dehydrated humulinic acid, a known gushing promoter (Laws and McGuinness, 1972), was not detected by Buffin and Campbell (2013), perhaps because advanced hops product technology is improving.

Brewers utilizing these hop products can perform analysis for oxidized hops acid species (Taniguchi et al., 2013), request that the vendor screen their product for gushing potential, pick a vendor that reduces α-acids prior to isomerization, or obtain Tetra from a vendor utilizing β-acids as the starting material. As a practical matter, large breweries dose hop products at controlled rates upstream of final filtration. Filtration should eliminate haze formation resulting from locally high concentrations of hop products, haze that could become a gushing enhancer.

SECONDARY GUSHING

Secondary gushing potentiators (Table 10.3) include hop resin degradation products, oxalate, filter aid breakthrough, metal ions, tensides, and uneven carbonation (Bamforth, 2011). Secondary gushing is not the result of increasing the package temperature, knocking, dropping, or agitating the product, but these abuses can lead to particulate formation that results in gushing. Although particulates and other secondary potentiators can cause gushing, they do not necessarily have to initiate gushing. This makes determining root causes of secondary gushing difficult. These potentiators can also cause a slower release of foam that might best be described as overfoaming. In this case, foam could collapse to a few millilitre of liquid flowing down the side of the bottle or even just to a cap of foam at the

Table 10.3 Factors Contributing to Secondary Gushing of Beer

Factor	Corrective Measure
Colloidal substances from inherent property of malt	Partner with maltster to modify malting parameters and/or use brewing enzymes. This problem is difficult to diagnose.
Calcium oxalate crystals	Ensure 50–100 ppm calcium in brewing liquor. Avoid introducing calcium salts after filtration, eg, no hard water to rinse filters, use low $CaCO_3$ diatomaceous earth (DE)/kieselguhr, prerinse bottles with soft water, rinse beer filter with weak acid. In cases in which calcium salt addition is not permitted, the adjustment of mash pH values is achieved by the use of biologically prepared lactic acid introduced into the mash either as acid malt or as acidified wort.
Oxidation products of hops and advanced hop products	Cold storage of hops and hop extracts. High-performance Liquid Chromatography (HPLC) diagnosis of hop oxidation products.
Metal ions (iron, nickel, and other heavy metals)	Treat incoming metal-containing water. Passivate vessels and pipes. Remove bimetallic junctions. Use low-iron DE/kieselguhr.
Polyphenols originating from hop pellets	Chillproof with PVPP.
Haze	Proper chillproofing. For difficult beer, this might require both hordein and polyphenol reduction, eg, chillproofing with silica hydrogel and PVPP. Silica xerogel gives a tighter filtration but shorter runs.
Turbulent bottle filling, cavitation causing flake formation	Bottle counterpressure should be the same pressure as the filler bowl.
High carbonation level	Proper carbonation control in package release tank. Lower the carbonation level if primary gushing is also a factor.
High air headspace in bottle; high level of oxygen and nitrogen	Ensure proper and consistent bottle fill height. Evacuate air from bottle or can several times before filling and/or create sufficient foam in the headspace (fobbing) to force air from the headspace before adding closure
Filter aid particles such as kieselguhr (diatomaceous earth), silica gel, Celite, PVPP	Proper filter setup. Check filter screen for integrity.
Foreign particles eg, impurities from bottles, dust from crowns	Improve returnable-bottle washing procedure. Ensure proper rinsing of cans and nonreturnable bottles before filling. Inspect new crowns, bottles, and cans before using. Inspect conveyer lines re GMP.
Cleaning agent (tenside, surfactant) residues	Proper rinsing, or switch to caustic or acid cleaning.

Continued

Table 10.3 Factors Contributing to Secondary Gushing of Beer—cont'd

Factor	Corrective Measure
Glass corrosion; rough bottle internal surface	Use good-quality bottles. Monitor returnable bottles for scratches and corrosion.
Bottle position and mechanical stress during storage and transportation	Bottles and cans should stand upright, not prone. Mechanical stress during transportation is difficult to control. In-package foaming, especially with tetra- and hexahydro-iso-α-acid-bittered beer, can cause foam flakes which lead to overfoaming.
Temperature and time of beer storage causing haze formation	Employ temperature-controlled warehousing. Plan deliveries to minimize shipping time. Work with retailers to control inventory.
Temperature at opening of packaged beer	Beer should not overfoam when opened at room temperature. If it does, check for presence of other secondary gushing factors, as well as for primary gushing.
Winter- and summer-type gushing	"Winter-type" gushing occurs after shaking or storing certain beers at cold temperatures and only occurs in beer which has residual papain or other cysteine protease activity. Winter-gushing factors are inactivated by acid proteases. Summer gushing is associated with certain malt batches.

Table significantly modified from Table 1, Sarlin, T., 2012. Detection and Characterisation of Fusarium Hydrophobins Inducing Gushing in Beer, Dissertation for Doctor of Science in Technology, Aalto University School of Chemical Technology, VTT Science 13. VTT, Espoo, Finland. http://www.vtt.fi/inf/pdf/science/2012/S13.pdf (accessed 13.09.13.). Additional material from Amaha, M., Kitabatake, K., 1981. Gushing in beer, In: Pollock, J.R.A. (Ed.), Brewing Science, vol. 2. Academic Press, London, pp. 457–489; Briggs, D.E., Boulton, C.A., Brookes, P.A., Stevens, R., 2004. Gushing. In: Brewing Science and Practice. Woodhead Publishing Limited, Cambridge, England, pp. 726–727. (Chapter 19.7); Casey, G.P., 2005. American Society of Brewing Chemists (ASBC) No. XXV Fishbone Diagrams. Process Control to Prevent Gushing in Beer. The Society, St. Paul, MN; Christian, M., Titze, J., Ilberg, V., Jacob, F., 2011. Novel perspectives in gushing analysis: a review. Journal of the Institute of Brewing 117, 295–313.

bottle mouth. The best defense against secondary gushing is a well-designed GMP program that includes specifications for ingredients, brewing, bottling, transportation, and point of sale.

GMP should include inline process photometers at a minimum of three locations: (1) after the lauter tun or mash filter to monitor wort clarity, (2) after the yeast centrifuge or filter to monitor effectiveness of yeast removal, and (3) after final filtration to ensure that yeast and filter aids are excluded from packaged beer. Dual-angle photometers are the norm. Measurements at 90° to incident light are sensitive to colloids, providing a quality check for the clarity. Forward-scattered light measurement at 25° or a lesser angle selectively detects filter aid particles and yeast cells. Low-angle detection also helps differentiate "invisible" or "pseudo" hazes, in which very small particles (<0.1 μm) cause high levels of light scatter when measured at 90° (Steiner et al., 2010). "Invisible haze" can be difficult to filter and to identify. However, it can be a predictive value of colloidal instability because very small proteo-tannic complexes, glucan particles, or yeast wall material can be precursors of visible haze that forms as the beer oxidizes (Malcorps et al., 2001).

DIAGNOSING SECONDARY GUSHING CAUSES

As a last resort, the beer inclusion or haze that caused the elevated turbidity and secondary gushing can be identified. There are numerous practical references for this purpose (ASBC, 2011a; Glenister and Paul, 1975; Steiner et al., 2010; Wiesen et al., 2011; Wiesen et al., 2012). In practice, identification is difficult, and the detected compound might not be the source of the primary problem. One challenge is removing a representative sample from a bottle that gushes when opened. The liquid that gushed may have concentrated the particulate source that triggered gushing, or new particles, especially protein precipitates, might form as the beer gushes.

WHAT ARE CONSUMER EXPECTATIONS?

Consumer studies show that preferences diverge between genders, race, and region (Bamforth, 2000; Evans and Bamforth, 2009). Both Italian and American consumers preferred a medium level of foam (Donadini et al., 2011; Smythe et al., 2002). "The perfect pour" may indeed be the classic two-finger (3 cm) foam head (Evans et al., 2012).

Men appreciate lacing more than women (Roza et al., 2006). For Italians, beers with medium foam levels were perceived as more alcoholic and fruitier, and beers with low foam levels were perceived as having off-flavors. Beers with high foam levels and heavy lacing were rated as more thirst-quenching than beers with high foam levels with poor lacing (Donadini et al., 2011). Contrarily, Scottish and American consumers did not associate differing levels of head and lacing with beer taste and olfactory sensations, with the exception that beers with low head levels were perceived as more likely to have off-flavors (Smythe et al., 2002).

In a novel approach, Czech pub consumers were asked a simple question "Is everything okay with the foam?" (Kosin et al., 2010). The results are perhaps a little disheartening to a brewer. The beer was problem free as long as there was sufficient foam to cover the beer surface. At this point, one quarter of the customers paid attention to the foam quality. The other three quarters were not concerned.

Some brewers are putting more emphasis on beer presentation. A few American craft brewers have introduced branded glassware that not only provides markings for foam height but is also designed to enhance their beer's flavor (Evans and Bamforth, 2009; Evans et al., 2012; Spiegelau, 2013). These brewers follow the proud tradition of European brewers (Evans and Bamforth, 2009; Delvaux et al., 1995), but perhaps take the concept a step further. They laser etch the bottom of the glass to provide nucleation sites for the continuous release of a stream of beer bubbles (beading). Beer presentation is also emphasized in programs designed to train servers. All of these steps will improve consumers' appreciation for beer's fine attributes, starting with the importance of "the perfect pour" for visual, tactile, and flavor appeal.

ACKNOWLEDGMENTS

Mr. Lusk thanks Kenneth Berg (PQ Corporation, United States), Brian Buffin (Kalsec, United States), Evan Evans (Australian Export Grains and Innovation Center, Australia), Nick Huige (United States), Mary-Jane Maurice (Malteurop North America, United States), Jay Refling (MillerCoors, United States), Tuija Sarlin (Technical Research Center of Finland), Paul Schwarz (North Dakota State University, United States), and Sandra van Nierop Stelma (Grolsch, Netherlands) for information and insightful suggestions.

REFERENCES

Aastrup, S., Legind-Hansen, P., Nielsen, H., 1996. Enzymatic reduction of gushing tendencies in beer. Brauwelt International 14 (2), 136–137.

Amaha, M., Kitabatake, K., 1981. Gushing in beer. In: Pollock, J.R.A. (Ed.), Brewing Science, vol. 2. Academic Press, London, pp. 457–489.

American Society of Brewing Chemists (ASBC), 2011a. Identification guides. In: Beer Inclusions: Common Causes of Elevated Turbidity, fourteenth ed. The Society, St. Paul, MN.

American Society of Brewing Chemists (ASBC), 2011b. Methods of Analysis, fourteenth ed. The Society, St. Paul, MN.

Anness, B.J., Reed, R.J.R., 1985. Lipids in the brewery – a material balance. Journal of the Institute of Brewing 91, 82–87.

Bamforth, C.W., 1985. The foaming properties of beer. Journal of the Institute of Brewing 91, 370–383.

Bamforth, C.W., 1995. Foam: method, myth or magic. The Brewer 81, 396–399.

Bamforth, C.W., 2000. Perceptions of beer foam. Journal of the Institute of Brewing 106, 229–238.

Bamforth, C.W., 2011. 125th anniversary review: the non-biological instability of beer. Journal of the Institute of Brewing 117, 488–497.

Beardmore, R., 2013. Bi-metallic Corrosion (Galvanic Corrosion). http://www.roymech.co.uk/Useful_Tables/Corrosion/Cor_bi_met.html (accessed 10.10.13.).

Besier, A., Müller, V., Fröhlich, J., Pätz, R., 2013. Reduce overfoaming. Method for the treatment of the phenomenon gushing in beer and malt beverages. Brewing and Beverage Industry International 46–49.

Blenkinsop, P., 1991. The manufacture, characteristics and uses of speciality malts. Technical Quarterly Master Brewers Association of the Americas 28, 145–149.

Boivin, P., Malanda, M., 1997. Improvement of malt quality and safety by adding starter culture during the malting process. Technical Quarterly Master Brewers Association of the Americas 34, 96–101.

Boivin, P., Malanda, M., 1999. Inoculation by *Geotrichum candidum* During Malting of Cereals or Other Plants. United States Patent US5955070.

Brewery Convention of Japan, 2004. Methods of Analysis of BCOJ (Revised Edition).

Brey, S.E., de Costa, S., Rogers, P.J., Bryce, J.H., Morris, P.C., Mitchell, W.J., Stewart, G.G., 2003. The effect of proteinase A on foam-active polypeptides during high and low gravity fermentation. Journal of the Institute of Brewing 109, 194–202.

Briggs, D.E., Boulton, C.A., Brookes, P.A., Stevens, R., 2004. Gushing. In: Brewing Science and Practice. Woodhead Publishing Limited, Cambridge, England, pp. 726–727 (Chapter 19.7).

Briggs, D.E., Hough, J.S., Stevens, R., Young, T.W., 1981. Malting & Brewing Science. In: Malt and Sweet Wort, second ed., vol. 1. Chapman & Hall, New York, pp. 134–141.

Buffin, B., Campbell, P.L., 2013. The influence of hop acid components on the phenomenon of gushing in various standard and non-alcoholic beers – evaluation of advanced hop products under induced gushing conditions. BrewingScience – Monatsschrift für Brauwissenschaft 66, 198–204.

Bühler, T., Michel, R., Kantelberg, B., Baumgärtner, Y., 2003. Dynamic low-pressure boiling – systematically optimized for top quality. Brauwelt International 21, 306–313.

Burger, K., Bach, H.-P., 2001. Mixtures of Enzymes Containing an Enzyme With Beta-glucanase Activity, to Be Used for Decreasing or Preventing Gushing. European Patent EP1164184B1.

Buttrick, P., 2006. A brewer's view on a modern brewhouse project. The Brewer & Distiller 2, 1–6.

Canadian Malting Barley Technical Centre, 2013. Recommended Malting Barley Varieties 2013–2014. Canadian Malting Barley Technical Centre, Winnipeg, Manitoba, Canada. http://www.cmbtc.com (accessed 15.09.13.).

Carlsberg, 2011. Press Release. Naturally-Bred Null-Lox, the Better Barley. Carlsberg Group. http://www.carlsberggroup.com/media/PressKits/Documents/Carlsberg/Null-LOX_the_better_barley.pdf (accessed 15.09.13.).

Carrington, R., Collett, R.C., Dunkin, I.R., Halek, G., 1972. Gushing promoters and suppressants in beer and hops. Journal of the Institute of Brewing 78, 243–254.

Casey, G.P., 1996. Primary versus secondary gushing and assay procedures used to assess malt/beer gushing potential. Technical Quarterly Master Brewers Association of the Americas 33, 228–235.

Casey, G.P., 2005. American Society of Brewing Chemists (ASBC) No. XXV Fishbone Diagrams. Process Control to Prevent Gushing in Beer. The Society, St. Paul, MN.

Catarino, S., Madeira, M., Monteiro, F., Rocha, F., Curvelo-Garcia, A.S., de Sousa, R.B., 2008. Effect of bentonite characteristics on the elemental composition of wine. Journal of Agricultural and Food Chemistry 56, 158–165.

Chandrashekar, J., Yarmolinsky, D., von Buchholtz, L., Oka, Y., Sly, W., Ryba, J.P.N., Zuker, C.S., 2009. The taste of carbonation. Science 326, 443–445.

Christian, M., Titze, J., Ilberg, V., Jacob, F., 2011. Novel perspectives in gushing analysis: a review. Journal of the Institute of Brewing 117, 295–313.

Coghe, S., Vanderhaegen, B., Pelgrims, B., Basteyns, A.-V., Delvaux, F.R., 2003. Characterization of dark specialty malts: new insights in color evaluation and pro- and antioxidative activity. Journal of Amercian Society of Brewing Chemists 61, 125–132.

Combe, A.L., Ang, J.K., Bamforth, C.W., 2013. Positive and negative impacts of specialty malts on beer foam: a comparison of various cereal products for their foaming properties. Journal of the Science of Food and Agriculture 93, 2094–2101.

Constant, M., 1992. A practical method for characterizing poured beer foam quality. Journal of Amercian Society of Brewing Chemists 50, 37–47.

Curtis, N.S., Martindale, L., 1961. Studies on gushing. I. Introduction. Journal of the Institute of Brewing 67, 417–421.

Deckers, S.M., Venken, T., Khalesi, M., Gebruers, K., Baggerman, G., Lorgouilloux, Y., Shokribousjein, Z., Ilberg, V., Schönberger, C., Titze, J., Verachtert, H., Michiels, C., Neven, H., Delcour, J., Martens, J., Derdelinckx, G., De Maeyer, M., 2012. Combined modeling and biophysical characterisation of CO(2) interaction with Class II hydrophobins: new insight into the mechanism underpinning primary gushing. Journal of Amercian Society of Brewing Chemists 70, 249–256.

Delvaux, F., Combes, F.J., Delvaux, F.R., 2004. The effect of wheat malting on the colloidal haze of white beers. Technical Quarterly Master Brewers Association of the Americas 41, 27–32.

Delvaux, F., Deams, V., Vanmachelen, H., Neven, H., Derdelinckx, G., 1995. Retention of beer flavours by the choice of appropriate glass. In: Proceedings of the European Brewery Convention Congress. Elsevier, Brussels, pp. 533–542.

Depraetere, S.A., Delvaux, F., Coghe, S., Delvaux, F.R., 2004. Wheat variety and barley malt properties: influence on haze intensity and foam stability of wheat beer. Journal of the Institute of Brewing 110, 200–206.

Dickie, K.H., Cann, C., Norman, E.C., Bamforth, C.W., Muller, R.E., 2001. Foam negative materials. Journal of Amercian Society of Brewing Chemists 59, 17–23.

Donadini, G., Fumi, M.D., De Faveri, M.D., 2011. How foam appearance influences the Italian consumer's beer perception and preference. Journal of the Institute of Brewing 117, 523–533.

Douglas, P.E., Flannigan, B., 1988. A microbiological evaluation of barley malt production. Journal of the Institute of Brewing 94, 85–88.

Engelmann, J., Walter, H., Illing, S., Jacob, F., 2012. Das phänomen gushing und rote körner im malz (teil 1). Brauwelt 32, 912–915.

Esterbauer, H., Schauenstein, E., 1977. Isomeric trihydroxy-octadecenoic acids in beer: evidence for their presence and quantitative determination. Zeitschrift für Lebensmittel-Untersuchung und -Forschung 164, 255–259.

European Brewery Convention (EBC), 2013. Analytica – EBC. Verlag Hans Carl Getranke Fachverlag, Nürnberg.

Evans, D.E., Bamforth, C.W., 2009. Beer foam: achieving a suitable head. In: Bamforth, C.W., Russell, I., Stewart, G. (Eds.), Handbook of Alcoholic Beverages: Beer, a Quality Perspective, first ed. Academic Press, New York, pp. 85–109.

Evans, D.E., Sheehan, M.C., 2002. Don't be fobbed off: the substance of beer foam – a review. Journal of Amercian Society of Brewing Chemists 60, 47–57.

Evans, D.E., Oberdieck, M., Redd, K.S., Newman, R., 2012. Comparison of the Rudin and NIBEM methods for measuring foam stability with a manual pour method to identify beer characteristics that deliver consumers stable beer foam. Journal of Amercian Society of Brewing Chemists 70, 70–78.

Faltermaier, A., Waters, D., Becker, T., Arendt, E., Gastl, M., 2014. Common wheat (*Triticum aestivum* L.) and its use as a brewing cereal – a review. Journal of the Institute of Brewing 120, 1–15.

Ford, Y.-Y., Westwood, K.T., Gahr, A., Ferreira, A.R., Wolinska, K., Lad, M., Wolf, B., Foster, T., 2012. Antifoam performance of hop extract emulsions in pilot brewing trials. Technical Quarterly Master Brewers Association of the Americas 49, 3–10.

Fritsch, A., Shellhammer, T.H., 2008. Relative bitterness of reduced and nonreduced iso-alpha-acids in lager beer. Journal of Amercian Society of Brewing Chemists 66, 88–93.

Garbe, L.-A., Hübke, H., Tressl, R., 2005. Enantioselective formation pathway of a trihydroxy fatty acid during mashing. Journal of Amercian Society of Brewing Chemists 63, 157–162.

Garbe, L.-A., Schwarz, P., Ehmer, A., 2009. Beer gushing. In: Bamforth, C.W., Russell, I., Stewart, G. (Eds.), Handbook of Alcoholic Beverages: Beer, a Quality Perspective, first ed. Academic Press, New York, pp. 186–212.

Gardner, R.J., Laws, D.R.J., McGuinness, J.D., 1973. The suppression of gushing by the use of hop oil. Journal of the Institute of Brewing 79, 209–211.

Glatthar, J., Heinisch, J.J., Senn, T., 2003. The use of unmalted triticale in brewing and its effect on wort and beer quality. Journal of Amercian Society of Brewing Chemists 61, 182–190.

Glenister, P., Paul, R., 1975. Beer Deposits: A Laboratory Guide and Pictorial Atlas. J.E. Siebel Sons' Company, Marshall Division, Miles Laboratories, Inc., Chicago.

Gruber, M.A., 2001. The flavor contributions of kilned and roasted products to finished beer styles. Technical Quarterly Master Brewers Association of the Americas 38, 227–233.

Hanke, S., Kern, M., Hermann, M., Back, W., Becker, T., Krottenthaler, M., 2010. Hop ingredients and their effect on gushing (part 1). Brauwelt International 28, 356–357.

Hanke, S., Kern, M., Hermann, M., Back, W., Becker, T., Krottenthaler, M., 2011. Hop ingredients and their effect on gushing (part 2). Brauwelt International 29, 24–26.

Hertrich, J.D., 2013. Topics in brewing: malting. Technical Quarterly Master Brewers Association of the Americas 50, 131–141.

Heymann, H., Goldberg, J.R., Wallin, C.E., Bamforth, C.W., 2010. A "beer" made from a bland alcohol base. Journal of Amercian Society of Brewing Chemists 68, 75–76.

Hii, V., Herwig, W.C., 1982. Determination of high molecular weight proteins in beer using Coomassie Blue. Journal of Amercian Society of Brewing Chemists 40, 46–50.

Hintz, M., 2011. A Spirited History of Milwaukee Brews & Booze. History Press, Charleston, SC.

Hippeli, S., Elstner, E.F., 2002. Minireview: are hydrophobins and/or non-specific lipid transfer proteins responsible for gushing in beer? New hypotheses on the chemical nature of gushing inducing factors. Zeitschrift für Naturforschung C: A Journal of Biosciences 57.

Hippeli, S., Hecht, D., 2009. The role of ns-LTP1 and proteases in causing primary gushing. Brauwelt International 27, 30–34.

Hirota, N., Kuroda, H., Takoi, K., Kaneko, T., Kaneda, H., Yoshida, I., Takashio, M., Ito, K., Takeda, K., 2006. Development of novel barley with improved beer foam and flavor stability-the impact of lipoxygenase-1-less barley in the brewing industry. Technical Quarterly Master Brewers Association of the Americas 43, 131–135.

Ishibashi, Y., Kakui, T., Terano, Y., Hon-no, E., Kogin, A., Nakatani, K., 1997. Application of ELISA to quantitative evaluation of foam-active protein in the malting and brewing processes. Journal of Amercian Society of Brewing Chemists 55, 20–23.

Jackson, G., Bamforth, C.W., 1982. The measurement of foam lacing. Journal of the Institute of Brewing 88, 378–381.

Johnson, D.D., Nganje, W., 2000. Impacts of DON in the Malting Barley Supply Chain: Aggregate Costs and Firm-Level Risks. Agricultural Economics Miscellaneous Report No. 187. Department of Agricultural Economics, Agricultural Experiment Station, North Dakota State University, Fargo, North Dakota. http://ageconsearch.umn.edu/bitstream/23103/1/aem187.pdf (accessed 31.01.14.).

Kaltner, D., Forster, C., Flieher, M., Nielsen, T.P., 2013. The influence of dry hopping on three different beer styles. Brauwelt International 31, 355–359.

Kaneda, H., Kobayashi, N., Watari, J., Shinotsuka, K., Takashio, M., 2002. A new taste sensor for evaluation of beer body and smoothness using a lipid-coated quartz crystal microbalance. Journal of Amercian Society of Brewing Chemists 60, 71–76.

Kieninger, H., Wiest, A., 1982. The utilization of caramel malt to improve beer quality. Technical Quarterly Master Brewers Association of the Americas 19, 170–174.

Kobayashi, N., Kaneda, H., Kano, Y., Koshino, S., 1994. Behavior of lipid hydroperoxides during mashing. Journal of Amercian Society of Brewing Chemists 52, 141–145.

Kobayashi, N., Kaneda, H., Kuroda, H., Watari, J., Kurihara, T., Shinotsuka, K., 2000. Behavior of mono-, di-, and trihydroxyoctadecenoic acids during mashing and methods of controlling their production. Journal of Bioscience and Bioengineering 90, 69–73.

Kobayashi, N., Segawa, S., Umemoto, S., Kuroda, H., Kaneda, H., Mitani, Y., Watari, J., Takashio, M., 2002. A new method for evaluating foam-damaging effect by free fatty acids. Journal of Amercian Society of Brewing Chemists 60, 37–41.

Kordialik-Bogack, E., Antczak, N., 2011. Prediction of beer foam stability from malt components. Czech Journal of Food Science 29, 243–249.

Kosin, P., Savel, J., Evans, D.E., Broz, A., 2012. How the method of beer dispense influences the served CO(2) content and the sensory profile of beer. Journal of Amercian Society of Brewing Chemists 70, 103–108.

Kosin, P., Savel, J., Evans, D.E., Broz, A., 2010. Relationship between matrix foaming potential, beer composition, and foam stability. Journal of Amercian Society of Brewing Chemists 68, 63–69.

Kühbeck, F., Back, W., Krottenthaler, M., 2006a. Influence of lauter turbidity on wort composition, fermentation performance and beer quality – a review. Journal of the Institute of Brewing 112, 215–221.

Kühbeck, F., Back, W., Krottenthaler, M., 2006b. Influence of lauter turbidity on wort composition, fermentation performance and beer quality in large-scale trials. Journal of the Institute of Brewing 112, 222–231.

Kunimune, T., Shellhammer, T.H., 2008. Foam-stabilizing effects and cling formation patterns of iso-alpha-acids and reduced iso-alpha-acids in lager beer. Journal of Agricultural and Food Chemistry 56, 8629–8634.

Kunz, T., Müller, C., Mato-Gonzales, D., Methner, F.J., 2012. The influence of unmalted barley on the oxidative stability of wort and beer. Journal of the Institute of Brewing 118, 32–39.

Kunzmann, C., 2011. Brewing with various adjuncts. In: Bangkok Brewing Conference. Thailand Beer Industry Guild (TBIG) and VLB Berlin. https://www.vlb-berlin.org/sites/default/files/Seite/Bangkok%20Conference%202011%20-%20Downloads/wed09-novozymes.pdf (accessed 20.03.14.).

Kuroda, H., Furusho, S., Maeba, H., Takashio, M., 2003. Characterization of factors involved in the production of 2(E)-nonenal during mashing. Bioscience, Biotechnology, and Biochemistry 67, 691–697.

Laitila, A., Sweins, H., Vilpola, A., Kotaviita, E., Olkku, J., Home, S., Haikara, A., 2006. *Lactobacillus plantarum* and *Pediococcus pentosaceus* starter cultures as a tool for microflora management in malting and for enhancement of malt processability. Journal of Agricultural and Food Chemistry 54, 3840–3851.

Langstaff, S.A., Lewis, M.J., 1993. Foam and the perception of beer flavor and mouthfeel. Technical Quarterly Master Brewers Association of the Americas 30, 16–17.

Laws, D.R.J., McGuinness, J.D., 1972. Origin and estimation of the gushing potential of isomerized hop extracts. Journal of the Institute of Brewing 78, 302–308.

Leather, R.V., 1998. The Cambridge Prize Lecture 1996. From field to firkin: an integrated approach to beer clarification and quality. Journal of the Institute of Brewing 104, 9–18.

Leiper, K.A., Duszanskyj, R., Stewart, G.G., 2002. Premixing of isinglass and silica gel to obtain improved beer stability. Journal of the Institute of Brewing 108, 28–31.

Lewis, M.J., Bamforth, C.W., 2006. Essays in Brewing Science. Springer Science + Business Media, LLC, New York, pp. 43–57.

Lowe, D.P., Arendt, E.K., 2004. The use and effects of lactic acid bacteria in malting and brewing with their relationships to antifungal activity, mycotoxins and gushing: a review. Journal of the Institute of Brewing 110, 163–180.

Lusk, L.T., Cronan, C.L., Chicoye, E., Goldstein, H., 1987. A surface-active fraction isolated from beer. Journal of Amercian Society of Brewing Chemists 45, 91–95.

Lusk, L.T., Cronan, C.L., Ting, P., Seabrooks, J., Ryder, D., 2003. An evolving understanding of foam bubbles based upon beer style development. Proceedings of the European Brewery Convention Congress, Dublin 29 (CD-ROM).

Lusk, L.T., Goldstein, H., Ryder, D., 1995. Independent role of beer proteins, melanoidins and polysaccharides in foam formation. Journal of Amercian Society of Brewing Chemists 53, 93–103.

Lusk, L.T., Goldstein, H., Watts, K., Navarro, A., Ryder, D., 2001. Monitoring barley lipid transfer protein levels in barley, malting, and brewing. Proceedings of the European Brewery Convention Congress, Budapest 28, 663–672 (CD-ROM).

Lynch, D.M., Bamforth, C.W., 2002. Measurement and characterization of bubble nucleation in beer. Journal of Food Science 67.

MacIntosh, A.J., MacLeod, A., Beattie, A.D., Eck, E., Edney, M., Rossnagel, B., Speers, R.A., 2014. Assessing the effect of fungal infection of barley and malt on premature yeast flocculation. Journal of Amercian Society of Brewing Chemists 72, 66–72.

Malcorps, P., Haselaars, P., Dupire, S., Van den Eynde, E., 2001. Glycogen released by the yeast as a cause of unfilterable haze in the beer. Technical Quarterly Master Brewers Association of the Americas 38, 95–98.

McMullen, M., Bergstrom, G., De Wolf, E., Dill-Macky, R., Hershman, D., Shaner, G., Van Sanford, D., 2012. A unified effort to fight an enemy of wheat and barley: Fusarium head blight. Plant Disease 96, 1712–1728.

Melm, G., Tung, P., Pringle, A., 1995. Mathematical modeling of beer foam. Technical Quarterly Master Brewers Association of the Americas 32, 6–10.

Michel, R.A., Vollhals, B., 2003. Dynamic wort boiling. Technical Quarterly Master Brewers Association of the Americas 40, 25–29.

Michel, R.A., Vollhals, B., 2002. Reduction of wort thermal stress – relative to heating methods and wort treatment. Technical Quarterly Master Brewers Association of the Americas 39, 156–163.

Mitteleuropäische Brautechnische Analysenkommission (MEBAK), 2012. Brautechnische Analysenmethoden: Würze, Bier, Biermischgetränke; Methodensammlung der Mitteleuropäischen Brautechnischen Analysenkommission. MEBAK, Freising, Germany.

Mitteleuropäische Brautechnische Analysenkommission (MEBAK), 2006. Brautechnische Analysemethoden (Band Rohstoffe). In: Anger, H.-M. (Ed.), MEBAK, Freising, Germany.

Müller, M.P., Schmid, F., Becker, T., Gastl, M., 2010. Impact of different hop compounds on the overfoaming volume of beer caused by primary gushing. Journal of the Institute of Brewing 116, 459–463.

Munar, M.J., Sebree, B., 1997. Gushing – a maltster's view. Journal of Amercian Society of Brewing Chemists 55, 119–122.

Neogen, 2012. Mycotoxin Handbook, third ed. Neogen Corporation, Lansing, Michigan http://www.neogen.com/FoodSafety/pdf/MycotoxinHandbook_12.pdf. (accessed 31.01.14.).

Oliveira, P., Mauch, A., Jacob, F., Arendt, E.K., 2012. Impact of *Fusarium culmorum*-infected barley malt grains on brewing and beer quality. Journal of Amercian Society of Brewing Chemists 70, 186–194.

Rehmanji, M., Gopal, C., Mola, A., 2005. Beer stabilization technology – clearly a matter of choice. Technical Quarterly Master Brewers Association of the Americas 42, 332–338.

Roberts, R.T., 1976. Some aspects of the stability of beer foam. In: Akers, R.J. (Ed.), Foams: Proceedings of a Symposium Organized by the Society of Chemical Industry, Colloid and Surface Chemistry Group, first ed. Academic Press, New York, pp. 257–271.

Robinson, L., David, P., 2008. Improving foam quality through wort boiling optimisation. In: Proceedings of the Institute of Brewing & Distilling. Asia Pacific Section, Auckland.

Ross, S., Clark, G.L., 1939. On the measurement of foam stability with special reference to beer. Wallerstein Laboratory Communications 6, 46–54.

Roza, J.R., Wallin, C.E., Bamforth, C.W., 2006. A comparison between instrumental measurement of head retention/lacing and perceived foam quality. Technical Quarterly Master Brewers Association of the Americas 43, 173–176.

San Martın, R., Briones, R., 2000. Quality control of commercial quillaja (*Quillaja saponaria* Molina) extracts by reverse phase HPLC. Journal of the Science of Food and Agriculture 80, 2063–2068.

Sarlin, T., 2012. Detection and Characterisation of Fusarium Hydrophobins Inducing Gushing in Beer. Dissertation for Doctor of Science in Technology. Aalto University School of Chemical Technology. VTT Science 13. VTT, Espoo, Finland. http://www.vtt.fi/inf/pdf/science/2012/S13.pdf (accessed 13.09.13.).

Sarlin, T., Nakari-Setälä, T., Linder, M., Penttilä, M., Haikara, A., 2005. Fungal hydrophobins as predictors of the gushing activity of malt. Journal of the Institute of Brewing 111, 105–111.

Sarlin, T., Vilpola, A., Kotaviita, E., Olkku, J., Haikara, A., 2007. Fungal hydrophobins in the barley-to-beer chain. Journal of the Institute of Brewing 113, 147–153.

Schaafsma, A.W., Nicol, R.W., Savard, M.E., Sinha, R.C., Reid, L.M., Rottinghaus, G., 1998. Analysis of Fusarium toxins in maize and wheat using thin layer chromatograph. Mycopathologia 142, 107–113.

Schneidereit, J.O., Kunz, T., Methner, F.-J., 2013. Application of Gallotannins to Prevent Gushing in Beer and Carbonated Beverages, Master Brewers Association of the Americas Annual Conference. Texas, Austin.

Schnitzenbaumer, B., Karl, C.A., Jacob, F., Arendt, E.K., 2013. Impact of unmalted white Nigerian and red Italian sorghum (Sorghum bicolor) on the quality of worts and beers applying optimized enzyme levels. Journal of Amercian Society of Brewing Chemists 71, 258–266.

Schnitzenbaumer, B., Kerpes, R., Titze, J., Jacob, F., Arendt, E.K., 2012. Impact of various levels of unmalted oats (*Avena sativa* L.) on the quality and processability of mashes, worts, and beers. Journal of Amercian Society of Brewing Chemists 70, 142–149.

Schwarz, P., Li, Y., 2011. Malting and brewing uses of barley. In: Ullrich, S.E. (Ed.), Barley: Production, Improvement, and Uses. Blackwell Publishing Ltd; Wiley InterScience (Online service), Ames, Iowa, pp. 478–521.

Shokribousjein, Z., Deckers, S.M., Gebruers, K., Lorgouilloux, Y., Baggerman, G., Verachtert, H., Delcour, J.A., Etienne, P., Rock, J.-M., Michielsa, C., Derdelinckx, G., 2011. Hydrophobins, beer foaming and gushing. Cerevisia 35, 85–101.

Shokribousjein, Z., Philippaerts, A., Riveros, D., Titze, J., Ford, Y., Deckers, S.M., Khalesi, M., Delcour, J.A., Gebruers, K., Verachtert, H., Ilberg, V., Derdelinckx, G., Sels, B., 2014. A curative method for primary gushing of beer and carbonated beverages: characterization and application of antifoam based on hop oils. Journal of Amercian Society of Brewing Chemists 72, 12–21.

Slack, P.T., Bamforth, C.W., 1983. The fractionation of polypeptides from barley and beer by hydrophobic interaction chromatography: the influence of their hydrophobicity on foam stability. Journal of the Institute of Brewing 89, 397–401.

Smythe, J.E., O'Mahony, M.A., Bamforth, C.W., 2002. The impact of the appearance of beer on its perception. Journal of the Institute of Brewing 108, 37–42.

Spiegelau, 2013. Spiegelau IPA Glass. Spiegelau GmbH, Neustadt an der Waldnaab, Germany. www.spiegelauusa.com (accessed 11.08.13.).

Steiner, E., Becker, T., Gastl, M., 2010. Turbidity and haze formation in beer – insights and overview. Journal of the Institute of Brewing 116, 360–368.

Stewart, G.G., Bothwick, R., Bryce, J., Cooper, D., Cunningham, S., Hart, C., Rees, E., 1997. Recent developments in high gravity brewing. Journal of the Institute of Brewing 34, 264–270.

Stewart, G.G., Mader, A., Chlup, P., Miedl, M., 2006. The influence of process parameters on beer foam stability. Technical Quarterly Master Brewers Association of the Americas 43, 47–51.

Taniguchi, Y., Matsukura, Y., Ozaki, H., Nishimura, K., Shindo, S., 2013. Identification and quantification of the oxidation products derived from α-acids and β-acids during storage of hops (*Humulus lupulus L.*). Journal of Agricultural and Food Chemistry 61, 3121–3130.

Taylor, D.G., 1990. The importance of pH control during brewing. Technical Quarterly Master Brewers Association of the Americas 27, 131–136.

Van Nierop, S.N.E., Evans, D.E., Axcell, B.C., Cantrell, I.C., Rautenback, M., 2004. Impact of different wort boiling temperatures on the beer foam stabilizing properties of lipid transfer protein. Journal of Agricultural and Food Chemistry 52, 3120–3129.

Vanderhaegen, B., Neven, H., Verachtert, H., Derdelinckx, G., 2006. The chemistry of beer aging – a critical review. Food Chemistry 95, 357–381.

Wallin, C.E., DiPietro, M.B., Schwarz, R.W., Bamforth, C.W., 2010. A comparison of three methods for the assessment of foam stability of beer. Journal of the Institute of Brewing 116, 78–80.

Wiesen, E., Gastl, M., Becker, T., 2011. Identification of hazes in beer – Part I. Brauwelt International 29, 374–379.

Wiesen, E., Gastl, M., Becker, T., 2012. Identification of hazes in beer – Part II. Brauwelt International 30, 41–46.

Wilson, J.H.R., Schwarz, H., Smith, R., 2011. Natural and Stable Solutions of Alpha-Acids and Their Use for the Improvement of Foam Quality of Beer. United States Patent Application US0287152A1.

Zapf, M.W., Theisen, S., Rohde, S., Rabenstein, F., Vogel, R.F., Niessen, L., 2007. Characterization of AfpA, an alkaline foam protein from cultures of *Fusarium culmorum* and its identification in infected malt. Journal of Applied Microbiology 103, 36–52.

Zapf, M.W., Theisen, S., Vogel, R.F., Niessen, L., 2006. Cloning of wheat LTP1500 and two *Fusarium culmorum* hydrophobins in *Saccharomyces cerevisiae* and assessment of their gushing inducing potential in experimental wort fermentation. Journal of the Institute of Brewing 112, 237–245.

Zufall, C., Tyrell, T., 2008. The influence of heavy metal ions on beer flavour stability. Journal of the Institute of Brewing 114, 134–142.

CHAPTER

11 Color

A.J. de Lange
Formerly McLean, Virginia, United States

INTRODUCTION

Ordering of color, that is, creating a system in which it is possible to specify the coordinates of each humanly visible color in some meaningful way, is a very challenging task that has engaged the minds of artists, philosophers, chemists, physicists, engineers, psychologists, and others from antiquity to the present day. The visible color of an object depends on two things: the way in which impinging light is modified by the object and second, the impression that the modified light makes on a human observer. This separates the problem of specifying a color (finding where it fits in an order system) into two parts. In the first, physical measurements are made of the relative energy at each visible wavelength in the illuminating spectrum and of the extent to which each wavelength of light is reflected or absorbed by the object. In the modern world, instruments are readily available which make these measurements to high accuracy and the first part is relatively straightforward.

In the second part of the problem measurement data are put into a model of human color vision. Because the mechanisms of human vision are not, even today, fully understood, because no two humans, even those with normal color vision, see in exactly the same way, and because the visual system is highly adaptive, the models are complex and are being constantly refined. Nevertheless, workable models do exist which serve as the basis for color photography, color printing, color television, and color control in industry.

In the brewing application the additional complexities of part two are enough to discourage many operators from using it, so today beer and wort colors are most often specified in terms of physical measurement alone, in particular the extent to which light of wavelength 430 nm is absorbed in passing through a 1-cm thick sample. This, as measured by a photometer or spectrophotometer, is the basis of the American Society of Brewing Chemists (ASBC) Standard Reference Method (SRM) (ASBC, 2006a) and European Brewery Convention (EBC)

Brewing Materials and Processes. http://dx.doi.org/10.1016/B978-0-12-799954-8.00011-3

[Central European Commission for Brewing Analysis (MEBAK, 2002)] color determination methods.

SRM and EBC values are points on one-dimensional color order scales. The Iodine, Brand, and Lovibond scales (DeClerck, 1958, Vol II, p. 332; Lovibond, 1915) which preceded the current SRM and EBC[1] methods are also one-dimensional scales. In these older methods, visual comparison was made, respectively, to a series of standard solutions of iodine, dyes, or colored-glass slides. Given a reported beer color in any of these scales, an experienced person can conjure up an approximate mental picture of what the beer will look like and based on this the one-dimensional scales can be considered at least partially successful as means of reporting beer color. The instrumental methods enjoy improved repeatability relative to visual methods by removing observer variability, but they, in common with the others, suffer from the limitation of any linear color order system: one degree of freedom. Beer and wort colors have more than one. If beer and wort colors had one degree of freedom all possible beer and wort colors would either be an exact match to one of a reasonably limited number of colored glasses or have a color that falls on a straight line between a pair in some perceptibly uniform color space. The deviations of actual beer and wort colors from those of the Lovibond glasses have long been noted, and better matching colors have been suggested (Bishop, 1950) and adopted, but even the improved glasses do not match all beers and worts exactly and best, but approximate, visual matches must be accepted in many cases. This means that it is entirely possible to have two beers with quite different color appearances that are both best matched by the same Lovibond glass. Sharpe et al. (1992) were motivated by this to develop an instrumental method based on the Commission Internationale de l'Eclairage (CIE) Tristimulus system. By considering more degrees of freedom (DOFs), color differences between beers with the same Lovibond rating can be resolved. The current ASBC Tristimulus method, Beer-10C (ASBC, 2006b) is similar to the method proposed by Sharpe. Both use the entire spectrum of the beer. Spectrum, in the context of this chapter, usually means a set of measurements of absorption or transmission made at wavelengths separated by 5 nm beginning at 380 and ending at 780 nm.

For the SRM and EBC scales to be completely adequate descriptors of color, all beer and wort absorption spectra as normalized[2] by the measurement at 430 nm would have to be the same. In fact, they are close to being so. We refer to the hypothesis that they are the same as the Constant Normalized Spectrum (CNS) hypothesis and will refer to it frequently throughout this chapter as it is

[1]Prior to adoption of the current 430-nm method the EBC used a single-wavelength measurement at 530 nm, and prior to that the Lovibond system which is still in use today especially by American maltsters.

[2]Every absorption measurement is divided by the absorption measurement at 430 nm. All normalized absorption spectra have value 1 at 430 nm.

FIGURE 11.1 Photograph Showing Effect of Path on Brightness of Beer Color. Hue and Chroma are Also Affected by Path.

close enough to being true to be a very useful device for exploring and explaining several aspects of beer color. An estimate[3] of the absorption CNS spectrum is

$$\underline{\widehat{a}}(\lambda) = 0.02465e^{-\frac{\lambda-430}{17.591}} + 0.97535e^{-\frac{\lambda-430}{82.122}}$$

in which λ is the wavelength in nm. The origins of this formula are discussed in the Implications of the CNS Hypothesis section later in this chapter.

The near-truth of the CNS hypothesis was noted during the development of the SRM method (Fig. 11.1, Stone and Miller, 1949) but as it is not entirely true, the SRM and EBC measures are not entirely adequate. As an example of this, the author measured two Pilsner beers the SRM values for which are the same (3.78), one of which when placed next to the other in 5-cm wide glasses appears about 5% brighter (relative brightness is measured on a scale of 0 to 100) and yellower (by about five color difference units) than the other. Shellhammer and Bamforth (Shellhammer and Bamforth, 2008, 2009) describe a similar pair that measured SRM 29.7. To resolve

[3]The fact that the CNS hypothesis is not absolutely true means that there is no CNS and that, therefore, the best we can do is estimate it. This is indicated by the hat over the a. The bar underneath indicates that the spectrum is normalized.

the differences between the members of these pairs, additional information (more DOFs) is needed. As the differences are proof of deviation from the CNS hypothesis, one approach might be to measure the deviations of the beers' normalized spectra from the CNS. Of the example Pilsners, one matches the CNS closely (deviation 0.058) and the other less so (deviation 0.535). The two beers are distinguishable by the deviation suggesting that SRM and deviation could serve as a two-DOF color descriptor. Though it requires the full spectrum to calculate deviation, the computation is a simple one[4].

With a bit more computational effort, the Principal Components (PCs) of the deviation can be computed. For example the first three PCs of the first Pilsner's deviation are −0.189, 0.140, and 0.039, whereas for the second they are −0.719, −0.117, and −0.009. Note that the sum of the squares of the PCs approximates the total deviation. The SRM (or EBC) and the set of PCs are candidates for a physical color specification.

Although an SRM-plus-deviation numbers system distinguishes beers with subtle color differences, one cannot look at the two sets of deviation components and tell that the second Pilsner is lighter and more yellow than the first. This is the obvious limitation of a system based purely on physical measurements. The entire justification for pursuing color as a quality control parameter is that the perceived color of a product influences the consumer's impression and acceptance of it and is a major reason for interest in tristimulus-based color specification, such as the ASBC Tristimulus (ASBC, 2006b) method which as, the name suggests, take beer spectra as input and produces triplets of numbers from which conclusions about visible color can be drawn. The ASBC method produces colors that would be seen were the beer viewed in a glass 1-cm wide. Beer is not ordinarily viewed in a glass that narrow, and so, again, we have the problem of a system which, although it adds two DOFs to the color model and is so able to distinguish similar beers, still does not tell us what the consumer is likely to see. Fortunately, the method is easily modified to compute colors under more realistic viewing conditions, but that leaves us with the problem of having to decide which set or sets of viewing conditions are appropriate.

An advantage of the deviation method is that if enough PCs are recorded with the SRM or EBC value, the spectrum can be reproduced from them with sufficient fidelity that accurate visible colors can be calculated under any viewing conditions. Of course, saving the original spectrum data also allows this to be done. The conservative approach, then, is to retain the physical measurement data, either as the complete spectrum or as SRM/EBC number and PCs and use them to compute visible color as and when that is necessary for a particular application.

[4]The CNS is subtracted from the normalized transmission spectrum of the beer and the elements of the difference squared and summed.

The remainder of the chapter expands on these ideas. We begin with a review of three-dimensional color spaces and then turn to methods of computing visible colors in them, pausing on the way to discuss methods for obtaining spectrum data including some practical aspects of spectrophotometer operation based on the properties of beer spectra. As part of that discussion, we review the Bouguer–Beer–Lambert law, its relationship to the CNS, the implications of the CNS, and its use in computing PCs. The discussions are illustrated with actual beer-spectrum data. Finally, some ideas related to the use of color data in a quality control/quality assurance (QC/QA) program are presented. As much of the broader general theory of color as is necessary to support these discussions is included, but readers wishing full understanding will have to consult some of the texts listed in the references.

Discussion extends beyond simple explanations of how to measure color using current protocols. Explanations of some of the shortcomings of the current methods and their underlying principles are given in the hope that further investigation will be stimulated. The science of beer color (nor of the color of things in general) is not settled.

As color reproduction is not possible in this volume, we cannot illustrate colors or color differences here. Readers are strongly encouraged to find a computer application that has what is typically called "color picker" capability. This widget allows the user to select colors in any of several color coordinate systems and displays a patch of that color on the screen. By moving the color controls in a picker, the reader will gain insight as to what the colors we are discussing look like and how much a specified change in color (color difference) changes their appearance.

HUMAN COLOR PERCEPTION—THREE-DIMENSIONAL COLOR SPACE

Colors as perceived by humans have three attributes. The first, hue, is associated with color names blue, green, yellow, red, and purple. These are usually envisioned as lying on the circumference of a circle. Intermediate colors are often given names based on combinations of the basic names such a blue-green or yellow-red. In the Munsell system, on which many of the current systems in use are based, five intermediate colors, blue-green, green-yellow, yellow-red, red-purple, and purple-blue are defined for a total of 10 which, in this system, are spaced at equal angles around the value (black, gray, white) axis. Purple itself is an intermediate color as it is a mixture of red and blue. The other unitary colors, blue, green, yellow, and red are unique in that no one of them contains any of the other three. Beers have hues which range from slightly greenish yellow to red.

The second attribute indicates, for example, how red a color is compared to an achromatic stimulus of the same brightness, or put another way, how intense or pure the color is, and is usually given a name such as purity, saturation, or chroma.

Pinks are lacking in this attribute, bright red is not. To see very pure (saturated, high chroma) colors, use a color picker in red, green, blue (RGB) mode with one of the three set to full on and the others to 0. The light from laser pointers is, being single wavelength, as pure as it is possible to be.

The third attribute is related to the brightness of a color, that is, where it falls on a scale between black and a reference white. It is given names like luminance or value.

This method of describing color suggests a three-dimensional space with a vertical axis of luminance (black at the bottom, white at the top) with the different hues arranged along the circumference of circles with chroma being the distance from the luminance axis to the point describing the color. Hue is then representable as an angle relative to a reference hue in a plane perpendicular to the luminance axis at the luminance level of the stimulus. A well-known system attributable to Munsell is in wide use today and was the basis for other systems such as the *Commission Internationale de l'Eclairage* $L^*a^*b^*$ (CIELAB) system which will be discussed extensively and in which we will perform color difference calculations. The coordinates in the Munsell system are hue, chroma, and value. Readers unfamiliar with the Munsell, or a similar system, should seek a picture of the Munsell "color tree" on the internet[5] or in a textbook.

The Munsell system clearly uses cylindrical coordinates. The CIELAB system uses Cartesian coordinates called Luminance (this is the same as Munsell value), a^*, and b^*. The relationships between a^* and b^* and the cylindrical hue and chroma coordinates will be developed later.

The color an observer sees when he looks at beer depends on five things:

1. The nature of the light (noon full-sun daylight, shadow, tungsten light, fluorescent light...) that illuminates the beer
2. The way in which the light interacts with the beer
3. The length of the path the light takes in passing through the beer
4. The angular subtense of the beer at the observer's eye
5. How the observer's brain processes the signals sent to it by the color sensors in his retina

Of these, only the second depends solely on the beer. By taking and recording absorption (or transmission) measurements at a sufficiency of wavelengths (spanning the visual spectrum of 380–780 nm) spaced sufficiently close to one another (spacing determined by nature of spectra), we completely capture the beer's color properties, and we could stop at this point. Once the physical color data has been captured and stored or suitably encoded, any of the color measurement methods described in this chapter can be implemented. Items 1, 3, 4. and 5 on the list need only be considered if visible color is to be computed.

[5]http://macboy.uchicago.edu/~eye1/images/Munselltree.jpg.

PATH

When the decision to compute visible color is made, the analyst must decide upon the viewing conditions (factors 1, 3, and 4 in the earlier list) for which the computation is to be made. The ASBC Tristimulus method uses CIE Illuminant C, the CIE 10 degree Observer (subtense greater than 4 degree, thus including rod vision to some extent) and a 1-cm path. Of these factors, the one that has the greatest effect on computed (and observed) color is the path. Fig. 11.1 is a black and white photograph of a Pilsner glass which tapers from about 8-cm diameter at the top to 3 cm at the bottom and which is filled with liquid[6] of SRM 5.9. Although the processes of photography and reproduction of the photograph in this book will doubtless have introduced distortion, the increase in brightness with shorter path should be plain. The CNS hypothesis predicts that as brightness decreases the color shifts from pale yellow to a more saturated orange, and the reader would observe that shift were Fig. 11.1 a color plate.

The typical British pub glass measures 8 cm at the top and 5.5 at the bottom, which, although not as dramatic a range as in the Pilsner glass, is enough difference to produce plainly perceptible color differences. Yet if we look at a glass of beer, we tend to see an object of single color. The human visual system is a marvel of adaptation. Knowing that the beer has constant physical color we tend to see it that way unless we are specifically looking for the color differences in a glass of varying width. That apparently changes the adaptation system, and we are now able to perceive the differences. More-sophisticated color systems attempt to account for some of these adaptive effects. We will not consider them here.

The obvious implied question is then as to what path should be used for computation of visible colors. Beer is never viewed by the consumer in 1-cm wide glasses but then we have argued that the purpose of tristimulus measurement may be simply to bring in the additional DOF required to distinguish beers with the same or near same SRM/EBC. With this purpose in mind, 1 cm is a good choice for path, as beers or worts with colors between 3 and 200 SRM produce luminances between 5 and 95%, whereas shorter or longer paths compress, respectively, the lighter and darker beers into the 5% regions at the ends of the scale.

The brewer concerned about the actual appearance of the beer can calculate tristimulus color for any path, any illuminant, and either observer. This is done by scaling an absorption spectrum to the desired path and choosing the appropriate tables for illuminant and observer from American Society of Testing and Materials (ASTM) Standard E 308 (ASTM, 1996), upon which the ASBC Tristimulus method is based (see following).

[6]The liquid is a dilution of Sinamar®, a commercial beer-coloring product made from dehusked, roasted malt, in DI water. Its normalized spectrum approximates the CNS.

OBTAINING VISIBLE BEER COLOR (TRISTIMULUS) DATA

To obtain the visible color coordinates for a beer or wort sample, we must do all of the following things:

1. Determine the fraction of light transmitted through a 1-cm wide sample at each of 81 wavelengths beginning at 380 nm and every 5 nm thereafter until 780 nm is reached. The resulting data set, the transmission spectrum, is complete in terms of the current requirement. It can be obtained by making 81 measurements on the sample or by calculating 81 transmission values from an encoded version of a previously measured spectrum.
2. Choose the light source under which the color information is desired. Determine the relative radiant power in that light source for each wavelength. The light source is usually chosen to be a CIE Standard Illuminant: A (tungsten), C (daylight), one of the D series (different phases of daylight), F series (fluorescent). The distribution is either tabulated or calculated from tabulated data as a function of its correlated color temperature (D Series). An arbitrary spectral distribution for which the analyst has tabulated distribution data can also be used.
3. Scale the transmission at each wavelength by the beer path length (glass width) for which visible color is to be determined and multiply the illuminant power at each wavelength by the scaled transmission. The result is the power spectrum of the desired illuminant as modified in passing through the beer in a glass of the specified width. This spectrum fully describes the physical stimulus that evokes a particular color sensation in the viewer.
4. Compute tristimulus values X, Y, and Z by calculating the dot product[7] of the modified power spectrum with each of three color-matching functions.
5. Map X, Y, and Z into a useful color space in which correlates of brightness, hue, and chroma can be computed.

Steps 1 through 5 summarize the prescription of ASTM E 308 (ASTM, 1966), the accepted practice for computing the visible color of reflecting and transmitting objects in the CIE color system. The document itself contains tables of the spectral distributions of the illuminants, values for the color-matching functions, and detailed instructions for making the computations (doing the projections) and formulae for mapping into the CIE $L^*a^*b^*$ (CIELAB) and CIE 1976 (L^*, u^*, v^*) (CIELUV) color spaces. Of these two color spaces, CIELAB is the more useful as it is frequently used

[7]"Dot product" is one name for the sum of point–wise multiplied pairs from two vectors (lists of numbers). If x_i represents the ith entry in the list of numbers representing the x color-matching function and S_i the ith element in the list of modified spectrum numbers, then the dot product of the x and S vectors is $\sum_{i=1}^{N} x_i S_i$. Each element of x is multiplied by the like-numbered element of S and the products summed. This may also be called the "inner product" or "projection of S onto x."

for comparison of the colors of objects. The ASBC Tristimulus method uses this space.

In the pages that follow, we discuss each of the five steps in detail. Before beginning, we introduce the Beer–Bouguer–Lambert law which explains the nature of beer's interaction with light and offer some examples of beer spectra.

BEER–BOUGUER–LAMBERT LAW

Consider a substance, labeled j, dissolved in water. The molecule of substance j includes a chromophore, a portion which interacts with light[8]. As a beam of photons of a particular wavelength passes through a solution of j, there is a very small but finite probability that a molecule of j will absorb a photon passing near it, thereby removing it from the beam. It is assumed that the probability that a particular molecule of j will absorb a photon is independent of the probability that any other one does and is independent of the probability that any molecule of any other substance, $i \neq j$, will absorb the photon. The number of molecules of j encountered in passing through a uniform solution along a path of length l clearly depends on the magnitude of l and on the concentration, c_j, of j in the solution. The probability that a photon makes it from one end of the path to the other depends on these two factors and on the probability of interception by a single molecule. Putting this probability into a factor, $\alpha_j(\lambda)$, which depends not only on the substance but also the photon wavelength, and performing the mathematics imposed by the assumptions, one quickly obtains the Beer–Bouguer–Lambert Law (Levine, 1995):

$$A_j(\lambda) = \log\left(\frac{I_i}{I_o}\right) = \alpha_j(\lambda)c_jl$$

in which $A_j(\lambda)$ is called the absorption at wavelength λ attributable to j; I_o, and I_i are the intensities of the light beams at wavelength λ, respectively, exiting and entering the beer; l is the length (cm) of the path through it; c_j is the molar concentration of substance j; and $\alpha_j(\lambda)$ is the molar absorption coefficient for substance j at wavelength λ.

Absorption is directly proportional to the length of the path along which the light beam travels (this is the Bouguer–Lambert part of the law). Given that the absorption of a sample measured in 1 cm, the standard path, is known the absorption in 2, 5, or 10 cm is easily obtained by scaling by those factors. Second, as the absorption is proportional to the concentration, one can, given an absorption measurement on a standard of known concentration, determine the concentration of a solution of unknown concentration based on an absorption measurement made upon it, or given

[8]Although the term "chromophore" technically refers to the portion of a molecule which produces color, we will use it here loosely to refer to a molecule which contains a chromophore.

the absorption at a particular concentration, obtain the absorption at any other simply by scaling by the ratio of the concentrations. This is the Beer part of the law.

Third, under the assumption that the probabilities of photon absorption by individual molecules are independent, we can calculate the absorption of a solution containing other color-absorbing substances from the sum of the individual absorptions:

$$A(\lambda) = l\sum_i c_i\alpha_i(\lambda)$$

In computing the visible colors of beer, we need the amount of light transmitted rather than its logarithm and so work with transmission spectra which are simply obtained from absorption spectra by taking antilogarithms

$$T(\lambda) = 10^{-A(\lambda)}$$

$0 \leq A(\lambda) < \infty$ implies that $0 \leq T(\lambda) \leq 1$. Given a transmission spectrum, $T(\lambda)$, the absorption spectrum is easily obtained from

$$A(\lambda) = -\log(T(\lambda))$$

IMPLICATIONS OF THE CNS HYPOTHESIS

The near validity of the CNS hypothesis implies that

$$A(\lambda) \approx lA(\lambda_n)\underline{a}(\lambda)$$

which, by comparison with the Bouguer–Beer–Lambert law, implies that beers and worts contain a single colorant material with molar extinction $\alpha(\lambda) = \underline{a}(\lambda)$ and molar concentration $A(\lambda_n)$. As the CNS is normalized to wavelength $\lambda_n = 430$ nm, the concentration of this fictitious colorant is

$$c = \left(\frac{\text{SRM}}{12.7}\right) = \left(\frac{\text{EBC}}{25}\right)$$

and a beer or wort spectrum is approximately:

$$A(\lambda) \approx l\left(\frac{\text{SRM}}{12.7}\right)\underline{a}(\lambda) = l\left(\frac{\text{EBC}}{25}\right)\underline{a}(\lambda)$$

which, in turn, implies that, at least approximately, given choice of illuminant and angular subtense, the characteristics of visible beer and wort color, both of which derive from the spectrum, depend only on the path length, l, and the SRM or EBC color rating of the beer. It is convenient to define the color depth as

$$D = l\text{SRM}$$

or

$$D = l\text{EBC}$$

in which l is the path length (cm). D is in units of SRM-cm or EBC-cm. Then

$$A(\lambda) \approx \left(\frac{D}{12.7}\right)\underline{a}(\lambda)$$

for D in SRM-cm. Divisor 12.7 is replaced by 25 if D is in units of EBC-cm.

The transmission spectrum of beer or wort is approximately

$$T(\lambda) = 10^{-A(\lambda)} \approx 10^{-\left(\frac{D}{12.7}\right)\underline{a}(\lambda)} = \left(10^{-\underline{a}(\lambda)}\right)^{\left(\frac{D}{12.7}\right)}$$

Define the transmission CNS as

$$\underline{t}(\lambda) = 10^{-\underline{a}(\lambda)}$$

Then

$$T(\lambda) \approx \underline{t}(\lambda)^{\left(\frac{D}{12.7}\right)}$$

is the approximate transmission spectrum of a beer, wort, caramel- or beer-coloring agent with color depth D in SRM-cm.

A more exact representation of the transmission spectrum of an actual beer is obtained from

$$T(\lambda) = \left[\underline{t}(\lambda) + \sum_{j=1}^{N} p_j \xi_j(\lambda)\right]^{\left(\frac{D}{12.7}\right)}$$

in which p_j are coefficients representative of the deviation of the normalized transmission spectrum of the beer from $\underline{t}(\lambda)$ and the $\xi_j(\lambda)$ are a set of eigenfunctions which depend, as does $\underline{t}(\lambda)$, on the general nature of the color of beer and similarly colored substances. To obtain an estimate of $\underline{t}(\lambda)$ and $\xi_j(\lambda)$, the absorption spectra of an ensemble of beers are measured; each is normalized by A(430) and converted to a transmission spectrum. These transmission spectra are averaged, and the average, symbolized by $\bar{\underline{t}}(\lambda)$, is subtracted from each normalized transmission spectrum. The resulting spectra are then loaded into a matrix which is decomposed using Singular Value Decomposition (SVD) (Press et al., 1986). The right eigenvectors are the $\xi_j(\lambda)$. Once a set of eigenvectors is available, the coefficients are obtained from the dot product of each eigenvector with the transmission spectrum. Details are in deLange (2008).

This $T(\lambda)$ equation shows equality, but equality is only satisfied when N, the number of eigenfunctions used, is equal to the number of wavelengths in the original spectrum data set. The value in using this technique is that an accurate, but not exact, approximation of a real beer spectrum is attainable for modest N with the actual number dependent on how well behaved the beer spectrum is, that is, how closely its normalized spectrum approaches $\underline{t}(\lambda)$. This is an application of Principal Components Analysis (PCA). PCA has previously seen limited application to color problems (Kuehni and Schwarz, 2008).

The p_j are referred to as the principal components (PCs) of the deviation of the beer transmission spectrum relative to $\underline{t}(\lambda)$ and can, taken with the SRM or EBC color number for a beer or wort, represent a full-color specification for the beer because its transmission, and hence absorption spectrum, can be reconstructed from them. The absorption spectrum is

$$A(\lambda) = -\log(T(\lambda)) = -\left(\frac{D}{12.7}\right)\log\left(\underline{t}(\lambda) + \sum_{j=1}^{N} p_j \xi_j(\lambda)\right)$$

The set of p_j represent, thus, a code for the spectrum. Taking $\bar{\underline{t}}(\lambda)$, based on 99 beers from (deLange, 2008), transforming it to absorption, and fitting a double-exponential function to the result, we obtain

$$\widehat{\underline{a}}(\lambda) = 0.02465\mathrm{e}^{-(\lambda-430)/17.591} + 0.97535\mathrm{e}^{-(\lambda-430)/82.122}$$

the estimated absorption form of the CNS presented in the introduction.

BEER SPECTRA

Fig. 11.2 shows the 1-cm absorption spectra of five beers, four of which are malt beers, and the fifth is a Kriek (cherry Lambic). The two dashed curves are $\widehat{\underline{a}}(\lambda)$ scaled by SRM/12.7 for the SRM values of the stout and the lambic. It is clear

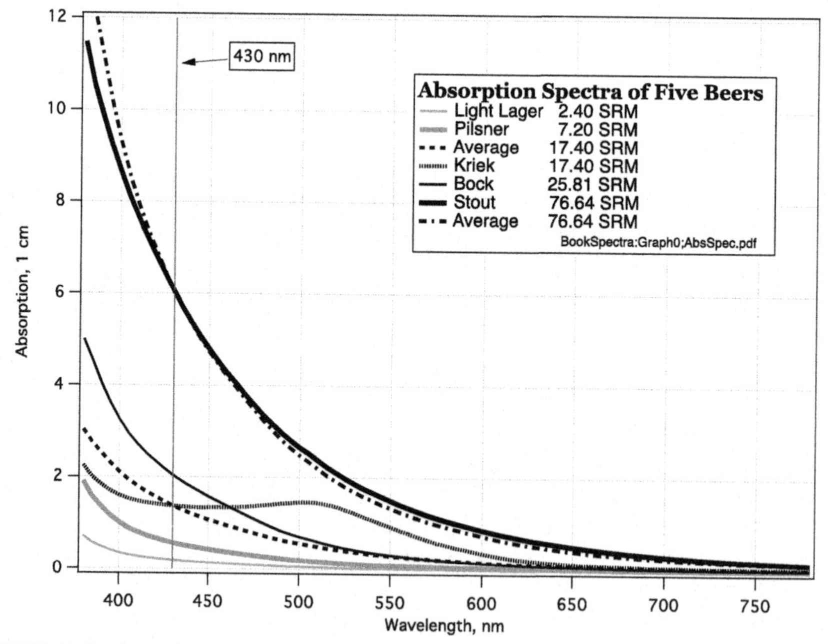

FIGURE 11.2 Absorption Spectra of Five Beers.

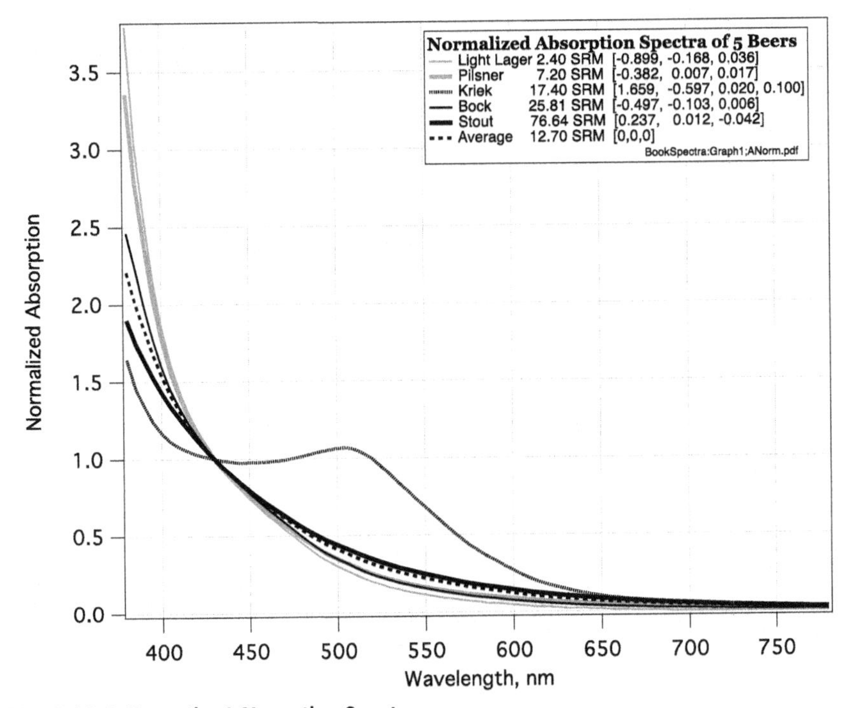

FIGURE 11.3 Normalized Absorption Spectra.

that scaled $\hat{\underline{a}}(\lambda)$ closely represents the stout spectrum but does not do so for the lambic.

In Fig. 11.3 the absorption spectra from Fig. 11.2 have been normalized by $A(430)$. The CNS hypothesis says that these curves will overly the dashed curve, $\hat{\underline{a}}(\lambda)$. Clearly, they do not do so exactly with the Kriek deviating from $\hat{\underline{a}}(\lambda)$ more than the others. The numbers in the brackets in the legends are the principal components of deviations measured relative to $\bar{\underline{t}}(\lambda)$ after transformation to transmission. The total spectral deviation is the sum of the squares of these numbers. The total deviation for the Kriek is 3.00. For the stout it is only 0.04.

Fig. 11.4 displays the transmission spectra of the same beers and Fig. 11.5 the normalized spectra of Fig. 11.3 transformed to transmission spectra. We refer to these as normalized transmission spectra but it is important to understand that they are really transformed normalized absorption spectra. As the normalized absorption spectra all have value 1.0 at 430 nm, the normalized transmission spectra all exhibit value 0.1 at that wavelength. Fig. 11.5 also displays three CIE color-matching functions which will be referred to later in the chapter.

As in the earlier plots, the dashed lines represent the CNS. The dashed line in Fig. 11.5 is $\bar{\underline{t}}(\lambda)$ measured as described previously. In Fig. 11.4, $\bar{\underline{t}}(\lambda)^{SRM/12.7}$ is shown for the Kriek and the stout. Be sure to compare the upper dashed curve to the Kriek spectrum and not the bock spectrum to which it lies closer.

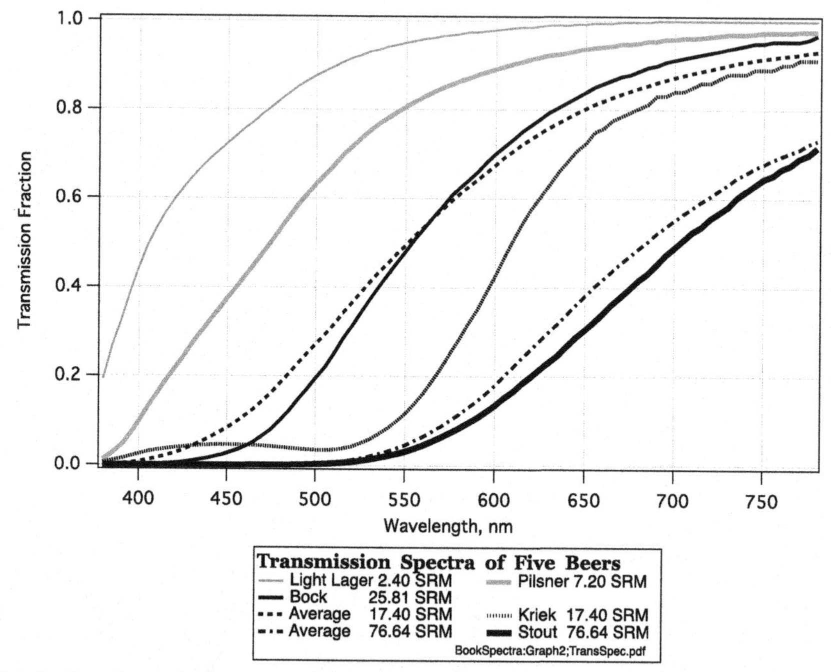

FIGURE 11.4 Transmission Spectra of the Beers of Fig. 11.1.

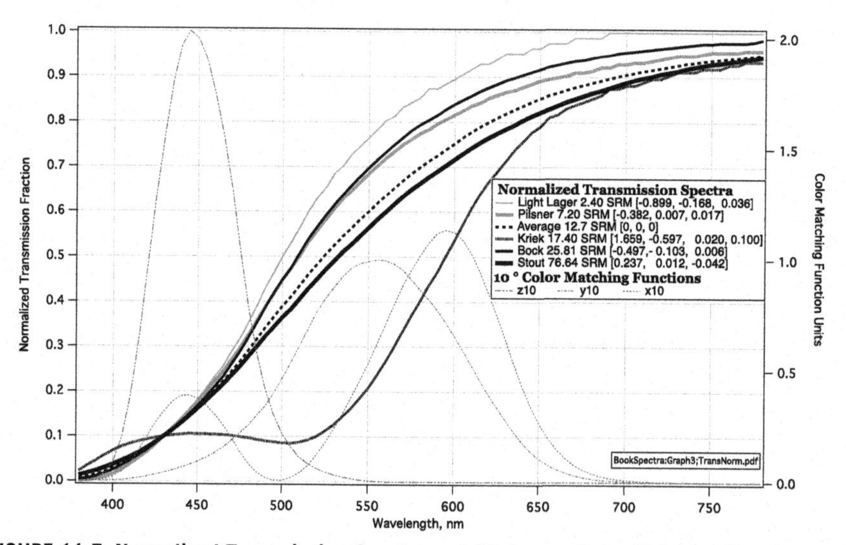

FIGURE 11.5 Normalized Transmission Spectra and 10 degree Color Matching Functions.

Figs. 11.4 and 11.5 tell the same story as the earlier figures: the CNS holds approximately for malt beer normalized transmission spectra. Fruit beer spectra depart appreciably. The deviation components are listed again on Fig. 11.5 for convenience.

DETAILS OF ASTM E–308 PRESCRIPTION

In the section Obtaining Visible Color (Tristimulus) Data, the five steps prescribed by ASTM E–308E for computation of visible beer color in the CIE system were briefly described. Each of these steps is described in detail in the material which follows.

STEP 1: OBTAINING BEER/WORT ABSORPTION/TRANSMISSION SPECTRUM DATA

Beer or wort spectrum data are easily obtained with a modern spectrophotometer or photometer. A spectrophotometer is an instrument containing a monochromator, a device which produces a light beam containing wavelengths in a narrow band around a selected wavelength, and a means of measuring the ratio of that beam's intensity as it enters and leaves a cuvette[9] placed in the beam which confines the sample to a path whose length is accurately known. A photometer is similar except that the narrow wavelength bands of light are obtained from filters or an array of light-emitting diodes. The cuvette can be of any convenient width as the Bouguer–Beer–Lambert law reveals that measurements obtained with any cuvette width are easily scaled to the standard 1 cm. A narrower cuvette will be needed for dark beers (see *Spectrophotometer Dynamic Range and Linearity* below).

A modern instrument is capable of making the requisite set of 81 measurements automatically adjusting the monochromator to each of the required wavelengths in turn, taking the measurements, and storing or transmitting them to a computer. If an available instrument is not capable of this level of automation wavelength setting, zeroing (see following), and data recording will have to be accomplished manually, and this is sufficiently labor intensive to greatly diminish the appeal of color methods which require the 81 point spectrum. E–308 has instructions and tables for color calculation using 21 measurements spaced 20 nm apart. It is possible, given the gentle variation of beer spectra, that even fewer might be taken and used with interpolation (discussed later). We have been assuming a laboratory environment in the discussion so far but must not forget that inline measurement is of interest

[9]This describes a single-beam photometer. In a double-beam instrument, one beam passes through the sample cuvette and the other through an identical cuvette filled with distilled water. The sample transmission is the ratio of the two beams' intensities after passage through the cuvettes.

as well. The need for a large number of automated measurements in a short period represents a challenge for designers of inline instruments.

Obtaining Spectrum Data: Some Practical Considerations

Zeroing the Instrument

We seek the absorption of the beer or wort, and thus instrument readings must be adjusted for the effective absorption[10] of the solvent (water) and the cuvette itself. This is done by first scanning a blank which is, ideally, the same cuvette that will be used for the sample but filled with deionized water. The absorptions measured from this blank are stored and subtracted, usually automatically, from the absorptions measured on the sample. In other words, the cuvette and solvent have effective absorptions of their own such that the observed absorption is

$$A_{\text{obs}}(\lambda) = A_{\text{blank}}(\lambda) + \sum_j \alpha_j(\lambda) c_j d$$

and subtracting off $A_{\text{blank}}(\lambda)$ gives us the absorption of the chromophores alone. Most instruments will not allow a scan to start until baseline-zeroing data has been collected.

When the sample and blank cuvettes cannot be the same, they should be a matched pair, that is, a pair that has been measured and determined very similar in optical properties. When the same cuvette is used, the deionized (DI) water from the blank scan should be completely removed by rinsing several times with sample.

Turbidity

Light that is scattered by the sample does not reach the photo detector and so causes the instrument to read higher absorption (or lower transmission) than it would without scatterers. It is, therefore, essential that beer or wort samples be free of bubbles, yeast, protein, protein-phenol, or other particles. Bubbles are particularly nettlesome as they can form in apparently degassed beer over the relatively long period which may be required for a full 380–780 nm scan. Sharply rapping the cuvette on the countertop before insertion into the instrument may be helpful here. The ASBC Tristimulus Method of Analysis (MOA) (ASBC, 2006b) suggests that beer be degassed by a combination of shaking and vacuum. The SRM and EBC single-wavelength color methods each specify turbidity requirements which are also applicable to full spectrum measurements.

Temperature and Exposure to Oxygen

Beer exposed to air, especially at elevated temperature, is subject to oxidation. This can lead to an increase in color depth. An example of this is shown on Fig. 11.20.

---

[10] Anything that prevents light entering the cuvette from exiting it contributes to effective absorption. This includes dirt or scratches and partial reflections from any interface between materials with different refractive indices such as the ones between air and glass and glass and sample.

Samples should, therefore, be kept cool and protected from air until just before spectrophotometric measurement. Cool samples should be brought to room temperature before placement in the instrument to avoid condensation on the cuvette. The physical treatment of samples should be as uniform as possible.

Spectrophotometer Dynamic Range and Linearity

Spectrophotometers and photometers are limited in their ability to measure highly absorptive samples. At $A = 3$ only one one-thousandth of the light beamed into the sample reaches the detector, and, in most instruments, the detector's self noise begins to compete with the signal produced by that light. At even higher absorption, the detector noise dominates the light-induced response completely and the instrument reports the noise—not the light passing through the sample. In addition to noise response, there is always some stray light within the instrument and the detector responds to this too. An ideal detector produces a response in which the reported absorption, $\mathscr{A}$, is equal to the actual absorption A.

$$\mathscr{A} = A$$

As the absorption approaches and passes the limits of the instrument's dynamic range, the response might be more like

$$\mathscr{A} = \mathrm{e}^{-\alpha A} A$$

in which the proportionality factor, $\mathrm{e}^{-\alpha A}$, becomes less than 1 as A increases. α is a small number.

In dealing with darker beers and worts, the analyst will eventually encounter samples that have spectra which, for some wavelengths, exceed instrument dynamic range. The spectrum labeled "1 cm no dilution" on Fig. 11.6 is an example of such a spectrum. Portions of that spectrum at wavelengths shorter than 400 nm are clearly dominated by noise and stray light. The analyst must be aware of the limitations of his instrument and when absorptions approach the limits of instrument linearity readings should be rejected.

The Beer–Bouguer–Lambert law suggests two ways to react when a spectrum such as the "1 cm no dilution" curve on Fig. 11.6 is obtained. One is to reduce the path and the other is to reduce the chromophore concentration by dilution. The other curves on Fig. 11.6 are spectra obtained from the same sample[11] using both these techniques. An estimate of the undistorted spectrum is then obtained by scaling the reduced absorption spectrum either by the path reduction or dilution factor. For example, if the path is halved or if the sample is diluted 1:1, the actual absorption values are estimated as twice the measured absorption values.

It has been suggested that when absorption scaling is necessary it is preferable to use a shorter path (Shellhammer, 2009) as dilution may change the chemistry of the

[11]The sample for Fig. 11.6 was a 1:99 DI water dilution of Sinamar, a product produced from dehusked roast barley that is sold to brewers for deepening the color of beer.

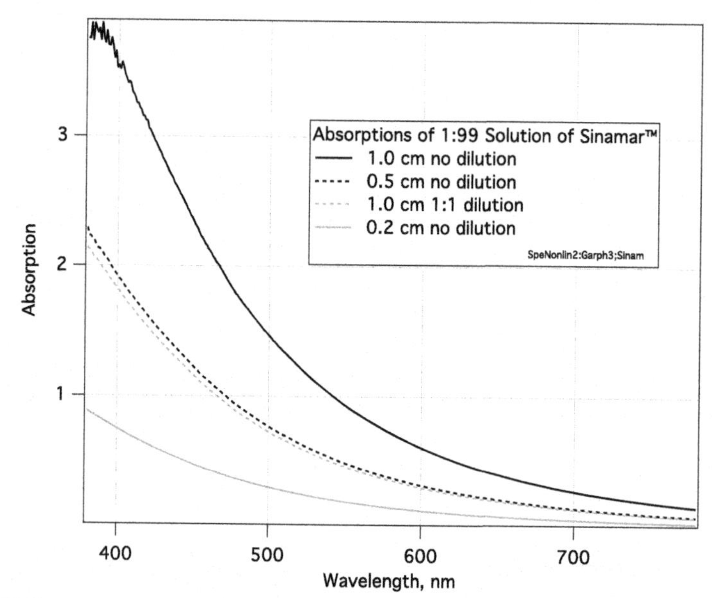

FIGURE 11.6 Measured Absorption Spectra of Sinamar in Various Paths and Diluted.

sample and thereby the $\alpha_j(\lambda)$. Although this may indeed be the case, both the ASBC and EBC color methods require that dilution be used when dynamic range reduction is required.

Whichever method is chosen, the analyst should scan both full strength and scaled aliquots from each sample and calculate absorption ratios at each wavelength. Even for very dark beers, there will be substantial portions of the spectrum at the longer wavelengths that will have absorptions in the linear range. It is informative to plot the ratio of the scaled spectrum to the full-strength/full-path spectrum using the full-strength/full-path value at each wavelength as the abscissa rather than the wavelengths themselves. Fig. 11.7 is such a plot. The topmost line on Fig. 11.7 plots ratios of absorptions obtained in a 5-mm cuvette to those obtained in a 1.0 cm cuvette. Ideally, each point would fall at value 0.50. The plot, however, shows that for absorptions of less than about 3.0 the ratio is 0.524. The constant ratio below $A = 3.0$ testifies to the linearity of the instrument up to about that level but shows that the ratios of the effective cuvette widths are not exactly 0.500. The effects of non-linearity and noise above 3.0 are also clearly demonstrated by this plot. The bottommost line on Fig. 11.7 is for the ratio between the 1-cm and 2-mm cuvettes. The portion of that curve below 3.0 indicates a ratio of 0.203 as the actual ratio of the paths rather than the expected 0.200. The ratio of absorption between 2- and 5-mm cuvettes is 0.388 rather than the expected 0.400. Thus, the three cuvettes used to obtain these data have paths in the ratio 1:0.524:0.203 rather than 1:0.5:0.2 as suggested by their labeling. It is not possible to determine which of the cuvettes is in error from these data; just that their ratios are not precise.

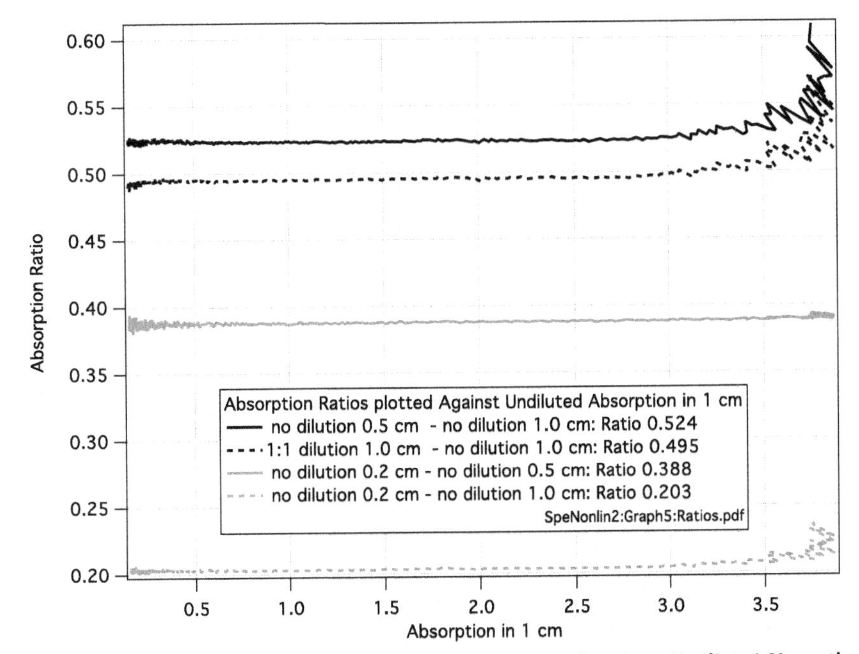

FIGURE 11.7 Ratios of Absorptions From Fig. 11.1 Plotted Against 1-cm Undiluted Absorption.

When dilution is used, scaling errors introduced by the geometric irregularities of the cuvettes are replaced by scaling errors from error in volume measurement. The 1:1 dilution curve on Fig. 11.7 shows a ratio of 0.495, suggesting that the dilution was 1: 1.020 rather than 1:1.

In summary: a plot like Fig. 11.7 shows not only where the linear region of absorption lies but gives the actual dilution or path ratios used to scale the reduced absorption measurements to the estimated full path with no dilution values. Small differences between expected and actual dilution ratios verify that, as in the example case, chemistry has not been significantly changed by this particular dilution.

The effects of spectrum distortions from nonlinearity at high absorption are mitigated by the fact that they occur at short wavelengths in which the color-matching functions (Fig. 11.5) are small. It does not much matter whether an absorption is 3.0 or 3.5 at $\lambda < 400$ nm as those values correspond to, respectively, 0.1% and 0.03% transmission, either of which, multiplied by any of the color-matching functions in that wavelength range, is close to 0. Errors of this type, thus, do not have a large effect on computed X, Y, and Z. For example, if it is assumed that the example bock spectrum is distorted such that all absorptions above $A = 3.0$ are clipped to 3.0, values of X, Y, and Z are affected in the fourth decimal place (Illuminant D65, 1-cm path). Although spectrum errors at $\lambda < 400$ nm do not much affect calculated color, it is clear that error in the measurement at 430 nm has a direct effect on calculated SRM and EBC values, and, as every point of the spectrum is normalized by A (430), the measurement at 430 nm must be in the linear range of the instrument for correct visible-color analysis.

Instrument Calibration

As is the case with any other piece of laboratory equipment, the spectrophotometer or photometer must be properly calibrated. If an instrument's absorption reading at 430 nm is in error by 10%, then SRM- or EBC-determined color values will be off by a like amount. Examination of $\underline{\hat{a}}(\lambda)$ shows its slope to be approximately $-0.013\ \text{nm}^{-1}$ [$\underline{\hat{a}}(\lambda)$ is dimensionless] at 430 nm so that an error in the wavelength reading at which the SRM absorption is measured will lead to an SRM error of approximately 0.013SRM per nm of wavelength calibration error.

Periodic maintenance and recalibration as recommended by the manufacturer should be carried out by an inhouse or contracted metrology service provider, but more-frequent performance checks should be carried out by the instrument user. Modern instruments tend to do at least some performance checks, for example, a wavelength calibration check, each time the instrument is turned on. Even so, and clearly in cases in which the instrument does not do self-tests, performance should be checked with external standards such as the National Institute of Standards and Technology (NIST) SRM (NIST, 2000) filters sold by various manufacturers. Wavelength accuracy, linearity, noise, and stray light can be easily checked in the laboratory by use of the proper filters and procedures which are set out in the instrument's user manual.

Instrument Spectral Bandwidth

Beer transmission spectra are smooth (Fig. 11.4). This has implications on both the number of measurements that must be taken to adequately and completely describe the spectrum and on the spectral bandwidth requirements of the spectrophotometer's monochromator. This, despite its name, does not produce light at a single wavelength but is rather spread over a small but finite band of wavelengths. The bandwidth of the instrument is usually specified as the width of the spectral energy distribution curve between its half-power points and may vary from 0.01 to 10 nm with the most expensive instruments featuring the narrowest bandwidths. In those instruments, bandwidth is often adjustable, but instruments found in a typical brewery's laboratory will have fixed bandwidths of from 1 to 10 nm. The implications of this are seen in the mathematical expression for the measured transmission at wavelength λ_0 which is the wavelength to which the monochromator is tuned:

$$\hat{T}(\lambda_0) = \frac{\int_{-w}^{w} S(\lambda)T(\lambda)H(\lambda_0 - \lambda)d\lambda}{\int_{-w}^{w} S(\lambda)H(\lambda_0 - \lambda)d\lambda} \xrightarrow{H(\lambda)\rightarrow\delta(\lambda)} \frac{S(\lambda_0)T(\lambda_0)}{S(\lambda_0)} = T(\lambda_0)$$

$T(\lambda_0)$ is the true transmission of the beer at λ_0, $H(\nu)$ is the monochromator spectral bandwidth power spreading function, and $S(\lambda)$ is the distribution of the light entering the monochromator. $\delta(\nu)$ represents the physically impossible but mathematically ideal spectral spread function which would pass light at only one wavelength[12]. $H(\lambda_0 - \lambda)$ is usually assumed to be a triangle with amplitude which is 1 at $\lambda = \lambda_0$, declines linearly to 0 at $\lambda = \lambda_0 \pm w$, and is 0 elsewhere where w (the width between the half-power points) is the instrument's specified bandwidth. The convolution in the numerator results in a blurring (distortion) of fine details in the observed spectrum which, if sufficiently large, will introduce errors into the color values calculated from it, but as beer spectra do not have fine details to blur (Fig. 11.4)[13], brewers need not adhere strictly to the requirements of E 308 (ASTM, 1996) which advises that data be corrected for instrument band pass, E1164 (ASTM, 2014) which requires that spectrum measurements be spaced no farther apart that 125% of the instrument band pass, or the SRM MOA (ASBC, 2006a) which requires instrument band pass of 1 nm or less.

It can be shown mathematically (Oleari, 2000) that the measured transmission spectrum does not depend on the width of the spectral distribution function window as long as the Taylor Series expansion of the transmission spectrum about each λ of interest has small third-order and higher terms. Beer transmission spectra have this property for triangular windows of total width of 100 nm or less (half-power point width 50 nm or less).

The Fourier transforms of beer absorption spectra are informative in this regard. Fig. 11.8 depicts an average of the power spectral densities (PSDs) of five malt beer transmission spectra[14]. PSD measures the rate of fluctuation of the spectrum with wavelength. Slowly varying functions, like beer transmission spectra, vary gradually with wavelength and have PSDs concentrated near 0 on the x-axis. Rapidly varying spectra would show appreciable density at higher frequencies (to the right on the plot). The x-axis is labeled in units of nm^{-1} (previously "cycles per nanometer" which form is somewhat helpful in conveying the idea that the value of the PSD is related to the rate of fluctuation in the spectrum per unit of wavelength). Fig. 11.8 demonstrates that beer transmission spectra do not have appreciable PSD (more than 60 dB down) at frequencies above 0.01 nm^{-1}. This suggests that any blurring of spectral detail caused by finite instrument spectral bandwidth which does not attenuate (blur detail) appreciably below 0.01 nm^{-1} will not be detrimental. Fourier transformation of the assumed spectral band pass shape (triangle) gives the effective gain of a filter that attenuates spectral variation at each spectral variation frequency. Fig. 11.8 depicts the effective gains of hypothesized triangular spectral

[12]This is the Dirac delta function which can be thought of as having infinite amplitude at $\nu = 0$, amplitude 0 for every other value of ν, and integral, over any finite range of ν that encompasses 0, equal to 1.

[13]Transmission spectra can show some detail close to transmissions of 1 as a consequence of quantization noise within the instrument which reads, typically, in steps of 0.1% transmission or 0.001 absorption. Blurring (removing) this noise is desirable.

[14]The transmission spectrum is transformed by the Fast Fourier Transform (FFT) into the "frequency domain." PSD is 20 times the log of the magnitude of the transformed data. The units are decibels.

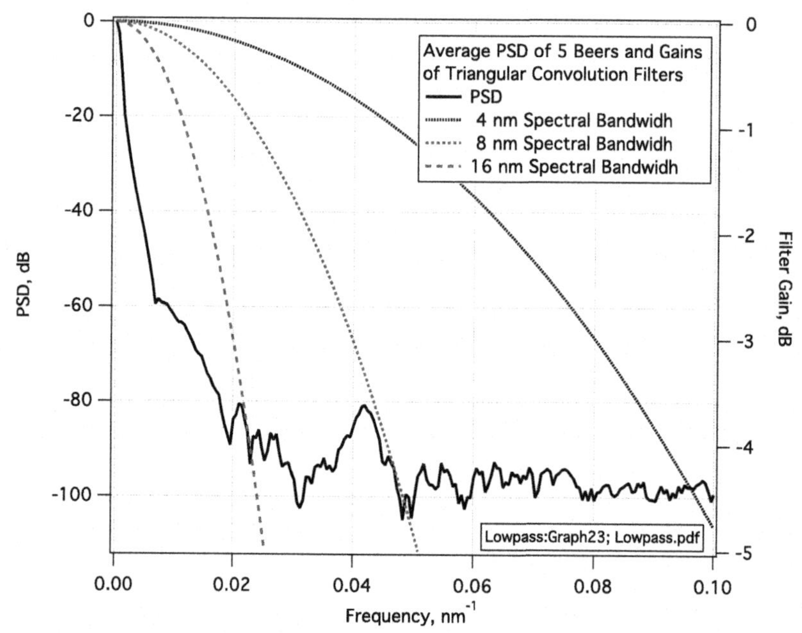

FIGURE 11.8 Power Spectral Density of Beer Transmission Spectra and Responses of Three Convolution Filters.

bandwidth distribution functions of half-power points separated by 4, 8, and 16 nm. The gains (right-hand vertical axis) of all three are within a decibel (dB) of 0 dB (unity) for frequencies below 0.01 nm^{-1} from which we conclude that instrument spectral bandwidths up to 16 nm should not produce appreciable spectral distortion nor significant error in color measurements.

It is, of course, reasonable to ask what "appreciable distortion" and "significant error in color" mean. Color error is the Euclidean distance in three-dimensional color space between the true color (computed from an error-free spectrum) and a color computed from a distorted or approximated one. In the section Measuring Color Differences, computation of CIELAB color difference will be discussed. If an observer presented with a pair of similarly colored patches cannot detect a difference between them, the color difference (error) is insignificant. The Just Noticeable Difference (JND) is the minimum difference that can be detected. It depends on, among other things, the direction of the color error (whether it is in value, chroma, or hue, with hue being the most sensitive and chroma the least) but on average in the CIELAB space, the space in which industrial colors are most often compared, the JND is about 1.06 units for light sources and about 2.3 units for surface colors (Mahy et al., 1994).

CIELAB color difference is symbolized by ΔE^*_{ab}. The magnitude, ΔE^*_{ab}, of color error attributable to spectral bandwidth effects depend, naturally, on a beer's spectrum, but, because of the near validity of the CNS hypothesis, Fig. 11.9 can

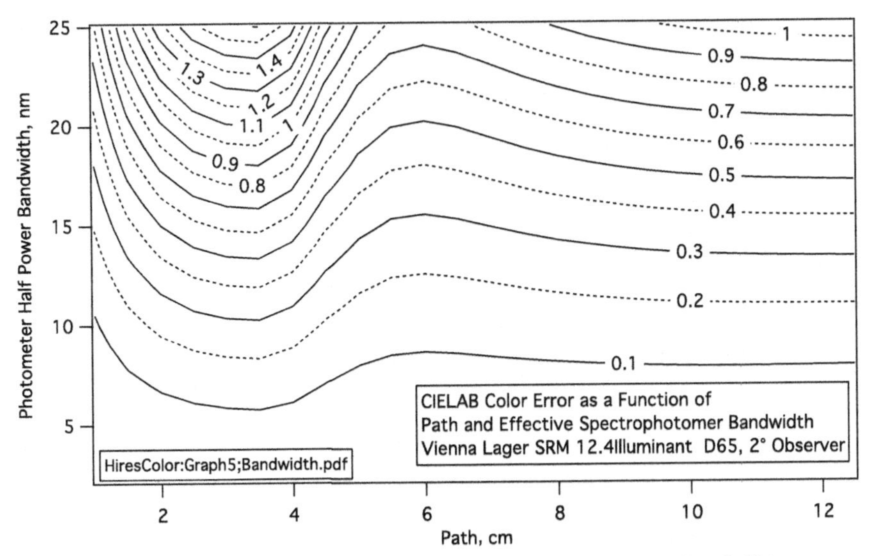

FIGURE 11.9 Color Error, ΔE^*_{ab}, Introduced by Finite Spectrophotometer Bandwidth.

be considered representative. To prepare this graph the transmission spectrum of a Vienna-style lager of 12.4 SRM was scanned from 360 to 790 nm in 0.5-nm steps using an instrument with spectral band pass of 2 nm. This spectrum was subsampled to 1-nm steps and CIELAB color computed, for several values of path length, using the 2 degree Observer color-matching functions and D65 illuminant[15]. The spectrum was then convolved with triangular band pass functions of various widths and the color recomputed. ΔE^*_{ab} between the unconvolved and convolved spectra were computed (see Measuring Color Differences) and plotted.

For broader interpretation of Fig. 11.9, the abscissas can be converted to color depth (SRM-cm) by multiplying each axis value by the SRM, 12.4, of the beer for which it was prepared. After this is done, Fig. 11.9 suggests that bandwidths less than 6 nm introduce color errors of less than $\Delta E^*_{ab} = 0.1$ at any color depth up to 155 SRM-cm (305 EBC-cm) and, judging from the asymptotic nature of the curves, beyond. The upper limit of 16 nm that we derived from consideration of spectrum PSDs earlier is supported by Fig. 11.9 which indicates errors of about $0.30 < \Delta E^*_{ab} < 0.60$ up to 155 SRM-cm and beyond for that bandwidth. The implication is that we need not correct for band bass of 16 or perhaps even 20 nm as failure to do so induces error of less than a JND. This is fortunate as band-pass correction is difficult and, if not done properly, can induce more error than it potentially removes. It also means we can use the uncorrected data tables in

[15]Readers wishing to experiment with 1-nm spaced color-matching functions will find them and the D65 spectrum in Wyszecki and Stiles (2000) or they can be downloaded from websites such as that of the Rochester Institute of Technology.

E−308 (Tables 11.1 and 11.2—these are the ones used in the ASBC Tristimulus MOA) rather than the ones to which corrections, based on assumptions that may not be relevant to beer, have been preapplied.

The ASBC Tristimulus method is silent on the subject of spectral bandwidth but the relevant ASTM practices, E−308 and E−1164, suggest that the instrument spectral bandwidth should be approximately equal to, but never less than, 80% of the spacing between readings, implying, for readings spaced at 5 nm, spectral bandwidth of 5 nm. This is valid advice for colored objects in general. We have just shown that any bandwidth less than 16 nm is adequate for beer.

Measurement Spacing

The low pass nature of beer absorption spectra suggests that 81 measurements (5 nm spacing) may be more than is necessary. Fig. 11.8 shows that the spectrum "signal" has apparent bandwidth[16] of 0.02 nm^{-1} which implies, from Shannon's sampling theorem, that it can be adequately represented by samples spaced every 50 nm (the reciprocal of the bandwidth). As we did when considering instrument bandwidth, we ask what "adequately" means. Fig. 11.10 is similar to Fig. 11.9 except that the ordinates are sample spacing, in nm, rather than spectral bandwidth. Fig. 11.10 was derived by subsampling the transmission spectrum for the same beer as in Fig. 11.9 and then reconstructing it from the reduced sample set using the well-known digital signal processing techniques (Crochiere and Rabiner, 1983) of zero insertion and low-pass filtering. As with Fig. 11.9 colors were computed from the original and reconstructed spectra using the CIE 1 nm, 2 degree matching functions for illuminant D65 and the CIELAB differences plotted.

Fig. 11.10 illustrates that subsampling at spacing up to 40 nm will introduce errors of 0.4 CIELAB units or less for paths up to 12.5 cm and beyond for this beer and for color depths of 155 SRM-cm and beyond in general. Spacing of 50 nm can produce errors of as much as 1.4 CIELAB units for color depths up to about 50 SRM-cm.

Although the techniques used to reconstruct subsampled spectra are well known in the communications industry, they are less well known outside, and it is therefore likely that a simpler reconstruction method would be greatly preferred by brewery analysts. Linear interpolation between widely spaced points is the simplest and most intuitive of these but unfortunately is relatively costly in terms of color error at wide wavelength spacing. An examination of typical transmission curves on Fig. 11.4 suggests why this is the case. Subsampling and reconstruction by linear interpolation replaces the original curves with a connected set of straight lines. These adequately follow the true curve in the mid-wavelength region but do not accurately reproduce it, unless the subsampling interval is small, where it first starts to rise in the blue wavelengths nor where it levels off in the red.

[16]Fig. 11.8 only shows half the PSD. The other half is a mirror image reflected about the vertical axis and so is not plotted.

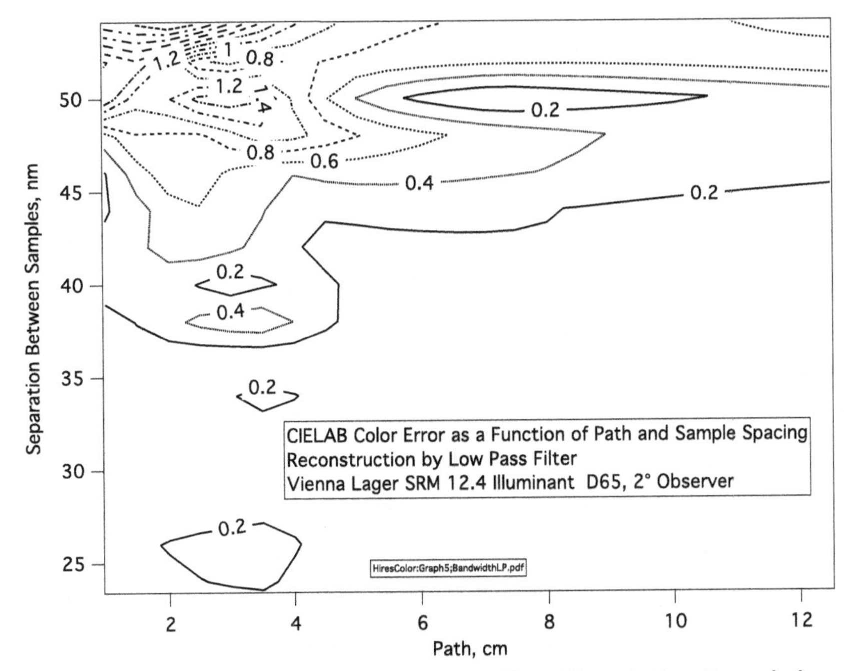

FIGURE 11.10 Color Errors ΔE^*_{ab}, Caused by Subsampling of Example Beer Transmission Spectrum With 0 Insertion and Low-Pass Filtering Reconstruction.

Fig. 11.11 shows computed color errors under the same conditions as for Fig. 11.10 except that the transmission spectrum used for color calculation is reproduced from the subsampled data set by linear interpolation. It is clear that there is heightened susceptibility to this problem at color depths of 36 to 40 SRM-cm and that to incur $\Delta E^*_{ab} < 1$ CIELAB unit sample spacing of no more than 19 nm should be used. Conversely, sample spacing of 8 nm gives $\Delta E^*_{ab} < 0.2$ implying that the 5-nm spacing recommended by E–308 is more than adequate.

Other Subsampling Schemes

Smedley (1992) experimented with reduced sampling by the use of Lagrange interpolation which expands the spectrum using Lagrange polynomials as basis functions. The method requires that transmission be measured at wavelengths which are not necessarily uniformly spaced but rather dictated by the necessity to suppress the oscillations to which polynomial expansions are subject. In fact, each spectrum has a best set of sampling wavelengths, but to determine what they are the spectrum must be measured on a fine spacing and an iterative process[17] used to determine what the best wavelengths are in terms of minimum distortion with respect to the full

[17]The author has experimented with simulated annealing (Černý, 1985).

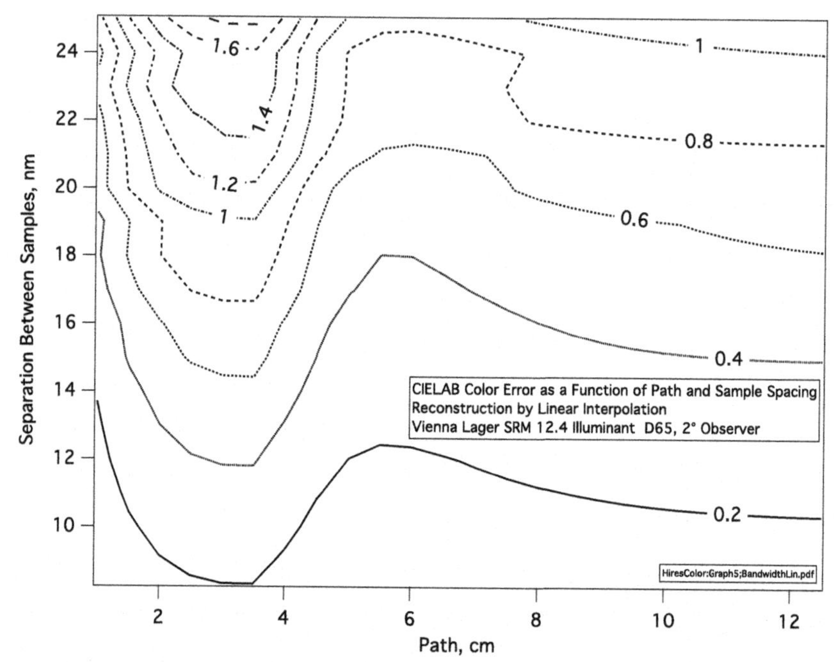

FIGURE 11.11 Color Errors Caused by Subsampling of Example Beer Transmission Spectrum With Linear Interpolation Reconstruction.

spectrum. Smedley found a set of wavelengths based on the spectra of several beers. This approach seems feasible using measurements at as few as seven or eight wavelengths. These, used with the Lagrange algorithm, produce a full set of spectral data (81 measurements) from which colors in error by less than the JND can be calculated.

Encoding the Spectrum

In Lagrange interpolation, the subset of spectrum measurements and the wavelengths at which they were taken constitute the code for a full spectrum. In general, the object of encoding is to represent the spectrum with as few coefficients as possible such that it can be well-enough reconstructed to produce calculated visible colors that do not differ by much (less than the JND) from the original spectrum. If visible color computation is not the goal, then enough coefficients are needed such that beers with the same or close SRM can be distinguished from one another. The codes then become the description of the physical beer color.

As was the case with Lagrange interpolation, the approach is usually to express the approximate spectrum as a sum of known functions. The general form for an expansion is

$$\widehat{T}(\lambda) = \sum_{i=0}^{N-1} b_i f_i(\lambda)$$

in which the f_i are the functions and b_i are coefficients which multiply f_i. The set of b_i is the code for the spectrum, and clearly we want accurate reproduction of the original spectrum (which can be the absorption spectrum as well as the transmission spectrum) with as few b_i as possible. For smoothly varying data, such as beer spectra, polynomials are often suitable. The simplest set of polynomials is

$$f_i(\lambda) = \lambda^i$$

and a related set, which produces a Taylor series expansion about λ_0, is

$$f_i(\lambda) = (\lambda - \lambda_0)^i$$

The obvious choice for λ_0 is 430 nm. Then $b_0 = T(\lambda_0)$ (or $A(\lambda_0)$ if the absorption spectrum is encoded) and is thus directly related to the SRM or EBC rating for the beer.

There is an art to polynomial fitting, and certain types of polynomials are more suited than others with efficiency depending on the degree to which they match the variations in the data being encoded. Experiments with $f_i(\lambda) = \lambda^i$, the Taylor series, Chebychev polynomials, and Legendre polynomials suggest that all are suitable and that seven or eight coefficients are needed to encode to less than the CIE JND. Space restrictions prohibit discussion of how coefficients are calculated beyond two comments. The first is that spreadsheet programs have curve-fitting routines which can determine them, and second is that if the polynomials are orthonormal as are, for example, the Chebyshev and Legendre polynomials, they can be precomputed and the coefficients determined as the dot product of stored polynomial vectors with the spectrum values, also easily accomplished in a spreadsheet.

The f_i can be functions other than polynomials. Astute readers may have noted the similarity between the current formula and the similar one given earlier for the expression of $T(\lambda)$ in terms of eigenvectors. Eigenvectors are orthonormal and have the additional property that they are computed based on the ways beer spectra differ from one another. This makes them efficient at encoding those differences and they become even more efficient if we take advantage of the CNS hypothesis by subtracting from a beer's normalized transmission spectrum the CNS and then encoding what is left over using eigenvectors attuned to the statistics of the differences exhibited by spectral deviations. The method was sketched earlier and full details and discussion are found in deLange (2008).

Physical Descriptions of Beer Color

Here, we assume that we have adequate beer spectrum data and consider how we might use it alone, that is, without regard for any aspect of human vision, as a basis for color reporting.

SRM and EBC Color

Both the SRM (ASBC, 2006a) and EBC (MEBAK, 2002) color specifications are derived from a single-spectrum datum: the 1-cm absorption at 430 nm. The SRM was originally defined as 10 times the absorption in a one-half inch cell. As one-half inch is 1.27 cm, this is equivalent to $12.7A_{1cm}(430)$. The EBC definition is simply $25A_{1cm}(430)$. The EBC color of a beer is thus 1.968 times the SRM color.

The potential problem of turbidity has been mentioned. The ASBC measurement procedure (ASBC, 2006a) requires that $A_{1cm}(700) \leq 0.039A_{1cm}(400)$ and prescribes filtration or centrifugation if that requirement is not met. The EBC method (MEBAK, 2002) requires that turbidity of the sample be less than 1 EBC turbidity unit [4 formazine nephelometric units (FNUs)] and prescribes treatment with kieselguhr and an 0.45 nm membrane filter when it is not. Some beers (eg, the stout of Fig. 11.3) can fail the ASBC turbidity test even when turbidity free because their normalized absorptions at 700 nm are higher than 0.039. These limitations aside, these two methods are in wide use today.

The Spectrum Itself

It is obvious that the complete absorption or transmission spectrum itself is a complete description of the beer's color. By studying the spectral shape relative to that of the CNS, an analyst should be able to learn to draw some conclusions about beer appearance beyond what the SRM and EBC measures tell him or her. For quality control purposes additional bounds on the deviation, such as the 700 nm one mentioned in the preceding paragraph, could be established. The augmentation coefficients, discussed next, effectively do this while automatically focusing on the parts of the spectrum in which most deviation is likely found.

Augmented SRM (EBC) Method

The augmented SRM and EBC methods begin with the 430 nm absorption measurement. Each point in the absorption spectrum is divided by $A(430)$, converted to transmission and $\underline{\bar{t}}(\lambda)$, the CNS, subtracted. The resulting vector represents the deviation of the normalized spectrum of the sample from the CNS. This deviation is characterized by its principal components (PCs). Each PC is computed by taking the dot product of the deviation vector with each of a set of eigenvectors[18]. SRM or EBC are then calculated from $A(430)$ as in the conventional SRM and EBC methods and the SRM or EBC value and the set of PCs become the code for the spectrum and the physical color of the beer. The numbers in brackets in the legends for Figs. 11.3 and 11.5 are example PCs for those beers.

This method has not been adopted by any standards body, and there is, therefore, no standardized set of eigenvectors or CNS ($\underline{\bar{t}}(\lambda)$) available. Readers wishing to experiment with this method can obtain example data from the author's website (deLange, 2014) or calculate their own based on measuring a set of beer spectra and processing them as described by deLange (2008). A set of four eigenvectors derived from an ensemble of 100 beers, including some fruit beers,

[18]Note the similarity to the ASBC Tristimulus method in which the dot product with color-matching functions is computed. Other differences here are that the spectrum is normalized and the CNS subtracted before the dot product is formed.

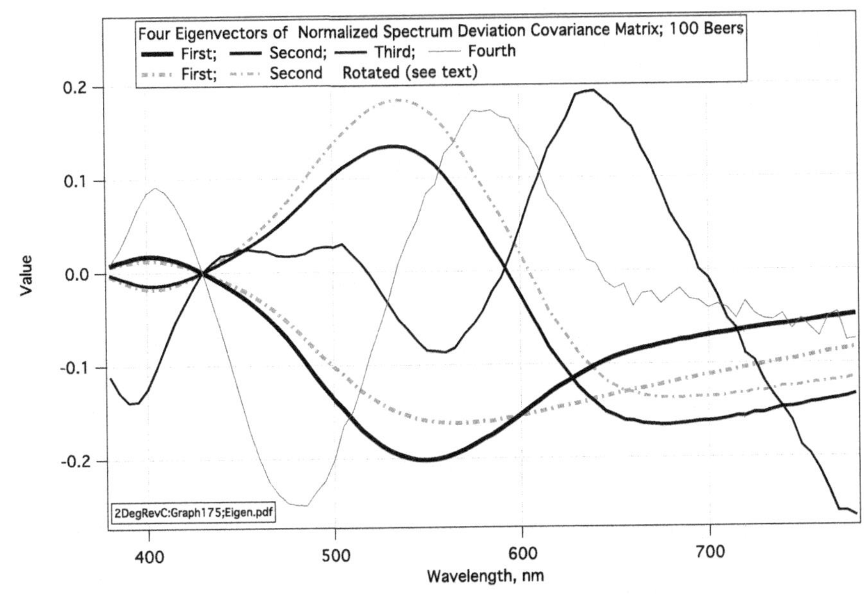

FIGURE 11.12 Four Eigenvectors From Normalized Transmission Spectra of 100 Beers.

is displayed in Fig. 11.12 as a set of four solid lines of varying weight. The dashed lines represent an alternate set of eigenvectors whose application will be explained shortly. They are potential replacements for the first and second normal eigenvectors.

The total deviation of a spectrum from the CNS is measured by the sum of the squares of the elements of the deviation vector. A highly deviant spectrum, such as that for a fruit lambic, will have total deviation of as much as 3, whereas most malt beers exhibit deviations of less than about 0.3 and only rarely above 1. One of the more interesting properties of the PCs is that the sum of their squares is equal to the sum of the deviation-vector element squares. As beer spectra have few degrees of freedom, the sum of the squares of the first few PCs is almost equal to the total with the first PC generally modeling most of the deviation, the second, most of what is left, and so on[19].

Fig. 11.13 is a plot of the first two PCs of 132 beers calculated using the eigenvectors of Fig. 11.12. The square of the distance from the origin to the point representing a beer is approximately equal to the deviation of that beer's spectrum because, as just stated, the first two PC's model most of the total deviation.

[19]The PC's are listed in the order in which they model the deviation of the beer ensemble from which the eigenvectors were derived. It is quite possible that a third or fourth PC may model more deviation of a particular beer spectrum than a lower-numbered one.

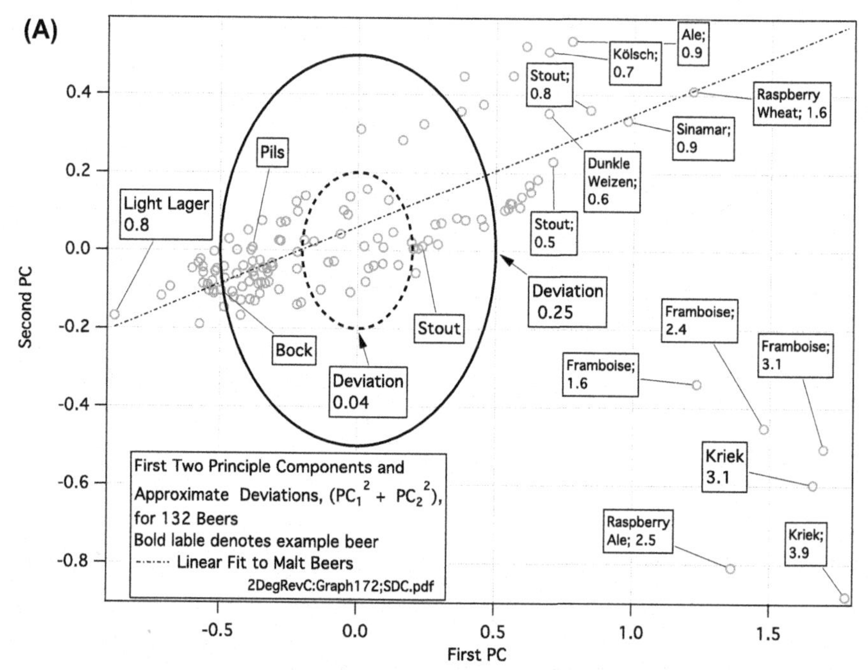

FIGURE 11.13A First and Second PCs of 130 Beer Normalized Spectra.

The ellipses[20] are loci of points for which the sum of the squares of the two PCs is, respectively, 0.04 and 0.25, and the majority of beers have deviations less that these. The high-deviation beers in the figure are labeled as to their types as are the more nominal beers which have been used as examples in this chapter. The most deviant beers are the framboises and krieks. The first eigenfunction and, to a lesser extent, the second have shapes somewhat similar to the dips in the normalized spectra of these beers between 500 and 550 nm. It is evident that all malt beers and fruit beers containing fruits such as peach and apple juice surround, in PC space, the dashed line on the figure. Fruit beers containing cherry and raspberry juice tend to lie at some distance from this line.

When confronted with a display such as Fig. 11.13A, the analyst is immediately tempted to rotate the axes such that the dashed line, which fits the malt beers, aligns with the horizontal. This is done in Fig. 11.13B and results in a new set of values for the first two PCs given by, respectively,

$$PC_{1,r} = PC_1 \cos\theta + PC_2 \sin\theta$$

and

$$PC_{2,r} = -PC_1 \sin\theta + PC_2 \cos\theta$$

[20]Had the two axes been scaled equally, these would plot as circles.

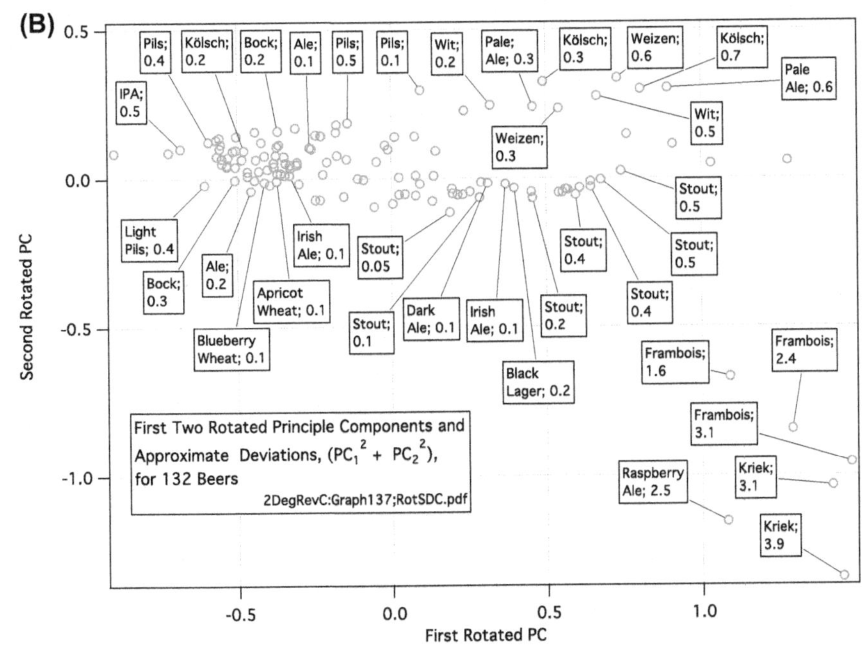

FIGURE 11.13B First and Second Rotated PCs of 130 Beer Normalized Spectra. Numbers in Label Boxes are Sums of Squares of the First two PCs.

PC_1 and PC_2 are the normal PCs computed using the solid-line eigenvectors in Fig. 11.12 and θ is the angle, 16.64 degree, between the dashed line in Fig. 11.13A and the PC_1 axis[21]. Doing this has the advantage that it makes the separation between beers as expressed in PC space easier to interpret. There are evidently three groups, and several beers in each group have been labeled to indicate the sorts of beers that fall into them.

$PC_{1,r}$ and $PC_{2,r}$ can be calculated from PC_1 and PC_2 using the rotation formulas given earlier, but it makes more sense to compute rotated eigenvectors

$$\xi_{1,r} = \xi_1 \cos\theta + \xi_2 \sin\theta$$

$$\xi_{2,r} = -\xi_1 \sin\theta + \xi_2 \cos\theta$$

and dot them ($\xi_{1,r}$ and $\xi_{2,r}$—the dashed lines on Fig. 11.12) with the normalized beer transmission spectrum instead of ξ_1 and ξ_2 when computing PCs. Note that $\xi_{1,r}$ and $\xi_{2,r}$ are still eigenvectors.

[21]The angle appears greater than 16.64 degree in Fig. 11.13A because the scaling of the two axes is different.

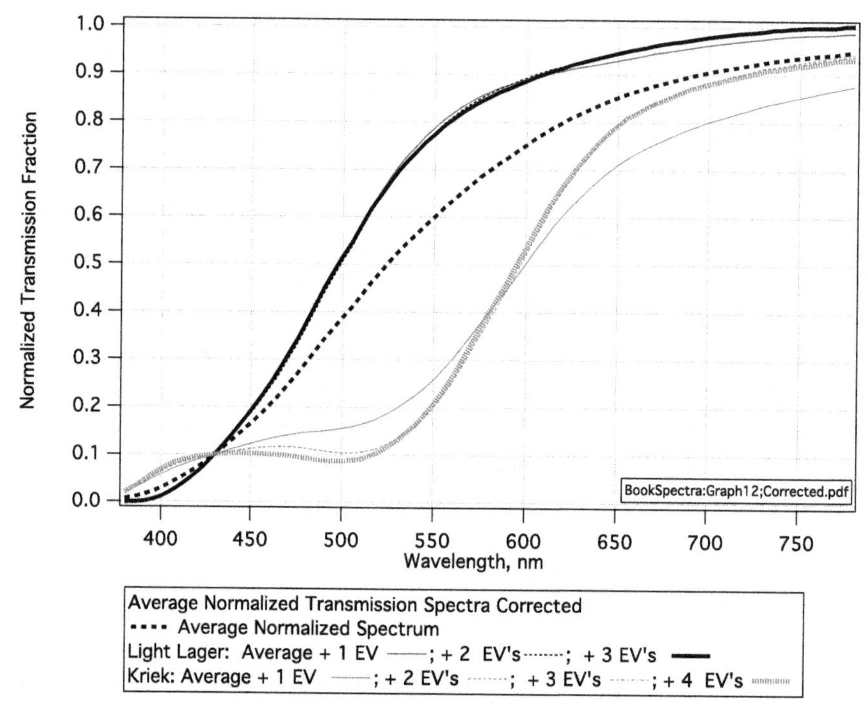

FIGURE 11.14 Average Normalized Transmission Spectrum of Light Lager and Kriek Corrected by Addition of Eigenvectors Weighted by PC Values From Fig. 11.4.

Reconstruction of Spectrum From Augmented SRM or EBC

Given a beer specification using the augmented SRM system, one can reconstruct the absorption and transmission spectra by using the PCs as weights for the eigenvectors[22] and adding these back to the CNS (Constant Normalized Spectrum in transmission form):

$$\hat{\underline{t}} = \bar{\underline{t}}(\lambda) + \sum_{j=1}^{N} p_j \xi_j(\lambda)$$

This is illustrated in Fig. 11.14 for the light lager and the lambic example beers (non-rotated EV's). The reconstructed spectra can be compared to the originals on Fig. 11.5. The accuracy of the reconstruction depends on the number of eigenvectors, N, used in the reconstruction. The analyst monitors the accuracy by comparing the sum of the squares of the utilized PCs to the total spectrum deviation (sum of the squares of the deviation vector elements).

[22]The first two of which, if rotation PCs were used, would be $\xi_{1,r}$ and $\xi_{2,r}$.

Because PCs can have either negative or positive values and because the eigenvectors have some negative elements, it is possible for the normalized, reconstructed transmission spectrum to have negative values. This happens at short wavelengths, is physically impossible, and represents imperfect approximation of 0 or very small positive numbers. It happens in the part of the spectrum in which the color-matching functions are small so that errors would not have appreciable effect on colors computed from the reconstruction, but as only integer powers of negative numbers can be computed (without resorting to complex arithmetic), it is necessary to replace any negative normalized transmission with 0. If the negative numbers are large and occur at any but the shortest wavelengths, that is evidence that too few PCs have been used to characterize the spectrum. The number of PCs needed for accurate reproduction of the spectrum need be considered only if the ultimate intent is to compute visible color. Fewer, eg, one or two, plus the SRM or EBC number, may be adequate if a physical color specification is the only goal.

The transmission spectrum used in CIE color computations is obtained from the reconstructed normalized transmission spectrum by scaling for SRM and path, l:

$$T(\lambda) = \left[\underline{\bar{t}}(\lambda) + \sum_{j=1}^{N} p_j \xi_j(\lambda)\right]^{\left(\frac{lSRM}{12.7}\right)}$$

The exponent is (lEBC/25) for augmented EBC.

STEP 2: OBTAINING ILLUMINANT SPECTRAL POWER DISTRIBUTION DATA

In going beyond Step 1, we begin considering the perceived colors of beer. These, of course, depend on the nature of the human visual system but also on the beer absorption spectrum and the distribution of energy within the illuminating light source. Fig. 11.15 shows the distributions for some of the standard CIE illuminants. Numerical values for the Fig. 11.15 (and some other) illuminants are tabulated in E 308 (ASTM, 1996) and an Illuminant C table is in the ASBC Beer 10C MOA (ASBC, 2006b) as that MOA calculates color for that illuminant[23]. D-series illuminant vectors are computed in the same way as reconstructed transmission spectra, ie, as the weighted sum of three eigenvectors the values for which can be obtained from (Wyszecki and Stiles, 2000) and elsewhere with the weights determined from the correlated color temperature[24] (CCT) using formulas also found in that reference. Thus, D65 is associated with a CCT of 6504 K and D50 with a CCT of 5003 K.

[23]Illuminant C has been replaced by Illuminant D65, to which it is similar, in most industries.

[24]These illuminants have spectra that are approximately those of the ideal black-body (Planckian) radiator at the specified temperature. See Fig. 11.16. K is the temperature in Kelvins on a scale with origin at absolute zero. Zero °C is equivalent to 273.15 K.

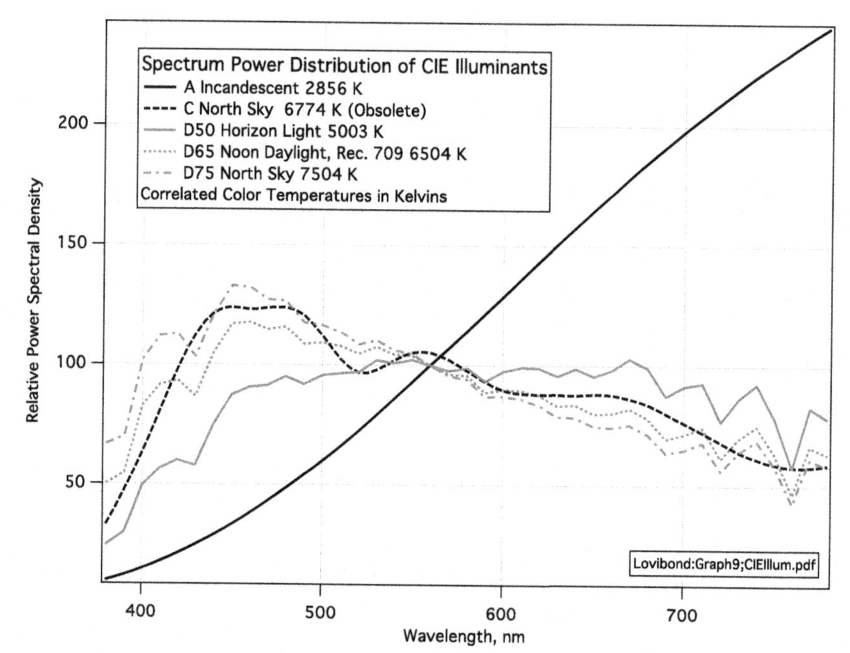

FIGURE 11.15 Energy Distributions, $S(\lambda)$, of Some CIE Illuminants.

Note that the higher the CCT, the more blue and less red light in a spectrum though lower CCT spectra are referred to as "warm" and higher CCT ones as "cool" by photographers and artists.

The eye tends to compensate for the color balance of the illuminant. In general, something that appears a particular color under a warm illuminant appears the same color under a cool one, but this is not exactly the case. Color spaces such as CIE 1976 (L^*, u^*, v^*) (CIELUV) and CIELAB account for this to some extent. Shifts due to illuminant change are small as is illustrated in Table 11.1 which shows, as an example, CIELAB color differences (ΔE_{ab}) expected between CIE illuminants

Table 11.1 CIELAB Color Differences (ΔE^*_{ab}) for 7.20 SRM Pilsner Beer Under Different CIE Illuminants in 5-cm Path, 10 degree Observer.

	A (2856 K)	C (6774 K)	D50 (5003 K)	D65 (6503 K)	D75 (7504 K)
A	0	8.69	5.01	7.97	9.75
C	8.69	0	5.57	2.00	1.54
D50	5.01	5.57	0	3.95	2.78
D65	7.97	2.00	3.95	0	2.16
D75	9.75	1.54	2.78	2.16	0

Values in Parentheses are illuminant correlated color temperatures

for a Pilsner beer measured in 5-cm path. Excluding Illuminant A (incandescent light), the color shifts of the table are all small (but larger than the JND).

STEP 3: SCALE AND MULTIPLY

In this step, the beer transmission spectrum measurements (or reconstructed spectrum), which are assumed to be in 1 cm, are scaled to the optical path length of interest and multiplied, point by point, by the spectral power distribution to give the spectrum of the illuminant as modified in passing through the beer. In ASBC MOA Beer 10-C (ASBC, 2006b), the color is computed for 1-cm path so that no scaling is necessary. As beer is not served in 1-cm wide glasses, other paths will be of interest to analysts concerned with what a consumer actually sees. Scaling of 1-cm absorption spectra is very easily done simply by multiplying each point by the new path length, l, in cm:

$$A_l(\lambda) = lA_{1\ \text{cm}}(\lambda)$$

Absorption measured in path l' is similarly converted (see Bouguer–Beer–Lambert Law) to path l

$$A_l(\lambda) = (l/l')A_{l'}(\lambda)$$

In transmission form

$$T_l(\lambda) = 10^{-A_l(\lambda)} = 10^{-lA_{1\ \text{cm}}(\lambda)} = T^l_{1\ \text{cm}}(\lambda)$$

Transmission form scaling can also be carried out by conversion to absorption, multiplication by path, and conversion back to transmission. Illuminant spectrum multiplication is done on scaled transmission spectra: $T_l(\lambda_j)S(\lambda_j)$.

STEP 4: COMPUTE CIE *X*, *Y*, AND *Z* (TRISTIMULUS)

Before computing tristimuli, a normalizing constant based on the illuminant is needed.

$$k_{I,\text{obs}} = 100 \Big/ \sum_{i=0}^{80} S_I(\lambda_i)\bar{y}_{\text{obs}}(\lambda_i)$$

in which $S_I(\lambda_i)$ are the elements of the spectral power distribution of chosen illuminant I and $\bar{y}_{\text{obs}}(\lambda_i)$ are the elements of the y color-matching function vector for the chosen observer (2 or 10 degrees). $\lambda_i = 380 + 5i$.

Given the value of $k_{I,\text{obs}}$ one then computes

$$X = k_{I,\text{obs}} \sum_{i=0}^{80} S_I(\lambda_i)T_d(\lambda_i)\bar{x}_{\text{obs}}(\lambda_i),$$

$$Y = k_{I,\text{obs}} \sum_{i=0}^{80} S_I(\lambda_i) T_d(\lambda_i) \bar{y}_{\text{obs}}(\lambda_i),$$

and

$$Z = k_{I,\text{obs}} \sum_{i=0}^{80} S_I(\lambda_i) T_d(\lambda_i) \bar{z}_{\text{obs}}(\lambda_i)$$

in which $\bar{x}_{\text{obs}}(\lambda_i)$ and $\bar{z}_{\text{obs}}(\lambda_i)$ are the other two color-matching functions for the specified observer. $\bar{x}_{10}(\lambda_i)$,$\bar{y}_{10}(\lambda_i)$ and $\bar{z}_{10}(\lambda_i)$ are plotted on Fig. 11.5. X, Y, and Z are the CIE tristimulus coordinates of the beer or wort for the chosen path, illuminant, and observer, respectively. The tristimulus of the white point, X_n, Y_n, and Z_n calculated as earlier with the values of the transmission equal to 1 at every wavelength, is needed for mapping into CIELUV and CIELAB spaces. For example,

$$X_n = k_{I,\text{obs}} \sum_{i=0}^{80} S_I(\lambda_i) \bar{x}_{\text{obs}}(\lambda_i)$$

It is clear from the definition of $k_{I,\text{obs}}$ that $Y_n = 100$. The CIE system is designed such that Y is the brightness of the stimulus and $\bar{y}_{10}(\lambda_i)$ is identical to the luminous efficiency function.

STEP 5: MAPPING INTO USEFUL COLOR SPACES

CIE xyY; CIE Chromaticity Diagram

Mapping of X, Y, Z into any of several color spaces enhances interpretation of tristimulus data. Projecting X, Y, Z onto the $X + Y + Z = 1$ plane by normalizing each of X, Y, and Z by their sum gives the chromaticities:

$$x = X/(X + Y + Z),$$

$$y = Y/(X + Y + Z)$$

and

$$z = Z/(X + Y + Z)$$

A plot of the x and y values of colors makes up a CIE Chromaticity Diagram, an example of which is given in Fig. 11.16. As neither X,Y, nor Z can be determined from x, y, nor z (though their ratios can), it is clear that the chromaticity diagram does not convey any information about the brightness of a tristimulus. To determine brightness X or Y must also be given. From either we can calculate

$$X + Y + Z = X/x = Y/y$$

and, as $z = 1 - x - y$, the set x,y,Y or x,X,y represents a complete specification of a tristimulus. xyY is customarily used.

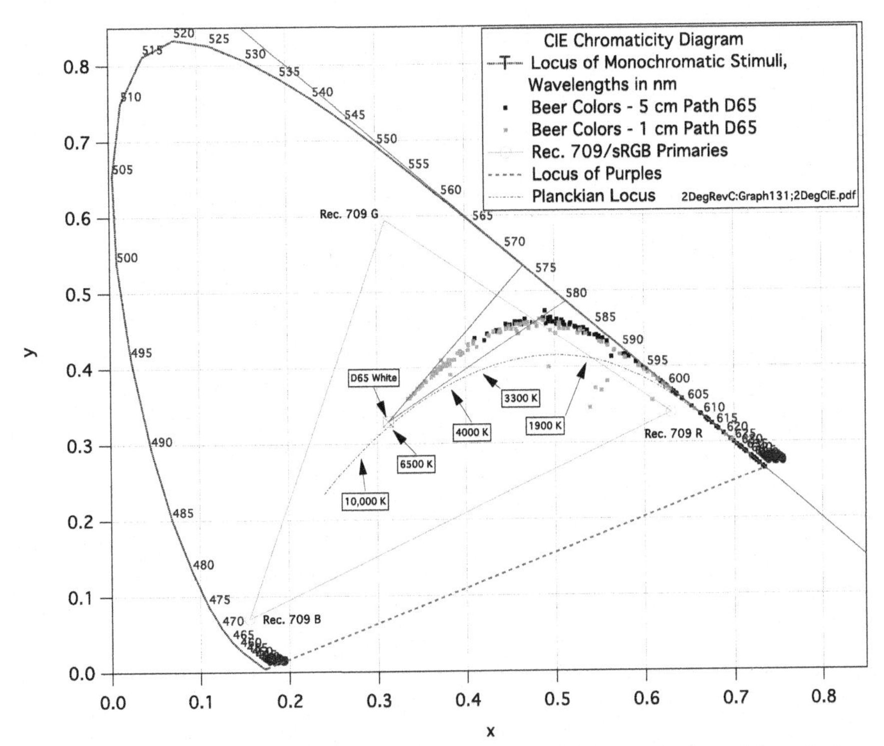

FIGURE 11.16 CIE Chromaticity Diagram Showing Loci of 124 Beers in 1- and 5-cm Paths, 2 degree Observer, Illuminant D65, D65 White Point, Planckian Locus, and Rec. 709 Primaries.

The CIE Chromaticity diagram reveals a great deal of information about color in general though it has its limitations when colors are quantitatively compared (see CIELAB System). If, for example, we have three light sources labeled[25] **R**, **G**, and **B**, measure the spectrum of light emitted from each of them and then plot the x and y coordinates of each. A triangle is formed like the one on Fig. 11.16 on which the vertices are labeled, respectively, "Rec. 709 R," "Rec. 709 G," and "Rec. 709 B." These are the loci of the three primaries recommended by the International Telecommunications Union (ITU) (ITU, 2002) for color TV broadcasting and are identical to those specified for the standard Red Green Blue (sRGB) space used by most computer monitors. To see these colors, open a computer application that has color picker capability and select RGB mode. Setting G and B to 0 shows the color of **R**; B and R to 0, **G**; and R and G to 0, **B**.

As a consequence of Grassman's Laws, any arbitrary color plotted on the interior of the triangle can be matched by a mixture of the **R**, **G**, and **B** light sources in

[25]The implications of the names given here are plain and indeed in most cases these sources will be red, green, and blue, but theoretically any colors that do not plot on a straight line can be used.

amounts proportional to the perpendicular distances of the color's plotted point from the lines which form the triangle. For example, the amount of **R** is proportional to the perpendicular distance of the point from the line joining Rec. 709 G and Rec. 709 B. Distances to points inside the triangle are considered positive, so that any point that lies outside the triangle will have at least one distance negative. As the sources cannot be configured to absorb light, such colors cannot be reproduced with these primaries[26]. The interior of the triangle is called the *gamut* of colors attached to the set of primaries. It should be clear that any three light sources which do not plot on a line can serve as a set of primaries. It should also be clear that a mixture of three primaries has a spectrum which is the weighted sum of the spectra of the individual primaries and is not the same as the spectrum of the light source the color of which is being reproduced in most cases. It should also be clear that an infinite number of spectra can map to a particular point (via the E-308 prescription) inside the triangle and which can, therefore, be matched by a mix of the primaries. This property, called *metamerism*, makes color photography, printing, and television possible.

To complete the definition of a gamut, the designer needs to specify the chromaticity of white. This is the x_w, y_w pair computed by E−308 when $T(\lambda) = 1.0$ at all wavelengths and is, thus, determined by the chosen illuminant. For the Rec. 709 color system, the chosen illuminant is D65. Its white point is plotted on Fig. 11.16. The units of R, G, and B are determined by scaling the perpendicular distances measured to this white point such that 1 unit of each of **R**, **G**, and **B** produces white.

The fact that a color which plots outside the Rec. 709 triangle cannot be reproduced using the Rec. 709 primaries because at least one of *R*, *G*, or *B* will be negative does not mean that this primary set cannot be used for specifying color. Primaries for color specification are arbitrary as long as their chromaticities form a triangle because simple transformations between tristimuli described with respect to a valid set of primaries and any other valid set of primaries exist.

The CIE chromaticity diagram is itself a color triangle based on fictitious[27] primaries, **X**, **Y**, and **Z**, which plot on the diagram at, respectively, (1,0), (0,1), and (0,0). These primaries were chosen to encompass all visible colors. Monochromatic stimuli (light sources of a single wavelength) define the boundary of this set of colors and that boundary is diagrammed as the sharkfin-shaped curve on Fig. 11.16. The corresponding wavelengths are marked adjacent to the curve. Any color inside or on the envelope defined by the monochromatic curve and the dashed line joining its ends is visible and can be reproduced by either a single monochromatic stimulus or a combination of two or three monochromatic stimuli. Moreover, as they are interior to the triangle defined by the three primaries **X**, **Y** and **Z**, they have all-positive tristimulus and chromaticity values.

---

[26]Though they can be matched by adding the negative primary amounts to the stimulus being studied.
[27]There are no physical light sources that exhibit the "colors" of **X**, **Y**, and **Z**.

The white point on Fig. 11.16 is seen to lie close to (but not on) a dashed curve called the *Planckian Locus*. This represents the chromaticities of light emitted by a black-body radiator at various temperatures. Illuminant D65 is so-called because its chromaticity is close to that of a black-body radiator at 6500 K (temperature labels are found at several points along the curve). The same is true for the other members of the D series.

Consider now the white point, the "color" that one would see if water were filled into a beer glass uniformly backlit by a light source with the distribution of D65. One can visualize this with color picker software with R, G, and B set to their maxima. If we gradually reduce the color picker blue slider, at first the color becomes a very pale yellow and as more are more blue is removed the yellow becomes more intense until blue is set to 0. In the computer's color space, reducing blue while holding R and G at maximum moves us along a line which starts at the white point and is perpendicular to the line which joins the G and R primaries. This line is not shown on Fig. 11.16. It would bisect the small triangle with vertex at the white point. When color picker blue is set to 0, the color on the screen has chromaticity on the bisecting line at the point where it cuts the line connecting G and R. Continuing that line until it intersects the sharkfin curve gives the locus of yellow colors that are more yellow than can be produced by the Rec. 709 primaries but are still visible colors. The point of intersection with the sharkfin curve defines the chromaticity of monochromatic light of wavelength 577 nm.

The saturation or excitation purity of a color is quantified by the ratio of the length of a line from the white point to the color's plotted chromaticity to the length of that same line extended to the sharkfin curve. Colors on the sharkfin curve are 100% saturated. The point of intersection of the extended line with the sharkfin curve defines the dominant wavelength of the color. A color which plots close to the spectrum locus line is pure and stimulates the eye in very nearly the same way as a monochromatic stimulus of the dominant wavelength.

The small gray and black squares on Fig. 11.16 are the colors of 124 beers as computed from their measured spectra for the 2 degree observer against the D65 illuminant. Gray squares correspond to the computed colors of the beers for a 1-cm path and the black square colors for a 5-cm path, considered more representative of the usual actual beer glass. The distribution of these squares reveals several important aspects of beer color which are predicted by the CNS hypothesis:

1. Beer chromaticities are very restricted. They fall close to the same curve in chromaticity space irrespective of whether the beer is light or dark or whether the path is 1 cm or 5 cm. Gray squares that are not close to the curve are associated with fruit beers. As the path lengthens, even the fruit beers start to exhibit colors that fall close to it. Anything that intensifies color (larger color depth, *D*) moves the color further along the curve away from the white point. Examination of Fig. 11.4 and the color-matching functions plotted on Fig. 11.5

shows that as D progressively increases, at first short wavelengths, then medium, and finally long wavelengths are attenuated so that at first the Z, then Y, and finally X tristimulus values decrease. Water would plot at the white point. Starting there and decreasing Z moves us along a line away from the **Z** primary location (0,0) through the white point until Y starts to decrease at which point the points on Fig. 11.16 are moving away from both **Z** (0,0) and **Y** (0,1) and the curve starts to bend to the right. Eventually Z gets close to 0 (the diagonal line from (0,1) to (1,0) is the locus of colors for which $Z = 0$) and further increases in D reduces the Y stimulus relative to X and the plotted points move further from **Y** along that diagonal line. At high values of D, a glass of beer is a red filter.

2. Light of wavelength 570–575 is called "greenish-yellow" and 575–580 is called "yellow" (Robertson, 1985) from which it is apparent that the dominant wavelengths for lighter-colored beers ($D < 25$ SRM-cm) are greenish-yellow to yellow. As D increases beyond 25 SRM-cm, they become at first yellowish-orange (580–585 nm), orange (585–595 nm), reddish – orange (595–620 nm), and eventually red (>620 nm)
3. As the depth of beer color exceeds $D = 25$ SRM-cm, the colors are too saturated to be depicted by a TV set or computer monitor using the Rec. 709 primaries nor can they be printed using typical cyan, magenta, yellow, and key (black) (CMYK) ink sets. This limitation needs to be kept in mind in planning print, TV, or web advertising that uses images of beer. It also frustrates attempts to produce printed or web-displayed color guides (color swatches labeled with SRM or EBC values).
4. At high beer-color depth, purities are very high being essentially those of monochromatic light sources of the dominant wavelengths. Verify this by pouring half an inch of stout into a glass and shining a powerful flashlight through it.

CIELAB Color Space

CIE chromaticity space has shown us a lot about beer color, but it has one major flaw which is significant to persons attempting to control or match color, and that is that it is not perceptually uniform which means that equal Euclidian distances between pairs of points on a chromaticity diagram are not representative of equally discernable color differences. As color comparison is important to industry, much work has been done and is ongoing in finding a mapping from X, Y, and Z into a space in which equal distances do represent equally perceptible color differences. Two mappings that are in common use are the CIELUV and CIELAB spaces. Neither is completely perceptually uniform and much of the ongoing work seeks modifications to make them more so. In both of these spaces, the brightness correlate, computed directly from the Y stimulus, is called L^* and is defined by

$$L^* = 116f\left(\frac{Y}{Y_n}\right) - 16$$

in which

$$f(\Theta) = \Theta^{1/3} \qquad ;\Theta > (6/29)^3$$
$$f(\Theta) = (1/3)(29/6)^2\Theta + (4/29) \quad ;\Theta \leq (6/29)^3$$

and Y_n is the Y stimulus for the white point which, as has been shown, is $Y_n = 100$. The two other color coordinates in CIELAB are:

$$a^* = 500[f(X/X_n) - f(Y/Y_n)]$$
$$b^* = 200[f(Y/Y_n) - f(Z/Z_n)]$$

from which hue angle[28]

$$h_{ab} = \tan^{-1}(b^*/a^*)$$

and chroma

$$C^*_{ab} = \sqrt{a^{*2} + b^{*2}}$$

are obtained. It is important to note that chroma is not a correlate of purity or saturation. There is no such correlate in the CIELAB system (Berns, 2000) although chroma normalized by L^* is often interpreted as purity.

It is also important to note that h_{ab} is not a perfect correlate of hue as planes through the L^* axis are not planes of constant Munsell hue (Sharma, 2003). The best that can be said in this regard is that colors with positive values of a^* are reddish whereas those with negative values of a^* are greenish. Similarly, $b^* > 0$ implies yellowish colors and $b^* < 0$ bluish. Again, we encourage readers to experiment with a computer's color picker to gain insight.

Fig. 11.17 plots a^* and b^* of 128 beers for 5-cm path viewed under Illuminant C by the 10 degree observer, the colors of the example beers in 1 cm (these are marked P1, B1, etc. and the 5-cm points for those same beers P5, B5, etc.) and the colors of hypothetical CNS beers for color depths from 2 to over 400 SRM-cm. Fig. 11.18 displays L^* for the same beers. The fact that the light lager in 1 cm, L1, plots close to the stout in 5 cm, S5, on Fig. 11.17 does not imply that these are the same color as they are at opposite ends of the L^* axis as shown on Fig. 11.18. It does demonstrate, however, that as both lie close to the L^* axis (which is perpendicular to Fig. 11.17 at the origin), neither is colorful one being almost white and the other almost black.

It is clear from the definitions of a^* and b^* that the white point is at the origin (0,0) of Fig. 11.17. In studying the CIE chromaticity diagram earlier, we noted that as beer color depth, lSRM, increases from 0 (water), first Z, then Y, and then X are progressively reduced. The formula for b^* suggests that it will at first increase as Z is decreased faster than Y with increasing lSRM but then, once Z is near 0, begin to decrease again as Y decreases at greater lSRM. Similar reasoning for a^* shows that

[28]For computation, a two-argument arctangent function is used so that radian values between 0 and 2π will be returned. These are converted to degrees.

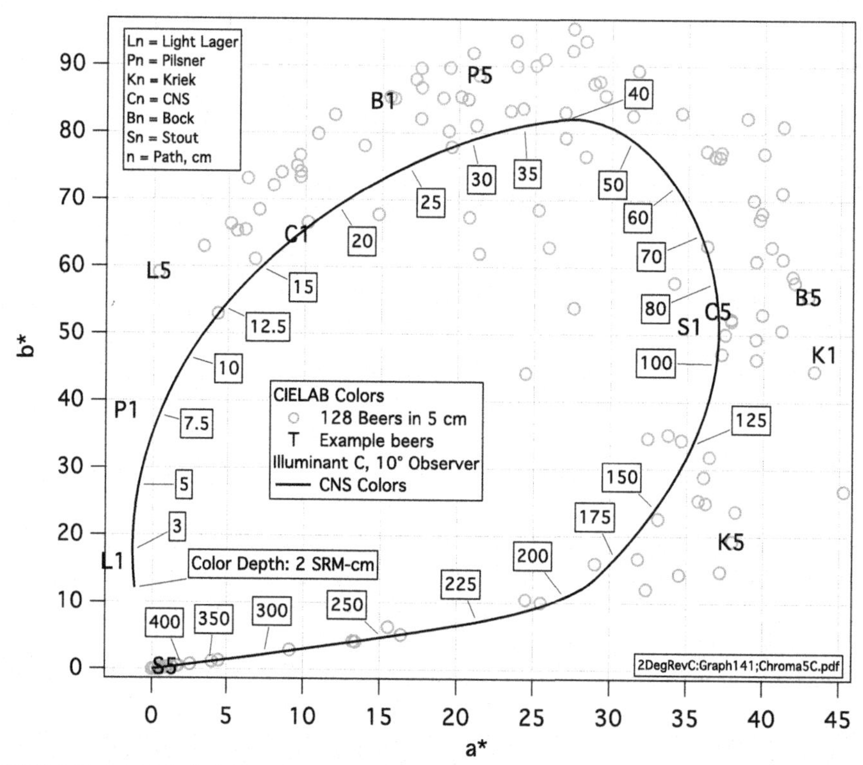

FIGURE 11.17 CIELAB Colors of 126 Beers in 5 cm, CNS Colors as a Function of Color Depth; Colors of the Five Example Beers in 1 and 5 cm.

it should similarly increase initially as *l*SRM increases because *Y* in being reduced faster than *X* and then at values for *l*SRM in which *Y* is very small, decrease again as the X color-matching function response is cut off. The solid curve on Fig. 11.17 which represents beers with the CNS and the open circles which represent measured beers both show these behaviors.

CIELAB space is good for measuring small color differences and that is the reason for our interest in it here: it is the space in which colors of objects are measured in industry, as opposed to CIELUV which is used for judging light emitters. It is not a chromaticity space.

Figs. 11.17 and 11.18 can be used to estimate the CIELAB color of a beer with known SRM or EBC in any size glass. First, convert an EBC color to the equivalent SRM number (divide by 1.9685) and then multiply SRM by the path to get the color depth. Enter Fig. 11.18 at the color depth and read L^* from the CNS curve. For example, $D = 70$ (eg, 10 SRM in a 7-cm wide glass) corresponds to $L^* = 39$. The spread of luminances attributable to deviations from the CNS can be judged from the vertical scatter of actual beer measurements about the CNS curve. Now read along the CNS curve in Fig. 11.17. $D = 70$ implies $a^* = 36$ $b^* = 63$. Once again, the scatter of the actual beer measurements perpendicular to the curve can be

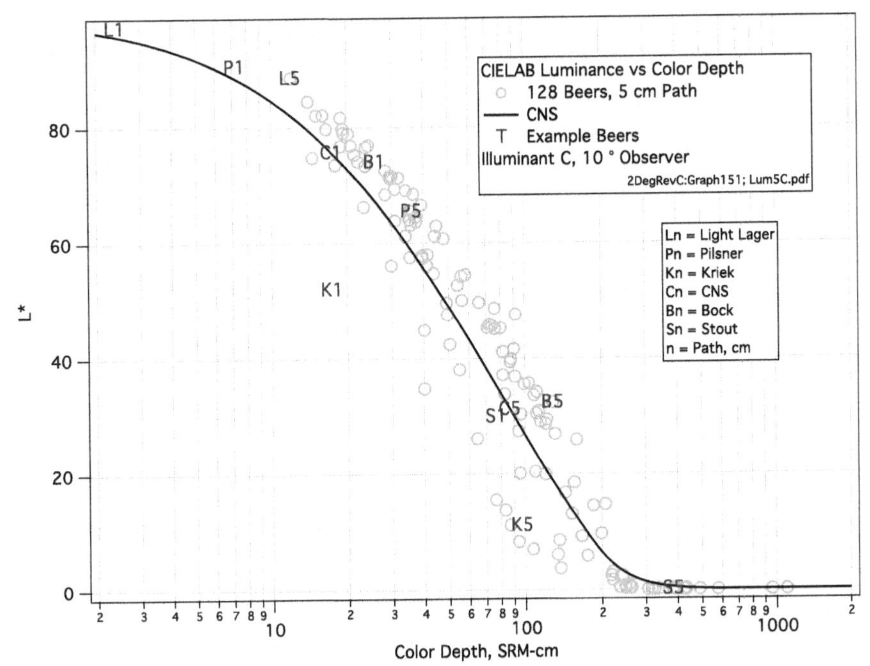

FIGURE 11.18 CIELAB/CIELUV Luminances of 126 Beers in 5 cm, CNS Luminances, Luminances of the Five Example Beers in 1 and 5 cm.

used to judge the dispersion of real beer colors. Using the CNS about 85% of beers' colors can be predicted to $\Delta E^*_{ab} < 20$ (Illuminant C, 10 degree Observer, 5-cm path) from the SRM or EBC color number alone; 44% can be predicted to $\Delta E^*_{ab} < 10$; 21% to; $\Delta E^*_{ab} < 5$ and 10% to less than the JND (2.3). Adding one or two PCs to the SRM or EBC brings the percentage less than the JND to 92%.

ASBC Tristimulus MOA

The ASBC Tristimulus Method, Beer 10C (ASBC, 2006b), is simply an implementation of Steps 1 through 5 using 1-cm cuvette width for spectrum measurement, Illuminant C, color-matching functions for the 10 degree observer and mapping into CIELAB space with L^*, a^* and b^* reported. As such, no further description is required.

Example Color Specifications

Table 11.2 lists the color specifications for the five beers we have been using as examples under each of the color specification methods that have been discussed so far. The PCs were computed without rotation. The table, in addition to the colors, lists the accuracy with which colors can be reconstructed from SRM or EBC and the PCs and the pseudosaturation of the colors computed as the 5-cm CIELAB chroma normalized by CIELAB L^*. CIELAB pseudochroma increases with increasing color depth (with the exception of the Kriek) as saturation (purity) does in CIExy space.

Table 11.2 Comparison of Color Specifications in Six[a] Color Measurement Systems

System/Beer		Light Lager	Pilsner	Kriek	Scaled CNS[b]	Bock	Stout
1.	EBC[c]	4.73	14.18	34.25	34.25	50.80	150.88
2.	SRM[d]	2.40	7.20	17.40	17.40	25.81	76.64
3.	Augmentation PC_1	−0.899	−0.382	1.659	0[e]	−0.497	0.237
	Augmentation PC_2	−0.168	0.007	−0.597	0	−0.103	0.012
	Augmentation PC_3	0.036	0.017	0.020	0	0.006	−0.042
4.	Pseudo-Lovibond[f]	2.76	7.31	55.42	18.15	24.51	–
5.	ASBC L^* (1 cm 10°, C)[g]	97.34	90.56	52.08	75.25	74.13	30.00
	ASBC a^* (1 cm 10°, C)	−2.49	−1.66	44.02	10.14	15.06	35.17
	ASBC b^* (1 cm 10°, C)	15.83	38.37	47.19	64.56	84.66	51.30
6.	L^* (5 cm 10°, C)	88.61	65.53	11.19	30.47	32.42	0.14
	A^* (5 cm 10°, C)	0.36	21.31	37.96	37.38	42.91	0.90
	B^* (5 cm 10°, C)	59.16	88.58	19.30	52.24	55.79	0.24
Pseudo-Saturation[h]		0.67	1.39	3.81	2.11	2.31	6.65
ΔE^*_{ab}(3 PC's, 5 cm, C)[i]		0.54	0.14	0.73[j]	0	0.97[k]	0.11

[a] *Of these, only four, EBC, SRM, Lovibond, and ASBC 1 cm, have, or have ever had, status as accepted standards.*
[b] *This is the CNS from deLange (2008) scaled to the same SRM as the Kriek.*
[c] *EBC (MEBAK, 2002).*
[d] *ASBC MOA Beer 10 –A (ASBC, 2006a).*
[e] *The PCs measure the deviation of the normalized beer spectrum from the CNS. A spectrum which is a scaled version of the CNS has a normalized spectrum identical to the CNS implying zero deviation.*
[f] *Data concerning the Lovibond glasses was obtained from Bishop (1950) and Judd et al. (1962) and used to compute their CIE colors as a function of Lovibond number. The given Lovibond color for each beer is the interpolated glass color that is closest in CIELAB space to the computed beer color in an 0.5-inch path.*
[g] *ASBC MOA Beer 10 –C (ASBC, 2006b).*
[h] $C^*_{ab}/L^* = \sqrt{a^{*2} + b^{*2}} / L^*$ 5 cm
[i] *This is the color difference between the original spectrum and a version of it as reconstructed from SRM and three PCs.*
[j] *Use of a fourth PC (0.100) reduces* ΔE^*_{ab} *to 0.11.*
[k] *Use of a fourth PC (0.032) reduces* ΔE^*_{ab} *to 0.32.*

Measuring Color Differences

Clearly, any quality-control program that intends to include color as a quality measure needs a means of measuring color differences so that the magnitude of deviation of a particular sample from a desired target can be quantified. This is the subject of ongoing research as no completely adequate metric has yet been discovered. It has been noted that the CIE chromaticity space is not perceptually uniform and that the CIELUV and CIELAB spaces were conceived to provide what were intended to be more so. Euclidean distance, as measured in either of these two spaces is more perceptually uniform but not entirely so. In $L^*a^*b^*$ (LAB) space, the difference between two colors has three components:

$\Delta L^*_{ab} = L^*_{ab1} - L^*_{ab2}$, $\Delta a^* = a^*_1 - a^*_2$, and $\Delta b^* = b^*_1 - b^*_2$ that is, it is a 3 vector. In our use of it thus far, we have only made use of the magnitude of this vector

$$\Delta E^*_{ab} = \sqrt{\Delta L^{*2} + \Delta a^{*2} + \Delta b^{*2}}$$

We have referred to this measure previously as the CIELAB difference. The average JND is stated as being about 2.3 (Sharma, 2003) units of ΔE^*_{ab} but the perceptible difference depends on direction as well as magnitude. The eye is, for example, more sensitive to changes in hue than in chroma. Working with chroma and hue angle as defined previously

$$\Delta E^*_{ab} = \sqrt{\Delta L^{*2} + C^{*2}_{ab1} + C^{*2}_{ab2} - 2C^*_{ab1}C^*_{ab2}\cos(h_{ab1} - h_{ab2})}$$

and we can define three components of difference as ΔL^*_{ab}, $\Delta C^*_{ab} = C^*_{ab1} - C^*_{ab2}$, and $\Delta h^*_{ab} = h_{ab1} - h_{ab2}$. The hue component of difference is often expressed as

$\Delta H^*_{ab} = \sqrt{\Delta E^{*2}_{ab} - \Delta L^{*2} - \Delta C^{*2}_{ab}}$, that is, as the part of color difference not due to luminance or chroma. It can be assigned the same sign as Δh^*_{ab} but as it is usually used squared in improved difference formulas such as

$$\Delta E_{(00)} = \sqrt{\left(\frac{\Delta L^{*2}_{ab}}{k_L S_L}\right)^2 + \left(\frac{\Delta C^*_{ab}}{k_C S_C}\right)^2 + \left(\frac{\Delta H^*_{ab}}{k_H S_H}\right)^2}$$

its sign is irrelevant. $\Delta E_{(00)}$ is intended be a metric which is more perceptually uniform than ΔE^*_{ab}. The k factors are set by the user (sometimes all to 1) depending on the perceived relative sensitivity to luminance, chroma, and hue and the S factors are determined, depending on the application and industry, by formulas which depend on the coordinates of the colors being compared in CIELAB space with this goal in mind. Details are found in color science texts such as Berns (2000) and Sharma et al. (2005).

BEER COLOR AS PART OF QUALITY CONTROL

Color is distinct from other quality-control parameters such as, for example, true extract in that it can be multidimensional and in that visible color depends on viewing conditions. A brewery using SRM alone would decide on a target SRM and tolerance for each product. A batch for which measured SRM differs from the target by more than the tolerance is sent back for rework. Similarly, a brewery using CIELAB color can accept any beer for which ΔE^*_{ab} or $\Delta E_{(00)}$ is less than some threshold ostensibly ignoring that color difference is three dimensional. Alternatively, limits can be specified for each color attribute separately, for example, tighter tolerance on hue than chroma, and each attribute separately compared to that attribute's target value. Relative importance of attributes can also be set via the k parameters in the $\Delta E_{(00)}$ formula.

Decisions concerning target color and tolerance ultimately derive from the visual impressions of human observers who examine multiple batches and decide which are acceptable with respect to color and which are not, and, if not, why they are not. Scatter plots of the instrumental measurements for these same batches are made using symbols which distinguish accepted and rejected points. An important consideration here is the choice of viewing conditions, in particular, path under which the plotted colors will be calculated. Increasing path magnifies color differences up to a point. Obviously, working with color differences at the path in which the beer is expected to be viewed leads to a better appreciation of how a particular color difference is likely perceived by the consumer.

A simple closed surface such as a circle, sphere, ellipse, or ellipsoid is sought which encloses as many of the acceptable points as possible while excluding as many of the rejected points as possible. Ordinary gradient search algorithms, such as the popular Excel Solver, are not suitable for this as the cost (number of unacceptable points surrounded plus number of acceptable points not surrounded) is not a continuous function of ellipsoid parameters. The concept is illustrated for two dimensions in Fig. 11.19 with color difference data randomly generated for purposes of illustration. Simulated annealing (Černý, 1985) does not have problems with discontinuous cost functions and was used to fit the ellipse in the figure to the data. The major and minor axes were added to the cost function thus insuring the smallest ellipse that satisfies the optimality criterion (fewest inclusion and exclusion errors).

There are other ways to obtain instrumental rejection criteria from visual history data. It should be clear from Fig. 11.19, for example, that if the acceptance ellipse (or ellipsoid) is not rotated about any of its axes (the ellipse in Fig. 11.19 is rotated about the ΔL^* axis perpendicular to the page), it can be transformed to a circle (sphere) by scaling the axes. Using that same scaling for the k parameters in the $\Delta E^*_{(00)}$ formula turns the acceptance criterion into the simple one of deciding whether $\Delta E^*_{(00)}$ is greater or less than the radius of the transformed circle. Berns (2000) is a good source of further insight.

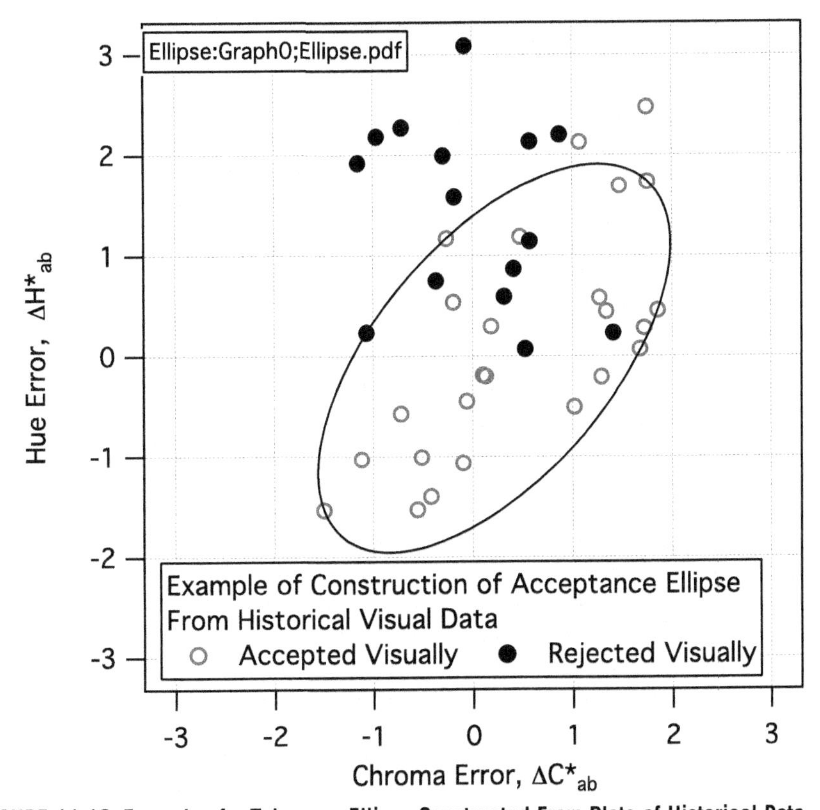

FIGURE 11.19 Example of a Tolerance Ellipse Constructed From Plots of Historical Data.

MODIFYING BEER COLOR

If a beer's color is deemed unacceptable, the cause of the unacceptability must be determined and corrective action planned. Use of lighter or darker or more or less colored malts in future gyles will move colors approximately parallel to the CNS curves in Fig. 11.17 and 11.18. This will not, of course, fix the current batch. A small operation such as a brewpub can serve that beer as is or designate it as a specialty under a different name. Breweries that wish to repair the out of tolerance beer have several ways to do so guided by the Bouguer–Beer–Lambert law which predicts that the absorption spectrum of two liquids mixed in the ratio $r{:}(1-r)$ with $0 < r < 1$ is

$$A_{\mathrm{m}}(\lambda) = rA_1(\lambda) + (1-r)A_2(\lambda)$$

This is easily solved for r,

$$r = \frac{A_{\mathrm{m}}(\lambda) - A_2(\lambda)}{A_1(\lambda) - A_2(\lambda)}$$

to give the proportion of liquid 1, with absorption $A_1(\lambda)$, to be added to liquid 2 to increase its absorption from $A_2(\lambda)$ to $A_m(\lambda)$. As SRM is proportional to absorption at 430 nm, this can be written as

$$r = \frac{\Delta \mathrm{SRM}}{\mathrm{SRM}_1 - \mathrm{SRM}_2}$$

If added liquid 1 is, for example, a *Farbebier* (Briggs, 2004) or other intensely colored product (SRM 3000 or more) $r \approx \Delta\mathrm{SRM}/\mathrm{SRM}_1$. At SRM 3000, only 33 mL of colorant would be required to add 1 SRM of color to a hectoliter of beer.

If 1 and 2 are close to one another in color, for example if they represent different batches of the same product, exactly the same principle applies, but blending ratios will be much more balanced. Blending multiple gyles can be used as a strategy for controlling beer color as blending N beers with the same average color and SRM standard deviation σ_{SRM} results in a mixture with SRM standard deviation $\sigma_{\mathrm{SRM}}/\sqrt{N}$. Blending, of course, mixes flavor as well as color compounds.

If $A_1(\lambda)$ and $A_2(\lambda)$ both closely adhere to the CNS hypothesis, the mix will too, so that blended and unblended colors will all appear on or near the CNS curves. For moderately deviant beers and colorants (caramels and *Farbebiers* appear to qualify) blended colors can be expected to lie on a curve connecting the unblended colors lying approximately parallel to the CNS curves. In the case in which it is desired to move color in a direction toward or away from the CNS curves, obviously highly deviant colorants must be used. Conceptually, a brewer can achieve any color he wants in the same way an artist mixing colors from his palette can. The blending formula allows prediction of the color of any mix by processing of the blended spectrum, $A_m(\lambda)$, in the same way (E−308) as any other spectrum. Some examples are shown on Fig. 11.20 which shows the measured colors of blends of an SRM 9.6 Pilsner beer with an SRM 12.2 Vienna in, respectively, 1, 5, and 8 cm as, respectively, triangles, circles, and squares. Blends of 100, 33.3, 50, 50, 66.5, and 0% Pilsner are shown with the most counterclockwise symbol in each group representing 100%.

The dual measurements at 50% were the results of first diluting Pilsner with Vienna and then Vienna with Pilsner and illustrate the kind of repeatability differences that can be expected from volume measurement errors. Color depth magnifies color differences so that the color difference between the 100 and 0% blends is $\Delta E^*_{ab} = 9.4$ (including ΔL^*) for 1-cm path; $\Delta E^*_{ab} = 13.2$ for 5-cm path, and $\Delta E^*_{ab} = 18.9$ for 8-cm path. The color difference attributable to a particular blend ratio will also increase with increased path.

The smooth curves near which the measured color symbols lie are the loci of theoretical colors computed from the blending formula and emphasize the point that the blending formula is not a perfect predictor. Deviations from the Bouguer–Beer–Lambert law will certainly cause deviations, but it is more likely that volume measurement errors are responsible for the ones seen on the figure.

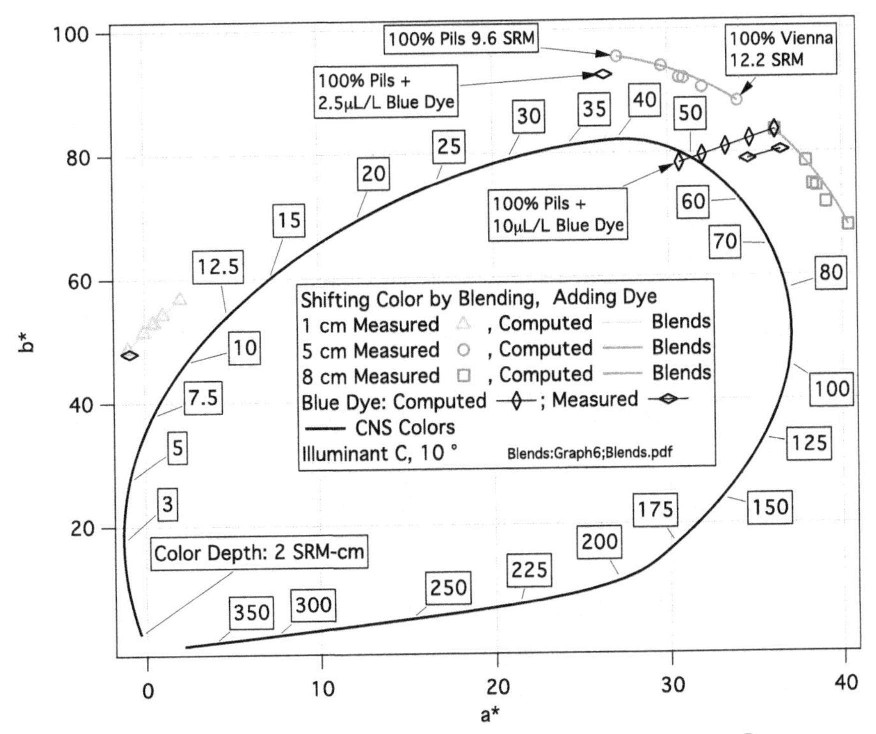

FIGURE 11.20 Effects of Blending Two Beers and of Adding Food Color Dye to Beer.

Fig. 11.20 also illustrates, as vertically oriented black diamonds, the theoretical colors in 8 cm expected from blending, from right to left, 0.0, 2.5, 5.0, 7.5, and 10.0 μL/L blue food-coloring dye with the Pilsner beer. The horizontal black diamonds show measured colors for the Pilsner beer with 2.5 μL/L added for 1-, 5-, and 8-cm (left-hand one of the pair) paths. The dye was added to a sample of the Pilsner which had stood at laboratory temperature in an open container for several hours. The right-hand horizontal diamond in the pair at 8 cm is the color of that beer before the dye was added and shows how far the exposure to air and elevated temperature caused the color to move from its original coordinates, those of the closest vertical black diamond. This enforces the earlier admonitions of the previous material on spectrum measurement practices.

SUMMARY AND CONCLUSION

Beer color is easily, reliably, and repeatably measured and specified in terms of the spectral absorption at 430 nm. These (SRM and EBC) measures are adequate for many purposes, including quality control, and give a rough indication as to the approximate visible color of beers $(\Delta E^*_{ab} < 20)$ but when better accuracy is needed more degrees of freedom are required in both data collection and the number of color descriptive parameters calculated from the data. When more degrees of freedom are considered, the complexity of data gathering and data processing both increase dramatically to the point that SRM and EBC methods remain the most commonly used methods today. Hidden within the complexities of higher dimensionality are opportunities to discover better methods of beer color specification. The ASBC Tristimulus method is a candidate as is the Augmented SRM method, both of which have been discussed in this chapter. There are certainly others.

REFERENCES

American Society of Brewing Chemists, 2006a. Methods of Analysis: MOA Beer 10 A Spectrophotometric Color Method. ASBC, St Paul MN.

American Society of Brewing Chemists, 2006b. Methods of Analysis: MOA Beer 10 C Spectrophotometric Color Method. ASBC, St Paul MN.

ASTM Standard E1164-12, 2014. Standard Practice for Obtaining Spectrometric Data for Object-color Evaluation. ASTM International, West Cohnshohocken, PA.

ASTM Standard E308-96, 1996. Standard Practices for Computing the Colors of Objects by Using the CIE System. ASTM International, West Cohnshohocken, PA.

Berns, R.S., 2000. Billmeyer and Saltzman's Principles of Color Technology, third ed. Wiley, New York.

Bishop, L.R., November-December. Proposed revisions of the Lovibond "52 series" of glass slides for the measurement of the colour of worts. Journal of the Institute of Brewing 56 (6).

Briggs, D.E., Boulton, C.A., Brooks, P.A., Stevens, R., 2004. Brewing Science and Practice. Woodhead Publishing, p. 45.

Černý, V., 1985. Thermodynamical approach to the traveling salesman problem: an efficient simulation algorithm. Journal of Optimization Theory and Applications 45 (1).

Crochiere, R.E., Rabiner, L.E., 1983. Multirate Digital Signal Processing. Prentice Hall.

DeClerck, J., 1958. A Textbook of Brewing Vol. II. Translated by Kathleen Barton −Wright. Chapman and Hall, London.

deLange, A.J., 2008. The standard reference method of beer color specification as the basis of a new method of beer color reporting. Proceedings, American Society of Brewing Chemists 66 (3), 143−150.

deLange, A.J., 2014. Color Calculation Spreadsheet. http://www.wetnewf.org/pdfs/Brewing_articles/MOAWorkbook.xls.

ITU, Recommendation ITU-R BT.709-5, 04/2002. Parameter Values for the HDTV Standards for the Production and International Programme Exchange.

Judd, D.H., Chamberlain, G.J., Haupt, G.W., July 1962. Ideal Lovibond color system. Journal of the Optical Society of America 52 (7).

Kuehni, R.G., Schwarz, A., 2008. Color Ordered, a Survey of Color Order Systems from Antiquity to the Present. Oxford University Press.

Levine, I.N., 1995. Physical Chemistry, fourth ed. McGraw Hill, New York.

Lovibond, J.W., 1915. Light and Colour Theories and Their Relation to Light and Colour Standardization. E. & F. N. Spon, Ltd, London.

Mahy, M., Van Eycken, L., Oosterlinck, A., April 1994. Evaluation of uniform color spaces developed after the adoption of CIELAB and CIELUV. Color Research and Application 19 (2).

MEBAK (Mitteleuropäischen Brautechnischen Analysenkommission), 2002. Brautechnische Analysenmethoden Band II 2.13.2 Spektralphtometrishce (EBC –Methode). MEBAK, Freising, p. 88.

NIST– "Technical Specification for Certification of Spectrophotometric NTRMS", 2000. NIST Special Publication 260 – 140 Standard Reference Materials. USGPO.

Oleari, C., Jun 2000. Spectral-reflectance-factor deconvolution and colorimetric calculations by local-power expansion. COLOR research and application 25 (3).

Press, W.H., Flannery, B.P., Teukolsky, S.A., Vettering, W.T., 1986. Numerical Recipes: The Art of Scientific Computing. Cambridge University Press, London.

Robertson, A.R., 1985. Color vision, representation and reproduction. In: Benson, K.B. (Ed.), Television Engineering Handbook. McGraw Hill, New York.

Sharma, G., Wu, W., Dalal, E.N., Feb. 2005. The CIE2000 color-difference formula: implementation notes, Supplementary test data and mathematical observations. COLOR Research and Application 30 (1).

Sharma, G., 2003. Digital Color Imaging Handbook. CRC Press.

Sharpe, F.R., Garvey, T.B., Pyne, N.S., 1992. The measurement of beer and wort color – a new approach. Journal of the Institute of Brewing 98, 321–324.

Shellhammer, T.H., Bamforth, C.W., 2008. Assessing color quality of beer. In: Culver, C., Wolstad, R. (Eds.), Color Quality of Fresh and Processed Foods, ACS Symposium Series, vol. 983. American Chemical Society, Washington, DC, pp. 192–202.

Shellhammer, T.H., 2009. In: Bamforth, C. (Ed.), Beer, a Quality Perspective. Academic Press, Amsterdam.

Smedley, S.M., November–December, 1992. Colour determination of beer using tristimulus values. J. Inst. Brew. 98, 497–504.

Stone, I., Gray, P., 1946. Photometric standardization and determination of color in beer. Proceedings, American Society of Brewing Chemists 41–49.

Stone, I., Miller, M.C., 1949. The standardization of methods for the determination of color in beer. Proceedings, American Society of Brewing Chemists 140–147.

Wyszecki, G., Stiles, W.S., 2000. Color Science Concepts and Methods, Quantitative Data and Formulae, second ed. Wiley, New York.

CHAPTER

Haze Measurement

12

C.W. Bamforth

University of California, Davis, California, CA, United States

THE NATURE OF THE PHYSICAL INSTABILITY OF BEER

Beer is inherently an unstable colloidal entity. It contains a diversity of materials than can lead to either

1. dispersed hazes (including "invisible" or "pseudo" hazes),
2. precipitates, or
3. "bits" (larger particles that are suspended in beer against a "bright" background; bits may arise from the disintegration of precipitates).

A detailed review of the various types of colloidal instability problems was offered by Bamforth (1999); see also Bamforth (2011).

Haze in beer is traditionally discussed in terms of

1. permanent haze: that which is continually present; and
2. chill haze: that which develops as beer is taken to 0°C but which redissolves when the temperature is returned to 20°C.

When discussing haze per se, notably permanent and chill haze, it is customary to be referring only to dispersed hazes. However, it is important that brewers be mindful that precipitates and bits are also manifestations of colloidal instability, and their presence may be induced by temperature shock, most notably the production of precipitates in beer that has been inadvertently frozen. This latter problem draws attention to the wisdom of including a cooling stage in any method employed to predict the physical instability of beer.

Colloidal instability is also exacerbated by shaking and agitation, and hazes have been linked to the motion of beer on board ship during lengthy journeys.

Haze is also characterized as

1. biological haze and
2. nonbiological haze.

The former refers to turbidity arising from the multiplication of living organisms in beer (see Chapter 14), whereas the latter concerns non-living materials that can cause clarity problems.

Brewing Materials and Processes. http://dx.doi.org/10.1016/B978-0-12-799954-8.00012-5

A diversity of materials can cause nonbiological haze (and precipitates and bits). The following list captures most of the known ones but may not be exclusive. In each case, more details are available from Bamforth (1999).

1. Proteins, reacting with
2. Polyphenols
3. β-Glucans
4. Arabinoxylans (pentosans)
5. Starch
6. Oxalate
7. Dead bacteria
8. Stabilizers (eg, propylene glycol alginate, silicates, papain, isinglass) reacting with one another under extreme conditions (eg, very high temperatures)
9. Can lid lubricants

THE MEASUREMENT OF HAZE

For the longest time, it has been customary to measure haze in beer as a function of light scattering. Historically, this has been by detecting the amount of light scattered at an angle of 90 degree to the incident beam, for example in a Radiometer haze meter or latter-day alternatives. However, there is a substantial scattering of light at this angle by extremely small particles that are not visible to the human eye. Such particles have been variously linked to entities such as tiny fragments of unconverted starchy endosperm and from material sloughed off the surface of yeast through aggressive treatment. These particles are too small to be removed by conventional filtration or centrifugation. The phenomenon is known as "invisible" haze or "pseudo" haze, and its presence obliges the operator to override the haze value delivered by the instrument and declare the product bright. The preferred alternative is to measure light scatter at a "forward" angle of scatter (13 degree or 30 degree), with which instrumentation there is no detection of invisible haze. However, if a full appreciation of all potentially problematic material is to be acquired, then it is advisable to measure at both forward and right-angle scatter angles.

MEASUREMENT OF BITS

Roy Cope, working in my laboratory in Bass Brewers in the 1980s, developed an unpublished procedure comprising the vacuum filtration of beer through filter paper (2 cm diameter) and staining retained bits on the paper with methylene blue. A series of reference papers was developed in which Cope had produced different quantities of stained bits and the severity of a bit problem in a beer was gauged in a semiquantitative way by comparison of the bits recovered with the reference portfolio of filter papers.

PREDICTION OF THE "HAZE STABILITY" OF BEER

Whilst proper processing, including filtration and/or centrifugation, should lead to "bright" beer in package, the concern is the extent to which turbidity develops in package over time.

Diverse methods have been proposed and used in an attempt to forecast the physical shelf life of beer.

In brief, they can be divided into methods that

1. measure specific haze components (see previous)
2. methods that "force" the beer, thereby accelerating the development of haze (and other elements of colloidal instability, notably precipitates and/or bits)

Clearly, the first type of method has serious inadequacies if only one or a relatively few materials are measured. For example, a method may reveal a beer not to have a worrisome level of haze-forming protein, but that says nothing about its content of polysaccharides, oxalate, and so on. For this reason, some brewers have based their predictive techniques on a combination of a pair of such methods, eg, measurements of protein and polyphenol, but even that may be inadequate.

The second type of method is more reasonable, as (depending on its precise nature) it should assess the tendency of **all** colloidally sensitive materials to "drop" out of solution. These methods can be divided into those that challenge the beer by extremes of heat or by hot–cold cycling and those that involve adding an agent (notably alcohol) that, allied to extreme chilling, will lead to any material that has a tendency to leave solution so to do.

In terms of the former type of method, we can include

1. for protein: the Saturated Ammonium Sulfate Precipitation Limit (SASPL) test and the tannic acid precipitation test
2. for polyphenol: the colorimetric determination of total polyphenol [American Society of Brewing Chemists (ASBC, 2015) method Beer-35], or high-performance liquid chromatography (HPLC) (Mitchell et al., 2005)
3. colorimetric or enzyme-based assays for protein (eg, Smith et al., 1985), total carbohydrate (eg, Dubois et al., 1956), β-glucan (eg, Bamforth, 1983), arabinoxylans (pentosan; eg, Kanauchi and Bamforth, 2003), starch (eg, Ahluwalia and Ellis, 1984), oxalic acid (Kanauchi et al., 2009)

Among the forcing tests are

1. The European Brewery Convention (EBC) (1963 method) in which beer is held at 60°C for 7 days, then cooled to 0°C for 24 h and the haze measured at a 90 degree angle.
2. The Harp method in which the beer is stored for 4 weeks at 37°C, followed by 8 h at 0°C and the haze measured at a 90 degree angle.

3. Various cycling methods, such as the one that holds beer for 24 h at 37°C, then 24 h at 0°C for 24 h prior to measuring haze at a 90 degree angle, this supposedly representing the equivalent of 1 month of natural storage.

Rather more valuable in this author's opinion are tests in which colloidally sensitive materials are **forced** out of solution. The most famous of these is the Chapon test in which a sample of beer is chilled to −8°C without freezing (added alcohol prevents freezing) and left for 8 h before the chill haze is measured at a scatter angle of 90 degree. This type of test is especially valuable because any material that displays a tendency to fall out of solution is likely to be detected in this test, which combines the very low temperature and the added precipitant (ethanol).

THE INVESTIGATION OF BEER HAZES

Clarity issues will arise, so it is important to have a strategy for investigating them in an ordered and logical way. A schematic for the sequence of operations is given in Fig. 12.1.

To recover haze by centrifugation, the beer should be spun at 10,000−15,000 g for 20 min at 20°C (permanent haze) or 0°C (chill haze). Lower speeds may be sufficient for certain hazes. As hazy beers might typically contain 2−3 mg haze per liter, some 5 L of beer is typically needed. Wash the pellets and combine them using a solution of ethanol of the same concentration [alcohol by volume

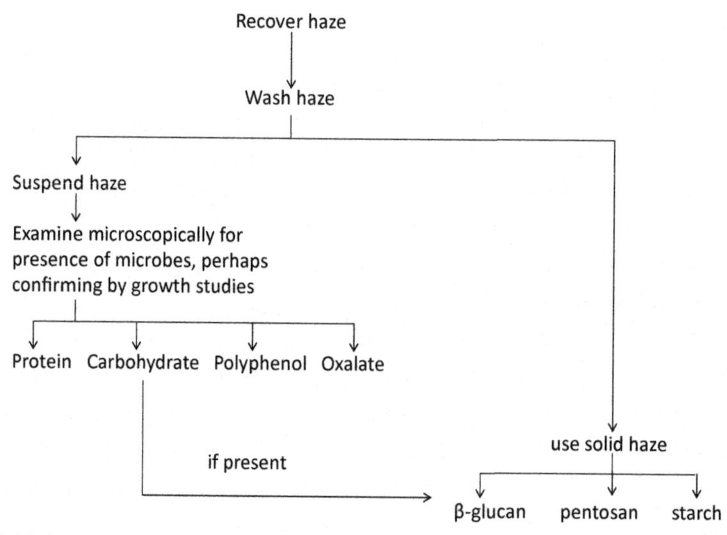

FIGURE 12.1

A systematic approach to analyzing beer haze.

(ABV)] as the beer. Resuspend in approximately 200 mL and recentrifuge. Dry the pellets in an oven at 96°C for 2–4 h, although it is better to use a desiccator, drying over phosphorus pentoxide for 24 h. Carefully weigh to one decimal place.

For invisible hazes, higher speeds of centrifugation are needed, eg, 100,000 g.

To recover haze by filtration, beer is passed through a membrane filter of pore size 0.45 μm and washed on the filter using ethanol solution of the same concentration as the beer. Haze material recovered from the filter is dried as previously.

To the dried haze from any of the earlier treatments is added deionized water at a rate of 0.2 mL per mg haze. The suspension is stirred well using a magnetic stirrer for approximately 1 h to evenly disperse the sample and any pipetting of the haze should always be when stirring is in progress.

Analysis of the various components according to the schematic in Fig. 1 is using methods referenced previously.

Readers are also referred to the ASBC Methods of Analysis. The current last entry under the Beer section is not numbered, but is titled "Beer Inclusions: Common Causes of Elevated Turbidity," and it does include a detailed summary of delving into turbidity issues.

A QA APPROACH TO MINIMIZING COLLOIDAL PROBLEMS IN BEER

Assessment of the quantity and nature of haze, invisible haze, bits, or precipitates in beer must be allied to a quality assurance mentality and the putting in place of raw material selection and process strategies that will ensure the required colloidal shelf life of a product. A holistic summary of the impact of raw materials and process stages on colloidal problems in beer has been offered elsewhere (Lewis and Bamforth, 2006).

REFERENCES

Ahluwalia, B., Ellis, E.E., 1984. A rapid and simple method for the determination of starch and β-glucan in barley and malt. Journal of the Institute of Brewing 90, 254–259.

American Society of Brewing Chemists, 2015. Methods of Analysis, fourteenth ed. http://methods.asbcnet.org.

Bamforth, C.W., 1983. *Penicillium funiculosum* as a source of β-glucanase for the estimation of barley β-glucan. Journal of the Institute of Brewing 89, 391–392.

Bamforth, C.W., 1999. Beer haze. Journal of the American Society of Brewing Chemists 57, 81–90.

Bamforth, C.W., 2011. 125th anniversary review: the non-biological instability of beer. Journal of the Institute of Brewing 117, 488–497.

DuBois, M., Gilles, K.A., Hamilton, J.K., Rebers, P.A., Smith, F., 1956. Colorimetric method for determination of sugars and related substances. Analytical Chemistry 28, 350–356.

Kanauchi, M., Bamforth, C.W., 2003. Use of xylose dehydrogenase from *Trichoderma viride* in an enzymic method for the measurement of pentosan in barley. Journal of the Institute of Brewing 109, 203–207.

Kanauchi, M., Milet, J., Bamforth, C.W., 2009. Oxalate and oxalate oxidase in malt. Journal of the Institute of Brewing 115, 232–237.

Lewis, M.J., Bamforth, C.W., 2006. Haze. Chapter 5. In: Essays in Malting and Brewing Science. Springer, New York.

Mitchell, A.E., Hong, Y.-J., May, J.C., Wright, C.A., Bamforth, C.W., 2005. A comparison of polyvinylpolypyrrolidone (PVPP), silica xerogel and a polyvinylpyrrolidone (PVP)-silica co-product for their ability to remove polyphenols from beer. Journal of the Institute of Brewing 111, 20–25.

Smith, P.K., Krohn, R.I., Hermanson, G.T., Mallia, A.K., Gartner, F.H., Provenzano, M.D., Fujimoto, E.K., Goeke, N.M., Olson, B.J., Klenk, D.C., 1985. Measurement of protein using bicinchoninic acid. Analytical Biochemistry 150, 76–85 (see also Erratum Analytical Biochemistry, 1987, 163, 279).

CHAPTER

Sensory Analysis in the Brewery

13

W.J. Simpson

Cara Technology Limited, Leatherhead, Surrey, United Kingdom

INTRODUCTION

Consumers today are more demanding than ever concerning beer quality. In the past beers contained numerous flavor defects, forgivingly referred to as "house flavors." Such flavors are now likely frowned upon by consumers.

To delight customers with their "secret sauce" of positive flavors, brewers have to rely on more than great raw materials, outstanding recipes, well-engineered equipment, good brewing practice, and instinct. Tasting is an essential skill for all successful brewers.

Only about 20% of the population have a genuine talent for tasting, allowing them, after thorough training, to identify and scale the intensity of flavors in beer. An equal proportion of individuals have a sense of taste and odor which is so underdeveloped that their perception of beer flavor has little relation to how their products are perceived by others.

INFORMATION REQUIRED BY THE BREWERY FROM SENSORY ANALYSIS OF BEER

OBJECTIVES OF SENSORY ANALYSIS OF BEER

There are many reasons for breweries to carry out sensory analysis of beer, including:

- Routine monitoring of beer in process to check for defects or decide on the optimal timing of process steps;
- Routine monitoring of beer prior to release to the packaging operation;
- Routine monitoring of beer prior to release to the market;
- Evaluation of samples from the market;
- Troubleshooting of problem beers, including flavor-stability issues, microbiological problems, and taint incidents;
- Evaluation of trial samples to support process or product improvement.

Many sensory test methods can provide this information (Fig. 13.1). In this Chapter, I will discuss how they are used to improve beer quality and consistency.

Brewing Materials and Processes. http://dx.doi.org/10.1016/B978-0-12-799954-8.00013-7

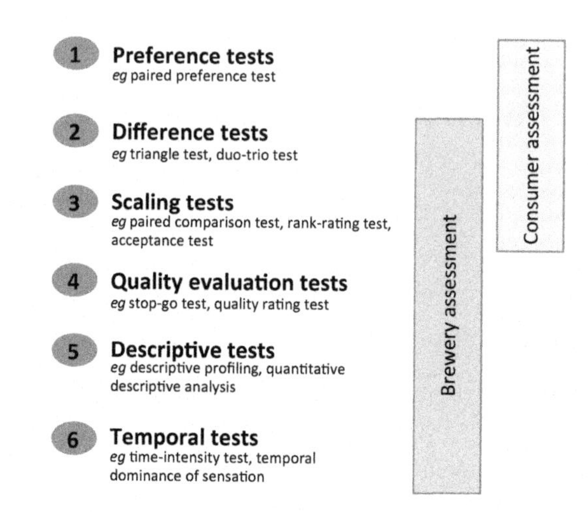

FIGURE 13.1

Sensory test methods which can be used to evaluate beer. Preference test methods belong to the domain of consumer testing, which should always take place outside of the brewery. All other methods can be used in the brewery to address specific testing issues.

SENSORY SPECIFICATIONS FOR BEER

Some beers have relatively simple flavor profiles, comprising 15–25 attributes. Others have profiles that are more complex. For all but the smallest of breweries, some form of formal flavor specification is needed. As a minimum, specifications should include:

- The positive flavors expected in a well-made example of the product;
- The relative intensity of each of these positive flavors.

Sensory specifications that are more rigorous have rules concerning the degree to which perceived flavor quality is affected by variation in the intensity of one or more positive flavors relative to the target, and the impact of off-flavors and taints. Reference standards for each flavor attribute in the product should be defined. The chemicals chosen as reference standards should be identical to those which impart the flavor characteristics naturally to the beer, rather than merely being similar to them. An example of a sensory specification for a pale lager beer is shown in Table 13.1.

ACTIONS WHICH CAN BE TAKEN IN RESPONSE TO THE RESULTS OF SENSORY TESTS

There is little point committing the resources needed for professional brewery tasting without a corresponding commitment to act on the results. Because of tasting

Table 13.1 Example Sensory Specification for a Mainstream Pale Lager Beer

Flavor Category	Flavor Attribute	Flavor Chemical	Target Flavor Intensity (Scale 0–10)	Impact of Variability in the Intensity of the Attribute to the Flavor Characteristics of the Product	Origin of Flavor
Positive flavors	Body	Various compounds	4.2	High	Various
	Carbonation	Carbon dioxide	4.0	High	Finishing and packaging
	Bitter	Hop bitter acids	2.8	Medium	Hops | brewhouse operations
	Isoamyl acetate	Isoamyl acetate	2.3	High	Yeast | fermentation process
	Sweet	Sugars	1.8	Medium	Yeast | fermentation process
	Malty–biscuity	2-Acetyl pyridine and others	1.5	Medium	Malt | brewhouse operations
	Caramel	Furaneol and others	1.3	Medium	Malt | brewhouse operations
	Citrus hop	Linalool and others	1.2	Medium	Hops | brewhouse operations
	Sour	Various acids	1.1	Medium	Brewhouse operations | yeast | fermentation process
	Astringent	Various polyphenols	1.0	Medium	Hops | brewhouse operations
	DMS	Dimethyl sulfide	0.8	Medium	Malt | brewhouse operations
	Ethyl acetate	Ethyl acetate	0.8	Medium	Yeast | fermentation process
	Solvent alcoholic	3- and 2-Methylbutanol	0.7	Medium	Yeast | fermentation process
	Floral hop	Geraniol and others	0.7	Medium	Hops | brewhouse operations
	Grainy	Isobutyraldehyde	0.6	Low	Malt | brewhouse operations
	Acetaldehyde	Acetaldehyde	0.5	Medium	Yeast | fermentation process
	Ethyl butyrate	Ethyl acetate	0.5	Low	Yeast | fermentation process
	Ethyl hexanoate	Ethyl hexanoate	0.5	Low	Yeast | fermentation process
	Sulfitic	Sulfur dioxide	0.5	Low	Yeast | fermentation process

Continued

Table 13.1 Example Sensory Specification for a Mainstream Pale Lager Beer—cont'd

Flavor Category	Flavor Attribute	Flavor Chemical	Target Flavor Intensity (Scale 0–10)	Impact of Variability in the Intensity of the Attribute to the Flavor Characteristics of the Product	Origin of Flavor
Off-flavors	Acetic	Acetic acid	0.0	Medium	Yeast | fermentation process
	Butyric	Butyric acid	0.0	High	Brewhouse hygiene
	Damascenone	β-Damascenone	0.0	Medium	Hops | beer flavor instability
	Diacetyl	2,3-Butanedione	0.0	Medium	Yeast | fermentation process
	H_2S	Hydrogen sulfide	0.0	Medium	Yeast | fermentation process
	Isovaleric	Isovaleric acid	0.0	Medium	Hops | brewhouse operations
	Leathery	Various compounds	0.0	Medium	Various
	Lightstruck	3-Methyl-2-Butene-1-thiol	0.0	High	Exposure of beer to light
	Mercaptan	Methanethiol	0.0	Medium	Yeast | fermentation process
	Metallic	Iron, copper, manganese	0.0	High	Various sources
	Methional	Methional		Low	
	Onion	Dimethyl trisulfide	0.0	Medium	Yeast | fermentation process
	Papery	*trans*-2-Nonenal	0.0	High	Beer flavor instability
	Phenolic-4-EP	4-Ethyl phenol	0.0	High	Brewery hygiene
	Phenolic-4-VG	4-Vinyl guaiacol	0.0	Medium	Malt | brewhouse operations | brewery hygiene
Taints	Bromocresols	Various bromocresols	0.0	High	External contamination of product or process
	Bromophenol	Various bromophenols	0.0	High	External contamination of product or process
	Chlorocresols	Various chlorocresols	0.0	High	External contamination of product or process
	Chlorophenol	Various chlorophenols	0.0	High	External contamination of product
	Inky	Various trihalomethanes	0.0	Medium	Contaminated water
	Musty	Various bromoanisoles and chloroanisoles	0.0	High	External contamination of product or process

individual samples, or recognition of trends emerging from a series of results, the following actions could be taken:

- Selection and purchase of better raw materials or yeast strains;
- Improvement of recipes to address flavor issues;
- Improvement of production and packaging methods to address product consistency;
- Improvement of brewery cleaning and sanitation practices;
- Blending of beer prior to packaging to ensure a more uniform product batch to batch;
- Reprocessing of beer, or extending the duration of key process steps, to deliver the required flavor profile and flavor life;
- Destruction of beer prior to or after packaging because of the presence of taints or off-flavors;
- Withdrawal of beer from the market;
- Introduction of remedial actions to prevent a recurrence of problems (for example in the case of taints), including improved training of production staff.

EVALUATION OF RAW MATERIALS AND PROCESS AIDS

MALT AND ADJUNCTS

Malted barley and wheat are evaluated prior to delivery to the brewery to ensure the absence of flavor defects. It is important to reinspect them on receipt and after storage because off-flavors or taints can develop.

Sensory evaluation usually consists of informal assessments by two or three individuals. A dry sample is checked for the presence of taints or off-flavors. Flavors of particular concern include musty, moldy, and earthy notes associated with growth of molds or other microorganisms. In the case of specialty malts, the presence of key flavor attributes such as bready, caramel, candy sugar, maple syrup, chocolate, or roasted characteristics can be confirmed during such evaluations. An alternative is to make small-scale mashes from the malt and evaluate the flavor of the resulting wort.

Solid adjuncts, including maize and rice, are evaluated in a similar way to malt. In addition to musty, moldy, and earthy characteristics, areas of concern include rancidity (resulting from oxidation of oils) and smokey characteristics (from drying over wood), especially in the case of rice.

Hygiene issues in the production and storage of sugar syrups can lead to development of butyric acid which imparts an odor of baby vomit to beer. This is produced by *Bacillus* spp. and *Clostridium* spp. during production or storage of sugar syrups (Hawthorne et al., 1991). This flavor cannot be detected directly in sugar syrup. Dilution and acidification of the sample is needed to release the butyric acid into the headspace. Samples prepared in this way are usually evaluated for the presence of such defects by two or three assessors.

BREWING WATER

Water is used in the brewery for three main purposes: (1) Brewing and dilution of beer after fermentation; (2) cleaning and rinsing of tanks and pipework; (3) cleaning of the brewery environment; and (4) supplying utilities, such as steam generators, tunnel pasteurizers, etc. Water used to make beer, or which is likely to come in contact with beer, must be free of undesirable tastes and odors, as well as complying with chemical, physical, and microbiological specifications.

Brewery water samples should include: (1) Incoming water; (2) dechlorinated water (if the incoming water is chlorinated); (3) deaerated water used for dilution. Typically, the procedures used are "informal." Two or three assessors evaluate samples to ensure samples are defect free. Defects should be described and identified.

HOPS AND HOP PRODUCTS

The main risk to beer flavor posed by hops and hop products is development of an isovaleric flavor, reminiscent of cheese. This arises from degradation of hop alpha acids. Use of large amounts of hop material late in the boil, or in fermentation, can also lead to development of a flavor character reminiscent of black tea or red wine, caused by β-damascenone. Astringent and metallic characteristics also pose a risk, together with lingering, harsh, or aspirin-like bitterness.

YEAST

Brewers' yeast of good quality has a neutral odor, being slightly estery and alcoholic with a distinctive B-vitamin character. However, yeast in various stages of degradation develops a number of distinctive odors and tastes. These include flavors imparted by volatile sulfur compounds such as H_2S, methanethiol, methional, and the breakdown products of water-soluble vitamins; aldehydes such as acetaldehyde; esters, in particular ethyl hexanoate; fatty compounds such as hexanoic, octanoic, and decanoic acids; and bitter-tasting peptides derived from enzymic breakdown of yeast proteins. Yeast which is contaminated with foreign microorganisms can develop phenolic flavors due to 4-vinyl guaiacol, 4-ethyl guaiacol, and 4-ethyl phenol, as well as sour notes from acetic and lactic acids.

Regular tasting of yeast slurries can warn of emerging yeast health issues which can be addressed by propagation of a new culture, acid washing, or adjustment of fermentation process parameters.

FILTER AIDS AND ADSORBENTS

Due to their porous nature filter aids such as diatomaceous earth and perlite, as well as adsorbents such as silica hydrogels, silica xerogels, and polyvinyl polypyrrolidone can adsorb odor-active materials from the environment and subsequently contaminate beer.

The taints most commonly associated with filter aids and adsorbents are airborne. They include musty notes caused by 2,4,6-trichloroanisole and 2,4,6-tribromoanisole; earthy notes caused by 2-methylisoborneol, geosmin, and 2-isopropyl-3-methoxypyrazine; and metallic notes from iron. Occasionally, disinfectant-like taints arise from chlorophenols, bromophenols, and chlorocresols.

To protect against such issues, filter aids and process aids should be transported and stored in clean odorless environments and stored in sealed bags.

Samples should be slurried in water prior to evaluation by two or three assessors. Diatomaceous earth powder should not be smelled directly in the dry state. The powder is carcinogenic when ingested via the lungs, so appropriate personal protective equipment must be worn when handling the dry filter-aid powder.

PROCESS GASES SUCH AS CARBON DIOXIDE, AIR, AND OXYGEN

Process gases represent a significant risk of taints. Carbon dioxide recovered from fermentation contains numerous undesirable flavors, including dimethyl sulfide (DMS), hydrogen sulfide, methanethiol, and acetaldehyde. Recovery systems for this gas are designed to remove such flavors through a combination of water scrubbing, oxidation, and adsorption. However, if the process is not carefully monitored, it can fail, resulting in contaminated carbon dioxide.

In the case of air and oxygen, the risk relates to traces of lubricant being contributed to beer. However, if beer or diluted beer is allowed to enter the air or oxygen supply lines, growth of microorganisms can lead to formation of musty, moldy, and earthy flavors which can contaminate wort during aeration.

Gases should be sampled for sensory evaluation by bubbling through sensory-pure water. Typically, the procedures used are "informal." Two or three assessors evaluate samples under blind conditions to assess whether the samples are free of defects. Defects should be described and identified.

EVALUATION OF IN-PROCESS SAMPLES

The key stages for evaluation of beer in-process are:

- Fermentation vessels at the end of fermentation;
- Maturation vessels at the end of maturation (or at various times during maturation in which a time-sensitive process has to be monitored);
- Filtered beer samples;
- Bright-beer tank samples.

Evaluation methods for in-process tasting are often empirical and do not meet the exacting standards of "officially-sanctioned" sensory procedures.

Examples of poor practice in such tests include the following:

- Sample coding and order of sample presentation are often poorly controlled;

- The number of assessors evaluating each sample is small;
- Assessors receive no formal training in evaluation of in-process samples;
- Assessments may be made by consensus rather than individually;
- Evaluations are carried out in the midst of the production environment rather than in a dedicated sensory laboratory—noise and ambient odors in such environments affect the ability of assessors to concentrate on the task in hand;
- Positive and negative control samples are often lacking.

Brewers would be well advised to consider these points to assure the performance of such tests.

EVALUATION OF PACKAGING MATERIALS

Primary packaging such as bottles, bottle caps, labels, cans, can lids, and kegs, and secondary packaging such as paper wraps, cardboard boxes, sleeves, and plastic keg spear caps are all potential vectors of taints in beer. Sources of taints include contaminated raw materials used to manufacture the packaging materials, inadequate curing of resins and inks, and adsorption of odors from the atmosphere.

Regular sensory evaluation of packaging materials is key to management of such risks. Materials should be assessed at the time of arrival at the brewery and close to time of use. Their odor can be evaluated directly. Alternatively, the materials can be rinsed with odor-free water which can then be assessed evaluated. The best way to perform such tests is to package a single batch of beer in both lots of packaging materials. Both types of samples should then be evaluated by difference testing when fresh and after storage.

EVALUATION OF PACKAGED BEER

Beer becomes increasingly valuable as it passes through the production process. The cost of each hectoliter of beer can double as it transfers from bright-beer tank into package, so quality defects are best identified and dealt with prior to packaging.

Most breweries taste all bright-beer tanks prior to release to packaging. Beer has to pass this "quality gate" prior to packaging. This tasting may be carried out by only two or three tasters. Use of only a few assessors carries with it a degree of risk, because one or more of the tasters may be blind to one or more flavor faults present in the product. This risk can be managed in one of two ways. Either the number of tasters can be increased (to 8–12) or tasters of higher skill level can be used to evaluate the samples. Most importantly, the selection of tasters for assessing such samples in a single session should be based on them having no overlapping anosmias, ie, if one taster is blind to diacetyl, the other two tasters should not be blind to this compound.

PRODUCT MANAGEMENT

MANAGEMENT OF PRODUCT CONSISTENCY

Brewers have to be confident of producing beer of consistent quality even in times when the raw materials used to make them may vary.

Difference tests can establish whether there are differences in the flavor of two or more samples. They include the triangle test, duo–trio test, paired-comparison test, and two-out-of-five test. Detailed descriptions of such methods can be found in textbooks on sensory analysis (Carpenter et al., 2000; Kemp et al., 2009; Kilcast, 2010; Lawless, 2013; Lawless and Heymann, 2010; Meilgaard et al., 2007; Stone and Sidel, 1993) and in relevant international standards [American Society of Testing and Materials (ASTM, 1997; ASTM, 2001b); International Organization for Standardization (ISO, 1983; ISO, 2003; ISO, 2004)].

Such methods are well established, but often misused. Details can be overlooked in an effort to save time. These methods are intended to detect differences between samples; different statistical measures must be applied if they are to be used to assess similarity between samples. For example, when attempting to approve a new supplier of bottle caps, the expectation (hypothesis) is that the new supply of caps will behave like the old one. Thus, similarity test formats are a more appropriate methodology in such cases.

MANAGEMENT OF PRODUCT CONFORMANCE

Descriptive analysis techniques are among the most widely used in the brewing industry. Several textbooks (Kemp et al., 2009; Kilcast, 2010; Lawless and Heymann, 2010; Meilgaard et al., 2007; Stone and Sidel, 1993), international standards (e.g., ISO, 2003), and practical workbooks (Carpenter et al., 2000; Lyon, 2002) describe how descriptive analysis can be carried out.

In many breweries, descriptive profiling methods deviate from published procedures. Elements of several methods are assimilated, with individual procedures differing in detail from one company to another. Arguably, the most obvious difference between descriptive analysis as practiced in breweries and implementations of the technique that are more rigorous is the lack of replication in brewery tests. Best practice typically involves three replicate measurements being made on each sample: most breweries make only one. Fig. 13.2 outlines the descriptive profiling procedure used in our company for evaluation of beer.

Use of an internationally agreed set of flavor terms is central to descriptive analysis techniques in breweries. First introduced in the 1970s (Meilgaard et al., 1979), this terminology system is today approved by several international bodies, including the *American Society of Brewing Chemists (ASBC)*, *Brewing Congress of Japan (BCOJ)*, and *Master Brewers Association of the Americas (MBAA)*. The "poster child" for this terminology system is the "Beer Flavor Wheel." The heart of the system, however, lies in the 122 individual flavor attribute names which cascade from

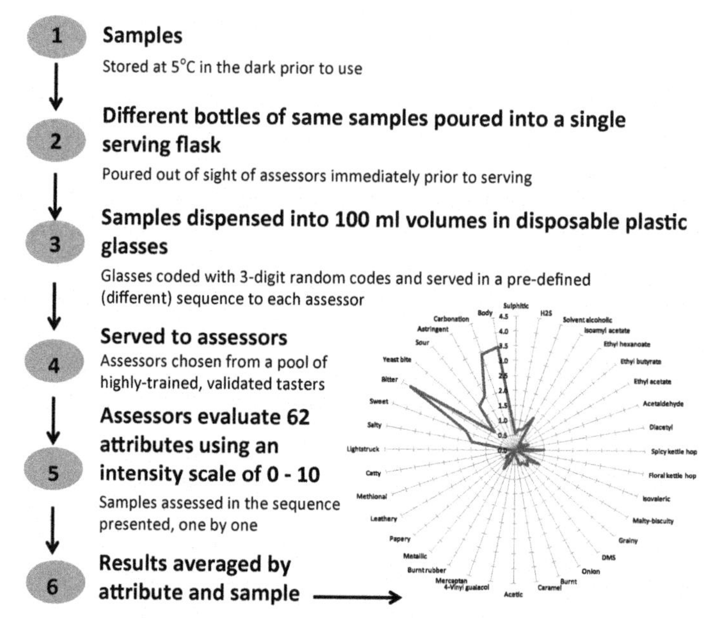

FIGURE 13.2

Example of a descriptive profiling procedure used to evaluate beer. In our company, we use this procedure, replicating each sample three times and including a variety of "decoy" and "reference" samples in the test set to minimize the risk of assessor expectation bias.

this top tier of 44 terms. Reference flavor standards to represent some of these attributes were developed in a large intercollaborative exercise some 40 years ago (Meilgaard et al., 1982). Today, these original references, and many more, are commercially available in stabilized form for routine use in breweries.

"Trueness-to-type" tests (Anon, 1995) are used by some breweries. Such tests assess the conformity of individual beer samples to tasters' views of the "ideal" example of the product. The technique involves carrying out descriptive analysis using a "just-about-right" scale for each attribute: the intensity of each attribute is scored relative to that of an "ideal" product. Off-flavors and taints are scaled using an absolute scale. To allow a comparison to "ideal," the identity of the product must be revealed to assessors prior to tasting. A single metric to represent product flavor conformance is calculated from the responses.

Scaling of attribute intensity is integral to many sensory procedures. Sometimes there is a need to estimate the intensity of single characteristics such as bitterness. In techniques such as Quantitative Descriptive Analysis, the intensity of many attributes has to be estimated on a single sample. Attribute intensity can be estimated using a variety of approaches (Fig. 13.3), all of which have their strengths and weaknesses. The rank-rating method (Kim and O'Mahony, 1998) is particularly useful in brewery applications. Assessors are tasked with ranking a set of samples in order of

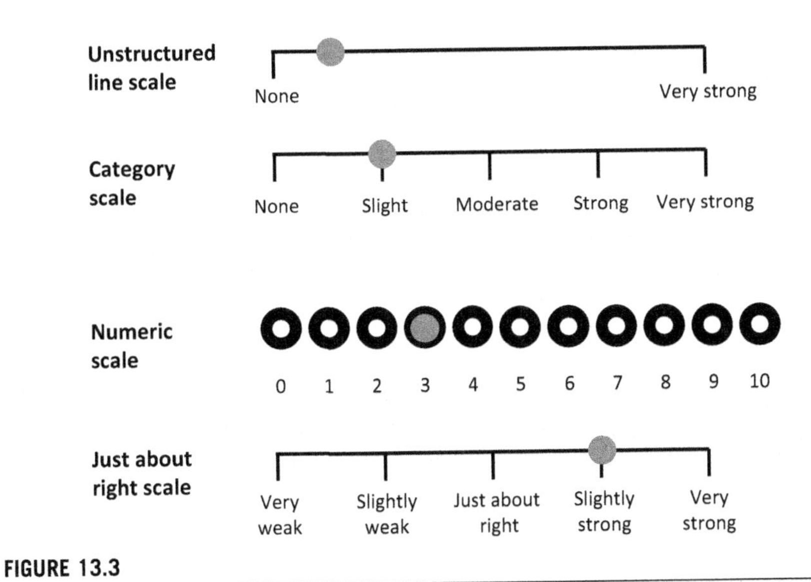

FIGURE 13.3

Examples of different types of response scale which can be used in sensory assessment of beer.

attribute intensity before rating attribute intensity. The advantages of this method are: (1) Assessors make fewer errors with this method compared to when conventional scaling procedures are used; (2) the method facilitates good discrimination and scaling of samples regardless of the type of scale used; (3) assessors have to repeatedly assess attribute intensity using this method, reducing errors, and increasing discrimination between samples.

In the difference-from-control test (Thompson, 1999), assessors receive a control sample together with one or more test samples. They are asked to rate the size of the difference between each sample, and the control for a chosen attribute. The mean difference-from-control values for each sample and for the blind-control sample are calculated and analyzed by ANOVA or with a paired *t*-test. This test has proven a useful way of estimating the degree of difference between the same product brewed and packaged in different breweries.

MANAGEMENT OF PRODUCT QUALITY

Objective assessment of "commercial quality" of beer is an important aspect of sensory analysis. Sensory specialists have to advise whether a batch of beer may lead to consumer complaints and, in the worst case, whether a product recall might be required.

Brewers use a range of methods to assign a quality grade to beer. In the simplest approach, assessors are asked to rate the quality of each sample, for example, using a

0–10 scale. This may be done in isolation, without assessment of other attributes. Quality may also be rated in parallel to assessment of individual flavor attributes, for example when carrying out Quantitative Flavor Profiling or Quantitative Descriptive Analysis.

A different approach combines allocation of a quality score for each sample together with an indication of nonconformances in the flavor profile of the sample. Specifically, the key nonconformances which have led to a less than perfect score should be indicated by the assessor. An example of a taste form used in such tests, together with example results, is shown in Fig. 13.4.

In an attempt to make quality assessment more objective, some breweries calculate their quality scores from descriptive profiling data (Axcell, 2008). High scores are awarded for beers which match the product flavor specification. Lower scores are allocated to beers which differ from the target profile or which contain off-flavors and taints. Specialist software handles the rules base required

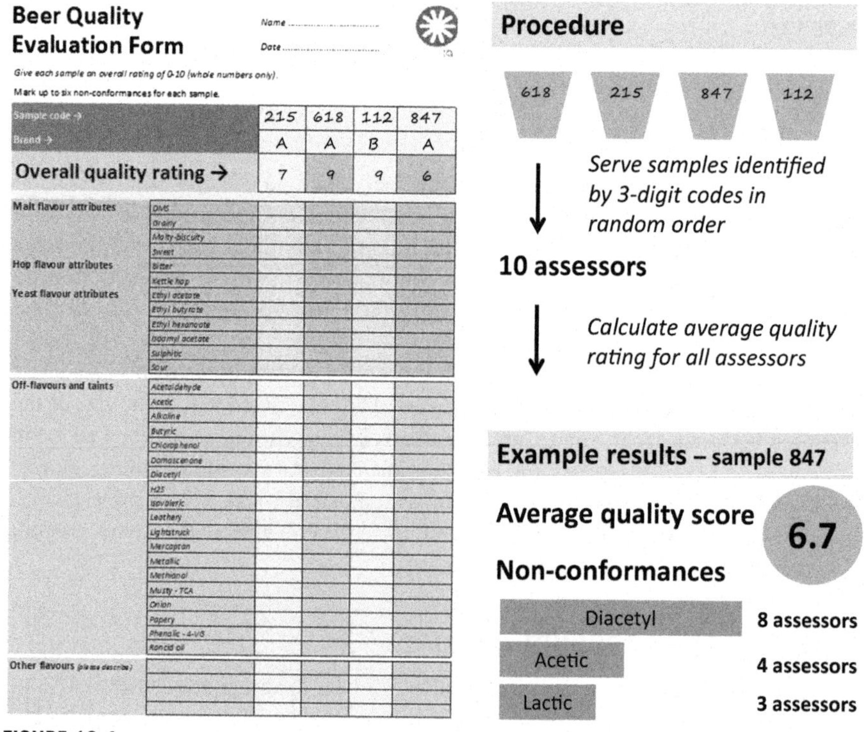

FIGURE 13.4

Taste form and procedure used to evaluate beer-flavor quality. The result from this test comprises a quality rating (scale of 0–10, in which 10 is excellent) and a ranked listing of the main flavor nonconformances.

for such sensory specifications together with comparisons of each test sample against the specification.

Point of Sale

The "moment of truth" for beer-flavor quality comes when the consumer drinks the product. The closer to this point that sensory evaluation can be conducted the better. For this reason, some breweries evaluate samples from the market, selecting samples which are typical of those purchased by consumers in terms of beer age and storage conditions. These are evaluated by the brewery taste panel, alongside "brewery-fresh" samples. Ideally, the flavor characteristics of both sets of samples should be indistinguishable.

Depending on the temperatures to which the beers are exposed, and the time spent at high temperatures, beers can develop a variety of age-related flavors. These include papery, leathery, aldehydic, caramel, metallic, stewed fruit, and astringent characteristics. Depending on the beer style, these changes may be viewed as negative or positive.

Rather than labeling products with arbitrary "best before" dates, it is arguably better for breweries to evaluate the flavor stability of their beers under market conditions using a trained taste panel. Having acquired reliable data, changes to beer production and packaging can be made to optimize the balance between declared shelf life and beer stability under market conditions.

TASTERS

All sensory tests rely on use of competent assessors. Competence may relate simply to following the test instructions without deviation. But for many tests, specific tasting skills are required. These include:

- Ability to discriminate samples of similar flavor on the basis of taste and odor;
- Ability to identify different flavor attributes in beer at commercially significant levels;
- Ability to rate the intensity of flavors in beer;
- Ability to rate the quality of beer samples, either in isolation, or by reference to a predetermined product or beer style sensory specification or reference sample;
- Ability to perform with a high degree of consistency.

An overview of the process by which tasting pools can be developed is shown in Fig. 13.5.

With the growth of craft beer, the requirement for tasters to be familiar with flavors associated with different styles has grown. In addition to self-study, knowledge and expertise can be acquired with the help of organizations such as the Cicerone Certification Program (www.cicerone.org) in the United States and Canada; Doemens Academy (www.doemens.com) in Germany; and the Beer Academy (www.beeracademy.co.uk) in the United Kingdom.

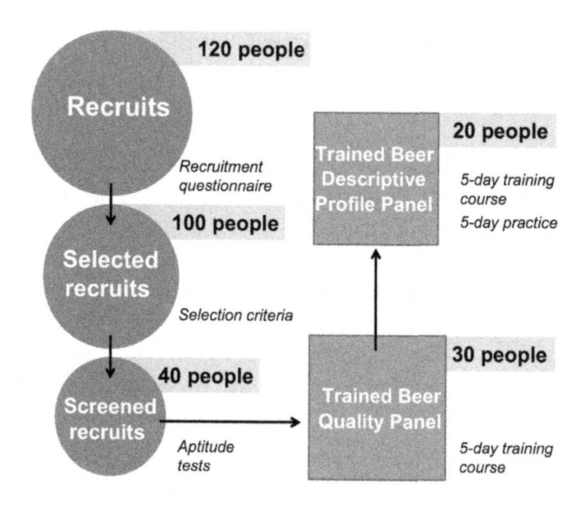

FIGURE 13.5

Overview of the process by which a pool of highly trained assessors can be developed to rate the quality of beer and describe its flavor.

RECRUITMENT AND SELECTION OF CANDIDATE TASTERS

Historically, breweries have recruited tasters from within their own ranks. An increasing number of breweries employ part-time workers, specifically selected, trained, and employed to taste beer.

The approach to selecting someone with the potential to be a productive and competent brewery taster is no different to that used to fill any skilled role. Rather than recruiting potential trainees at random, the "chosen few" should be selected from a larger group using a number of relevant, well-defined criteria. A larger talent pool inevitably delivers better quality recruits. Thus, the best 10 tasters chosen from a pool of 200 will always outperform the best 10 tasters chosen from a pool of 20.

The advantages associated with recruiting from a larger pool include:

- Assessors are easier to train and less dependent on having a skilled teacher to achieve good results;
- They perform well in assessments during training;
- They perform well in routine tasting after training;
- The higher levels of motivation and enjoyment they experience from their success usually results in high levels of taste-panel attendance.

What types of people make good beer tasters for routine deployment in breweries?

- They should be routinely available for tasting duties—heavy travel commitments or other limitations on attendance make it difficult to justify the commitment in time and money needed to develop a candidate taster's skills;
- Because the samples contain alcohol, assessors should be of legal drinking age;
- They should like beer;

- They may be regular beer drinkers, but not necessarily so;
- They should have an interest in flavor, food, and beer;
- They should have good mental abilities, able to focus their attention on the task at hand;
- They should be in good health and, in particular, not predisposed to, or suffering from, alcohol-related health problems or major impairments to their sense of smell or taste;
- They should not be taking medicines that impair their sensory capabilities;
- They should not suffer from allergies or sensitivity to beer components such as sulfur dioxide or gluten.

Questionnaires used to recruit candidate tasters should:

- Encourage candidates to become professional beer tasters;
- Provide a basis for selection of candidates based on interest, motivation, and availability;
- Provide a first-level screen to identify those whose health might be placed at risk by participation in professional beer tasting due to allergies, food intolerances, or other medical conditions.

Recruitment questionnaires should comprise questions that elicit quantitative responses rather than comments. This allows selection decisions to be made in a consistent, transparent way.

Tools such as CAGE and AUDIT screening questionnaires (McCusker et al., 2002) can be used to estimate an individual's risk of developing alcohol-related health disorders. Although such tests have not yet been widely used in recruitment of beer tasters, they may have a place in protecting tasters from the risk of alcohol-related health disorders, and breweries from the risk of litigation.

SCREENING OF CANDIDATE TASTERS

After selection, candidate tasters should be screened to identify those most likely to perform best in routine tasting. Screening allows those with inherent defects in their sensory acuity [including anosmia (specific smell blindness) and ageusia (specific taste blindness)] to be identified. The focus should not be on current competence. In fact, in a well-devised screening test, assessors who have already received some training should have no advantage over novices.

For such tests to have the predictive power expected of them, the way in which sessions are run during screening must be similar to how they will be run during training. An example of a screening protocol used to assess the potential of candidate beer tasters is shown in Fig. 13.6.

TRAINING OF TASTERS

Training of professional beer tasters has undergone a radical transformation over the last 20 years. This is perhaps best understood by analogy, relating the process of learning to taste beer to that involved in learning a new language.

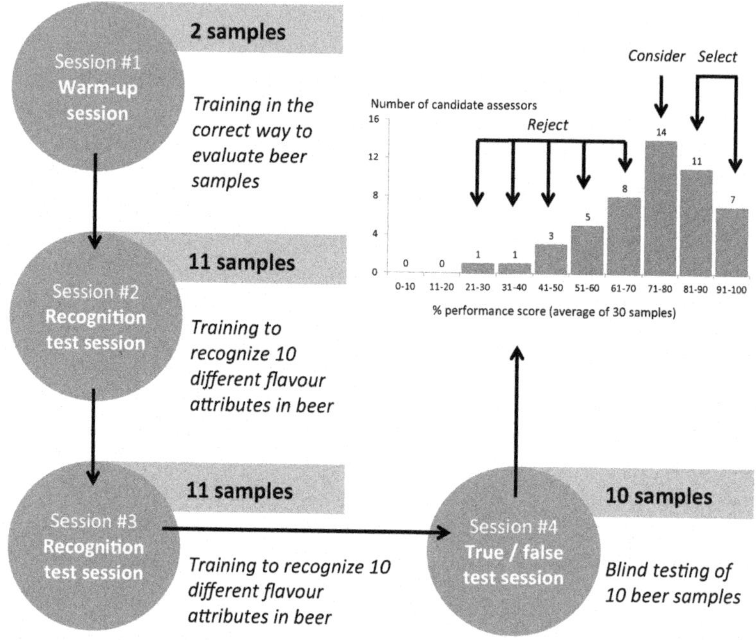

FIGURE 13.6

Example of a half-day module for aptitude assessment (screening) of candidate beer tasters. Only the highest-scoring individuals are chosen for further training.

There are two possible approaches to learning a new language:

- Live among users of the language and learn to speak, read, and write it by looking, listening, and copying what is seen and heard—we can term this method "passive learning";
- Take an intensive course in the language, learning the vocabulary and grammatical rules, practicing what has been learned, and continually taking tests and receiving feedback on the progress of learning—we can term this method "active learning."

Both approaches work. Millions of adults learn new languages in both ways every year. But there are differences in the time needed to learn languages in each way and in the level of skill which can be developed.

Passive learning takes a long time, typically years to develop the skill needed for conversation and paid work. Development of written communication skills are likely to lag behind the spoken word.

Active learning is usually faster. An intensive language course can allow someone to achieve a good level of spoken and written competence within weeks or months.

In terms of the skill level developed, and the potential to build on that skills base in the future, most would agree that active learning provides a better foundation, especially for written language, because the rules of grammar and spelling are likely better understood.

Another important difference is the issue of dialect. When assimilating language skills from native speakers, it is inevitable that a local dialect will form part of the learner's use of the language. In contrast, when the language is learned actively in a systematic way, dialect is less likely to feature.

Finally, we can compare the two methods in terms of proven competence and test results. In passive learning, the opportunity for feedback on performance from friends, colleagues, and strangers diminishes over time. In fact, it may well be considered rude to correct someone's grammar or their choice of one word over another. The objectivity of this feedback can also be questioned. In contrast, in active learning, tests and assessments provide objective feedback on competence at all stages of the learning process.

The "traditional" way that people learned to taste beer had much in common with the passive learning of languages referred to previously. Trainee assessors tasted alongside experienced ones. Over time, trainees learned to mimic the behavior of experienced colleagues. Patterns were recognized by trainees—beers referred to by experienced tasters as being bitter would, over time, become recognized by trainees as having a different taste from other samples. Eventually, tasters would learn to describe that taste as "bitter." Beers described as having hoppy characteristics would be recognized as distinctive, and use of the term "hoppy" would be reinforced by the opportunity to smell real hops in the hop store or in the brewhouse.

But the difficulties that beset people who learn a new language from their peers also affect trainee beer tasters:

- The process takes a long time; and
- The trainee can only be as good as the people whose behavior they learn to mimic.

In contrast, systematic, active learning allows people to develop their beer-tasting skills faster and to reach a level of skill which exceeds that of the people they have learned from, provided they are capable of doing so.

So how can we actively train professional beer tasters? The use of standardized training session formats is central to the active learning approach (Larsen et al., 2005). In our own taster training activities we use a wide range of training session types. Some of these are listed in Table 13.2.

Sessions can be customized by including different beer styles or brands, and by use of different beer flavors at a range of flavor intensities. Individual sessions can then be sequenced to form training modules. Because each training session takes 60–90 min, regardless of whether the number of trainees is 5 or 25, typically five or six sessions can be covered in a single day's training. For example, with novice trainees who have been selected and screened as described earlier, a 25-session program which aims to train assessors how to objectively evaluate the flavor quality of

Table 13.2 Examples of Types of Training Session Used to Develop Beer Tasting Skills in Breweries

Session Type	Session Description	Purpose
Recognition test	Ten beer samples, each containing a single-added flavor are presented to assessors one by one together with a reference sample. Assessors are coached extensively in how to detect and identify each flavor in beer. After the training session, the performance of each individual assessor with respect to identification of each flavor is evaluated in a blind test. Points are awarded for each correct answer.	To develop assessors' skills in recognition of individual flavors in beer. To teach assessors the most appropriate specific physical detection methods for each flavor.
Beer style matching test	Four different beer styles are presented to assessors to allow them to familiarize themselves with the flavors associated with each of the samples. Samples are removed, then six blind-coded samples are presented. Assessors have to identify the beer style of each sample in a blind test. Points are awarded for each correct answer.	To develop assessors' skills in discrimination of different beer samples and recognition of different beer styles.
Flavor matching test	Beers flavored with four different reference standards are presented to assessors to allow them to familiarize themselves with the flavors associated of each of the samples. Samples are removed then six blind-coded samples are presented. Assessors have to identify the beer flavors in a blind test. Points are awarded for each correct answer.	To develop assessors' skills in discrimination of different beer samples and recognition of different beer-flavor attributes.
True–false test	Ten beer samples are presented to assessors. Some are control beers, others contain one or more quality defects introduced using reference flavor standards. Assessors are invited to agree or disagree with a statement about each sample using a response form. For example, "this sample has an estery flavor." Points are awarded for each correct answer.	To allow assessors to make use of the skills they have developed in other training sessions. To develop assessors' skills in relation to beer-quality assessment.

Table 13.2 Examples of Types of Training Session Used to Develop Beer Tasting Skills in Breweries—cont'd

Session Type	Session Description	Purpose
Flavor recognition test	Ten beer samples, each containing a single-added flavor are presented to assessors all at once together with a reference sample. Assessors have to identify the flavor in each sample from a pick list of 30–40 attributes. Points are awarded for each correct answer.	To evaluate assessors' skills in identification of individual flavor attributes at the chosen intensity. To identify common confusions arising in such evaluations so that they can be corrected in subsequent training sessions.
Stop–go test	Ten beer samples are presented to assessors. Some samples are control beers, others contain one or more quality defects introduced using reference-flavor standards. Assessors have to decide whether each sample is good enough to package or release to market (go), or whether it should be put on hold (stop) pending a more detailed evaluation of its flavor quality.	To develop assessors' skills in assessment of product quality and detection of flavor defects.
Quality rating test	Ten beer samples are presented to assessors. Some samples are control beers, others contain one or more quality defects introduced using reference-flavor standards. Assessors have to rate the quality of each sample in relation to their expectations for the product or style using a scale of 0–10, in which 10 represents a perfect example of the product.	To develop assessors' skills in assessment of product quality and detection of flavor defects.
Quality evaluation test	Five beer samples are presented to assessors. Some samples are control beers, others contain one or more positive flavors or flavor defects introduced using reference-flavor standards. Assessors have to rate the quality of each sample in relation to their expectations for the product or style using a scale of 0–10, in which 10 represents a perfect example of the product. In addition, assessors have to mark up to five nonconformances from a list of 20–40 flavor attributes. These nonconformance flavors provide an explanation for less than perfect quality scores.	To develop assessors' skills in assessment of product quality and detection and identification of flavor defects.

Continued

Table 13.2 Examples of Types of Training Session Used to Develop Beer Tasting Skills in Breweries—cont'd

Session Type	Session Description	Purpose
Rank-rating test	Five, eight, or ten samples are presented to assessors. These contain an identical base beer but differ in the concentration of a single-flavor compound introduced using reference-flavor standards. Some concentrations of the flavor are replicated, others are not. Some samples contain no added flavor. Assessors have to first rank the samples in order of the intensity of the added flavor. They then have to scale the intensity of the added flavor attribute in each sample, using a scale of 0–10, in which 0 indicates the absence of the attribute. Performance scores for each individual assessor are calculated from a regression analysis of intensity scores against the concentration of the added flavor compound.	To develop assessors' skills in assessment of attribute intensity and discrimination between samples of similar flavor.
Descriptive profile test	Four or five samples are presented to assessors. Some samples are control beers, others contain one, or more positive flavors, or flavor defects introduced using reference-flavor standards. Assessors have to rate each of 20–50 flavor attributes in the descriptive profile form using a scale of 0–10, in which 0 indicates the absence of the attribute.	To develop assessors' skills in descriptive profiling of samples.

beer can be completed in 5 days. Once assessors have reached this standard, an additional 25 sessions delivered during 5 days of training can develop the advanced skills needed to allow the tasters to carry out descriptive profiling of beer. Having acquired the core skills, further practice is needed for assessors to consolidate these skills. This helps them to evaluate samples more quickly and with a greater degree of certainty.

The second component of this type of training is the use of reference flavor standards to develop skills in attribute recognition and scaling. Perhaps the greatest deficiency associated with passive learning of beer-flavor evaluation was the ambiguity inherent in learning activities. The fact that the "teacher" might describe a beer as a having a "fruity" character was no guarantee that: (1) The trainee would be find the same sort of fruity character in the beer; or (2) they would be genetically capable of detecting the compound responsible for the fruity character detected by the first assessor. In addition, different teachers, and therefore different trainees might use different terms for the same flavor characteristic, or the same term for two different characteristics.

The most important aspect here is the fact that assessors identify and rate the intensity of flavor notes at the level of specific, individual chemicals. In many cases, assessors can discriminate between optical isomers of the same chemical—one of the best-known examples is our ability to discriminate D-carvone and L-carvone, which smell of peppermint and spearmint, respectively. Thus, whereas we may think we are detecting a "banana-like" note in beer, we are in fact detecting isoamyl acetate, a specific flavor compound. As explained earlier, it is essential that assessors are trained using the actual compounds that impart such flavor characteristics to beer, rather than compounds which smell similar to them. Stabilized reference flavor standards provide a way of achieving this.

Notwithstanding the importance of the chemical aspects of training, it is important not to overlook the psychological aspects. Constant feedback to assessors on their performance builds confidence, and confidence builds performance.

MANAGEMENT OF TASTE PANELS

The role of the taste-panel leader involves:

- Recruitment and selection of candidate assessors;
- Screening of candidates to assess their aptitude;
- Training and coaching of assessors;
- Management and administration of routine taste sessions;
- Analysis and reporting of test results;
- Responsibility for taster welfare.

Management of brewery taste panels requires a mix of skills and behaviors. Panel leaders need to run sessions in the same way every time, with no deviation in procedure from one day to the next. They need to make every effort to eliminate sources of bias during assessments. This requires strength of character to suppress conversation or the answering of telephone calls during taste sessions. But panel leaders also have to be great motivators, providing encouragement and feedback to assessors, if they are to get the best out of the panel.

Above all, panel leaders need to understand the importance professional beer tasting plays in the success of the brewing enterprise and the assessors in their care have to share that belief.

It is not surprising that breweries which perform best in sensory analysis activities are those which have the most able taste-panel leaders. When standards of tasting fall below expectations, the problem often lies with the attitude, skills, and credibility of the person occupying this important role.

MANAGEMENT OF TASTER WELFARE

Guidance on ethical practices for sensory analysis has been published by the Institute of Food Science and Technology (Anon, 2010). A guidance note from the ASTM (ASTM, 2001a) addresses issues specific to alcoholic beverages. The main points for breweries to be aware of are:

- Sensory tests involve human subjects, so the scope of tests and authorization to sanction them should be defined in a written policy;
- Participation in tasting should be voluntary, although the specific arrangements between taster and taste session organizer can be contractual or ad hoc;
- Organizers of taste sessions have a legal liability toward their tasters and to the public;
- Tasters should be made aware of the risks associated with tasting;
- Individuals for whom alcohol might be harmful, including pregnant women, should be excluded from tasting activities;
- Sample volumes should be minimized as far as possible and volumes consumed recorded;
- In some jurisdictions, government agencies may have an interest in serving and tasting of alcoholic beverages—it is essential that you keep up-to-date on any rules and regulations which may apply.

Routine health checks on tasters can include tests on nonspecific indicators of liver function or measurement of fatty acid esters found in hair (Simpson, 2006). The latter allows alcohol consumption patterns of individuals to be monitored over time, even after a period of abstinence, and can provide a bank of evidence for later analysis in cases in which self-declared fitness for tasting is called into question. Ultimately, genetic screening may provide the best protection against tasting-related health risks. Tests to identify at-risk individuals are available (Tabakoof et al., 2004), but, in common with other forms of genetic testing, obstacles to their use remain.

ASSESSMENT OF TASTER COMPETENCE

Twenty years ago, the competence of professional beer tasters was expressed in terms of their experience. A taster with many years' tasting experience was considered more competent than one with several months' experience. Of course, the assumptions on which such generalizations were based are flawed. Experience and competence are not synonymous.

Before considering how best to assess competence, let us consider specific aspects in which taster performance can be found wanting. Poor taster performance can result from:

- Confusion of one flavor with another;
- Differences in how assessors use intensity scales;
- Differences in how assessors perceive one or more flavor attributes (including complete blindness to some flavors);

- Inconsistencies in sample assessments;
- Susceptibility to distraction and bias.

Several "ring analysis" or "intercollaborative" schemes are available to measure beer-taster competence. Some involve supply of samples of one or more batches of beer to different brewery sites. Taste-panel members evaluate these samples and the results are then statistically evaluated, comparing those generated by different tasters and different taste panels.

Other taster performance-assessment schemes rely on the use of reference flavor standards to prepare beers with consistent, defined sensory profiles from those made locally. They differ from the "reference" beer from which they have been prepared in the level of one or more flavor attributes. Assessors have to identify the attribute(s) present in different samples, choosing from a pick list of 30–40 attributes (described using the official terms of the ASBC, BCOJ, or MBAA). The results of each test are entered into a secure online database. The resulting data can be analyzed in a variety of ways (Fig. 13.7).

Statistical procedures are available which allow the competence of individual tasters to be established from analysis of the results of replicate samples in routine

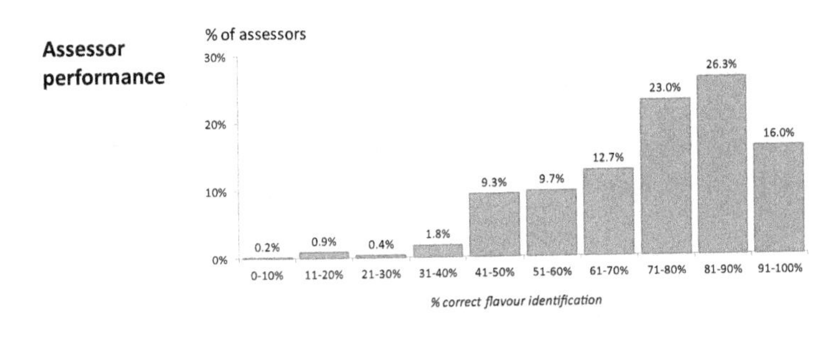

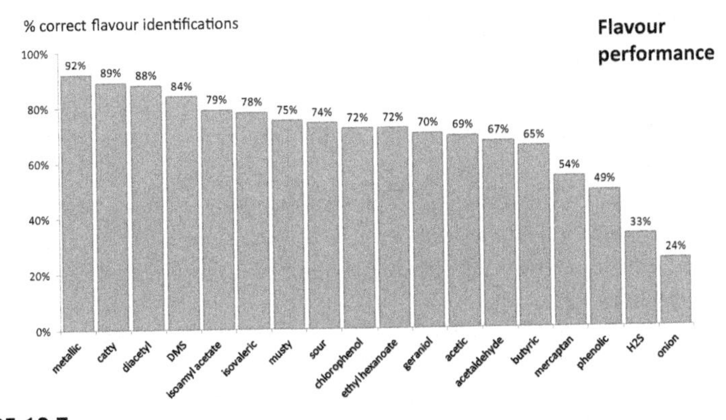

FIGURE 13.7

Example results from the validation of assessor competence using beers prepared with reference beer-flavor standards.

taste sessions (Cliff and Dever, 1996; Piggott and Hunter, 1999). The aim of such analyses is to measure:

- The repeatability of the results of each assessor for each attribute (at least two replicates are needed to carry out such analyses);
- The degree of agreement between the mean attribute score for each assessor compared to the mean value for the panel;
- The degree of discrimination that an assessor achieves between closely separated levels of individual attributes;
- The way in which the assessor uses the intensity scale compared to how the panel uses the scale.

A high-performing assessor shows a high degree of repeatability for assessment of all attributes, together with a high degree of discrimination. They use the scale in a way that is representative of the panel as a whole. By constructing a graph of degree of discrimination against repeatability for all attributes, the relative performance of each assessor for each attribute can be compared (Naes, 1990; Naes and Solheim, 1991). By plotting such data in the form of control charts, trends in performance can be identified. "Eggshell" plots can be used to gain further insight into assessor performance. These make use of ranking data or continuous data converted to ranks. They have the advantage that, being based on nonparametric statistics, they do not rely on underlying assumptions concerning the distribution of the data.

In my view, such techniques should be part of the routine work of all brewery sensory analysts. An excellent free-software tool to carry out such analyses is available (www.panelcheck.com).

PRACTICAL ASPECTS OF BEER TASTING

TASTE ROOM FACILITIES

The taste room environment should allow assessors to concentrate on the task in hand with no risk of distraction. It should provide everything needed by the taste-panel leader to obtain, store, prepare, serve, and dispose of samples, and execute a range of sensory tests.

So where do you start? Regardless of whether the taste room is part of a new building or a modification of an existing one, it is important from the outset that the sensory analyst works closely with architects and building contractors to define the current and future needs.

To aid this working relationship, the development of a detailed body of information is needed, including details on:

- The amount and location of available space required;
- The volume of current and anticipated tasting on a daily, weekly, or monthly basis;
- Current and anticipated types of products which need to be tested;
- Current and anticipated test methods; and
- Current and future availability of tasters.

Such details assist in planning of a new facility, or upgrading of an existing one. The key requirements for the brewery taste room are listed in Table 13.3 with an example of assessors evaluating samples within taste booths shown in Fig. 13.8. Further details can be found in the ISO standard relating to this topic (ISO, 1988) and in textbooks on sensory analysis (Lawless and Heymann, 2010; Meilgaard et al., 2007; Stone and Sidel, 1993).

PLANNING OF TASTE SESSIONS

Taste sessions require meticulous planning if they are to produce valid results as efficiently as possible. Considerations include:

- The samples to be tasted;
- The type of test which has to be carried out to achieve the objective of the analysis;
- The number of replicates to be tasted;
- The number of samples which will be presented to the assessors in each test session;
- The number of test sessions;
- The number of assessors required for the test;
- The assessors who have to be invited to the test;
- The date and time at which the test will take place;
- The order in which the samples are to be presented to each assessor;
- The codes which will appear on each glass presented to each assessor;
- The temperature at which the samples will be presented to assessors;
- The volume of sample which will be presented to each assessor;
- The taste form to be used to record the results (including the attributes to be evaluated and the type of response scale to be used);
- The time allowed for the test;
- The amount of information about the samples to be shared with the assessors (eg, beer style, beer brand, etc.);
- Whether the results of the test will be shared with assessors after the taste session and, if they are to be shared, how this is to be done;
- How the results will be analyzed and reported.

Some of the previous aspects can be decided on a day-to-day basis. However, many can be decided on in advance as policy and applied to all brewery taste sessions.

Scheduling of Taste Sessions

Many breweries hold their taste sessions at a fixed time of day. This can be advantageous from the perspective of scheduling other activities. In choosing the best time for tasting, the following factors should be considered:

- The risk of schedule conflicts with other activities, such as production meetings, time-sensitive work, or lunch breaks;
- The risk of assessors consuming alcohol within a short time of having to drive or operate hazardous machinery;
- The risk of poor tasting performance due to tiredness, the after-effects of eating, or the risk of poor concentration due to having conducted a mentally demanding activity just prior to the session.

Table 13.3 Key Requirements for the Brewery Taste Room

Requirement	Feature	Options
Hygienic, easily cleaned facility	The facility should be designed with hygiene and cleaning in mind. Appropriate security measures, such as door access control systems, can help minimize the risk of cross-contamination with odors, or other contaminants. Use of "lean" practices, such as "5S" can help assure taste-room hygiene.	The basic requirements are: (1) A reception area; (2) an area for delivery, storage, and preparation of samples; (3) a sample assessment area, containing tasting booths; and (optionally) (4) a postsession discussion area.
Facilities for sample delivery and storage	Space for delivery, booking, long-term and short-term refrigerated-sample storage is required. To minimize the risk of bias, samples should not be visible to assessors under any circumstances.	Warm storage facilities (typically 30–32°C) for shelf-life evaluation.
Facilities for sample preparation and serving	Refrigerated storage space (<5°C). Adequate space for sample preparation, labeling, and serving. Glass-washing facilities will be needed unless disposable glassware is used (beware of noise!). Separate bins for disposal of glass, metal, and plastic containers. Samples should not be visible to assessors under any circumstances.	Cool boxes packed with ice can be used to maintain sample temperature prior to serving. If draught beer samples are to be served, appropriate dispensing systems together with equipment to clean them are required.
Assessor access to tasting facility	A ground floor location, near to the site entrance can be beneficial for access by "external" assessors. If "internal" assessors are used it may be more appropriate to choose a location for the taste room which provides easiest access for the greatest number of assessors. To prevent physical or line of sight access to information that may bias their responses, assessors should be allowed to enter and leave the test and discussion areas only via the reception area.	If "external" assessors are to be used, an additional "reception area" equipped with lockers for storage of outdoor clothing and personal effects is required. Comfortable seats should be provided—the area should be well lit and kept tidy.
Taste room co-ordinator access to tasting area for sample serving	Samples can be served on trays to individual assessors through a hatch. Alternatively, booths can be located along one or more walls of the sample preparation area, allowing samples to be provided to and removed from individual assessors directly from the preparation area.	Serving of samples must be done in such a way that distraction of assessors is minimized.

Odor-free environment for sample assessment	High-specification air conditioning to maintain the tasting environment at 20–22°C with a relative humidity of 50–55%. The entire booth area should have a slight positive pressure relative to other areas.	Noise from air conditioning units must be minimized. Carbon filters can be used to remove odors from air which might emanate from the brewery environment.
Appropriate lighting for sample assessment	Lighting should be under the control of sensory staff, not assessors. Lighting should be indirect to minimize the risk of lightstruck-character development in samples—shadow-free incandescent light, sufficient to provide 300–350 lux, controllable with a dimmer switch to a maximum of 700–800 lux at the counter surface, is recommended within booths. Low-level white lighting is suitable in most circumstances and preferable to daylight.	Red or amber lighting can be used to obscure differences between beer samples of different colors.
Separate booths for assessment of samples by isolated assessors	Individual booths should be about 1 m square and separated by opaque dividers that extend 50 cm beyond the edge and a 1 m above the countertop. Number of booths should allow all assessors for beer-quality evaluation and descriptive analysis sessions to attend together—typically, 10–12 booths are satisfactory. Taste room facilities should preferably not have windows. Assessors should be comfortably seated—all booths and seats must be identical. Computer-based data acquisition is preferable to paper-based forms for most tests.	Booths must be easy to clean and hygienically designed. Individual sinks and drains in booths are best avoided.
Minimal opportunities for assessor bias during sample assessment	Quiet location with no extraneous noise from refrigerators and other equipment. Posters or other visual materials should not be displayed on taste room walls as these represent a potential source of assessor bias. The temptation to display information within taste booths should be resisted as this represents a potential source of assessor bias.	To minimize the risk of bias, it is best that all assessors arrive and leave the taste session at the same time.
Group area for postsession discussion and feedback	Assessors are typically seated around a centrally located table. Decor and furnishings should be neutral to minimize distraction. Feedback on results collected using paper forms can be via a white board or flipchart. Feedback on results collected using computer-based data acquisition is best done using a PC or tablet and projector.	Alternatively, feedback can be provided to individual assessors while they remain in booths using data generated by the sensory software used to acquire the tasting data.

FIGURE 13.8

Assessors evaluating samples within booths in a taste room. Control of air purity within the room is of particular importance, together with the absence of any source of distraction for assessors.

Sample Storage

Prior to evaluation, it is important that samples are stored under conditions which minimize flavor change. Typically, that means storage in the dark at a temperature of 2–5°C for no longer than 2 or 3 weeks prior to testing. The storage area must be odor-free, because some odors (particularly, "musty" notes) can enter sealed beer bottles through plastic liners in crown corks.

Samples should be clearly labeled and allocated a unique code, specific to a batch of beer and a particular set of storage conditions. Records should tie this sample code to the product name, production location, packaging date and time, sample storage conditions, and any other relevant information.

It is particularly important that information about the samples is concealed from assessors prior to tasting. Knowledge of the samples, or the reason for testing them, is likely to bias the assessments. To reduce this risk, samples can be placed in sealed cardboard boxes and labeled with the date of the scheduled taste session. Bottles and cans can be wrapped in aluminum foil to obscure labels from view.

Sample Preparation

Samples should be poured into glasses or serving flasks prior to taking them into the taste room. They should not be poured directly from the original bottles or cans as sight of them by tasters can bias assessors.

Unless all assessors can be served from a single package [for example, in the case of a 30 L keg or 2 L polyethylene terephthalate (PET) bottle], it is better to combine several bottles or cans in a single-serving flask, because the flavor of beers from the same batch can vary from package to package, particularly if the samples have aged prior to evaluation. There is a small risk of loss of carbon dioxide in this procedure, but this is of minor significance relative to package-to-package variation.

Sample Coding

The way in which glasses are identified to assessors is of great importance. Glasses should not be labeled with single-digit numbers such as 1, 2, or 3, or letters such as A, B, or C. This is because assessors usually have expectations related to such labels. These expectations will bias their responses. Instead, glasses should be labeled with three-digit codes, such as 319, 638, or 017. Each sample should be labeled with a unique code. Codes should not be reused from one session to another.

Various online sample-code generators are available. A good example can be found at the website of the University of Oregon (https://registrar.uoregon.edu/faculty-staff/random-number-generator).

A further enhancement is to avoid particular three-digit codes which may have significance to assessors. For example, in the United States, the number 911 should be avoided as it is the telephone number for the emergency services. Similarly, the numbers 666, 888, and 999 should be avoided in some cultures.

Sample Serving

In many breweries, samples are evaluated by all assessors in the same sequence. For example, all assessors might start with sample "612" before moving into sample "593." This is bad practice and a significant source of bias. The effect of this bias depends on the type of test being conducted. For example, if a quality-rating test is being carried out, the first sample evaluated usually scores higher than it otherwise would if its position was randomized within the set. This could lead to a sample of "poor" flavor quality being graded as "satisfactory," or a sample of "satisfactory" quality being graded as "excellent."

Ideally, each sample presented to assessors should have an equal chance of occupying any one of the serving positions. For two samples coded "612" and "593," the samples can either be presented in the sequence "612" and "593," or in the sequence "593" and "612." With only two assessors, all possible sample sequences can be covered. However, an even number of assessors is required to avoid an imbalance in the order of presentation.

When there are three samples, there are six possible serving sequences. For a sample set coded "442," "975," and "118," the six possible serving sequences are: "442," "975," and "118"; "442," "118," and "975"; "975," "442," and "118"; "975," "118," and "442"; "118," "442," and "975"; and "118," "975," and "442." Thus, six assessors, or multiples of six, will be required to cover each sample-serving sequence an equal number of times.

When the number of samples exceeds three, the number of potential sample combinations rises considerably. Determining an order of sample presentation is a more complex task in such circumstances. To help with this problem, various techniques are available (see Lawless and Heymann, 2010 for details). These include:

- Random sample allocation;
- Balanced test designs;
- Complete block designs;
- Incomplete block designs.

The dual issues of sample-code allocation and sample-serving order can be conveniently addressed through use of sensory-evaluation software or software specifically designed for this purpose.

Sample Evaluation

There are several options available with respect to evaluation of samples by assessors.

- Monadic assessment, in which a single sample is presented for evaluation;
- Sequential monadic assessment, in which a series of samples is presented to assessors, each of which is evaluated individually in a specific sequence without reverting to previous samples;
- Semimonadic assessment, in which a series of samples is presented to assessors, each of which is evaluated individually in a specific sequence but with reverting to previous samples permitted;
- Comparative assessment, in which several samples are presented together and assessors are instructed to compare them to one another as part of the evaluation procedure.

It is important not to leave the method of evaluation to chance. The way in which it is to be done must be specified in the test instructions, and assessors must be clear on what is required of them.

STATISTICAL ANALYSIS OF SENSORY DATA

For some sensory test procedures, calculations of the average responses for individual attributes or quality ratings suffice. In the case of quality assessment methods in which nonconformances are recorded, counting of the number of times individual nonconformances are indicated provides a measure of their importance.

However, tests of statistical significance can be important in understanding the extent to which differences in mean responses can be relied upon when comparing results from two or more samples. These statistical tests fall into two categories:

- Parametric tests, which assume that the data fits a predefined distribution pattern;
- Nonparametric tests, which make no assumptions about the distribution of the data.

Common tests include the Chi-squared test, *F*-test, *T*-test, Kramer's rank-sum test, Friedman test, and Analysis of Variance (ANOVA). A discussion of the various methods used to analyze the statistical significance of data is beyond the scope of this chapter; the reader is referred elsewhere (Stone and Sidel, 1993; Carpenter et al., 2000).

Basic statistical testing can be carried out using spreadsheet applications (such as Microsoft Excel). More advanced procedures require the use of specialist software.

Multivariate analysis techniques can be used to explore relationships within complex data sets. Useful techniques include principal component analysis, Generalized Procrustes Analysis, and cluster analysis.

REPORTING OF SENSORY DATA

Sensory reports in breweries should be generated with one question in mind: "What action is required by the recipient in response to this set of results?" Often the answer may be "none." Such inaction can be, in itself, a valuable outcome. If the flavor of the samples tested is a good fit to specification, it is important for production staff not to tinker with the recipe and process. If they do, they may make a good product bad.

When samples fail to comply with the specification, reports should provide sufficient guidance to brewers to allow them to make targeted improvements. There is little point in grading a product or set of products as "unsatisfactory" if no specific quality defects are identified. A preferable approach for samples of suboptimal flavor quality is to indicate up to three nonconformances which, if addressed, would result in an improvement in the quality score. For example, one set of products scoring an average of 7.1 on a scale of 0–10 (in which 10 is perfection) might benefit from optimizing bitterness and kettle hop character—flavors which require intervention in the brewhouse. Another set of products achieving the same average quality score might require a reduction in the intensity of H_2S, mercaptan, and onion characteristics—flavors which require intervention in the maturation cellar.

It is important not to overreact to results of assessment of individual samples. It is preferable to analyze those of a number of samples, ideally from one product stream, and tested over a period of a few weeks. Construction of a Pareto chart (Fig. 13.9) can then aid identification of the main nonconformances associated with that product, allowing preventative and corrective actions targeted where they matter most.

The issue of data management should not be taken lightly. A typical large brewery generates close to 300,000 data points each year derived from about 6000 to 9000 samples. Collecting that data in the most efficient way and extracting the greatest possible business intelligence from it are significant challenges that cannot be adequately addressed using paper forms and spreadsheets. Computer-based data acquisition, analysis, and reporting is a fundamental prerequisite for efficient sensory analysis in today's breweries.

Finally, it cannot be overemphasized that sensory results should not be released to production staff unless they are valid. Thus, for release of results, tests should have been carried out with a sufficient number of well-trained assessors, of known

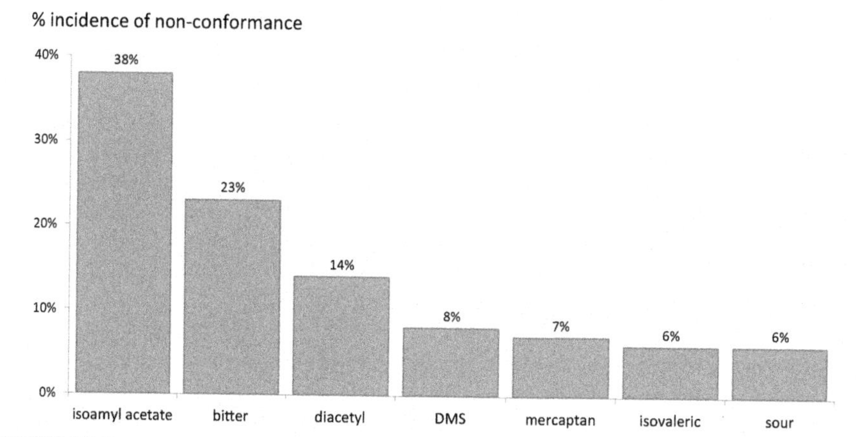

FIGURE 13.9

Pareto chart used to identify the main nonconformances in a single brand of beer produced in one brewery during a 4-week period.

competence, under conditions specified by the test procedure (Etaio et al., 2010). Practices described in international standards such as ISO 17025 (ISO, 2005) provide a good basis for assessment of test method validity in this regard.

REFERENCES

Anon, 1995. Sensory Analysis Manual. Institute of Brewing, London, 44 pp.

Anon, 2010. IFST: policy statement on ethical and professional practices for the sensory analysis of foods. Available from: http://www.ifst.org/knowledge-centre/information-statements/ifst-guidelines-ethical-and-professional-practices-sensory (issued January 2010. accessed 07.04.15.).

ASTM, 1997. Standard Test Method for Sensory Analysis – Triangle Test. ASTM E1885–97.

ASTM, 2001a. Standard Guide for Sensory Evaluation of Beverages Containing Alcohol. ASTM E1879–00.

ASTM, 2001b. Standard Test Method for Directional Difference Test. ASTM E2164–01.

Axcell, B., 2008. A new global approach to tasting. MBAA Technical Quarterly 45, 348–351.

Carpenter, R.P., Lyon, D.H., Hasdell, T.A., 2000. Guidelines for Sensory Analysis in Food Product Development and Quality Control, second ed. Aspen Publishers Inc., Maryland.

Cliff, M.A., Dever, M., 1996. A proposed approach for evaluating expert wine judge performance using descriptive statistics. Journal of Wine Research 7 (2), 83–90.

Etaio, I., Albisu, M., Ojeda, M., Gil, P.F., Salmeron, J., Perez Elortondo, F.J., 2010. Sensory quality control for food certification: a case study on wine. Panel training and qualification, method validation and monitoring. Food Control 21, 542–548.

Hawthorne, D.B., Shaw, R.D., Davine, D.F., Kavanagh, T.E., Clarke, B.J., 1991. Butyric acid off-flavors in beer: origins and control. Journal of the American Society of Brewing Chemists 49, 4–8.

ISO, 1983. Standard 5495: Sensory Analysis – Methodology – Paired Comparison Test. International Organization for Standardization, Paris.

ISO, 1988. Standard 8589: Sensory Analysis – General Guidance for the Design of Test Rooms. International Organization for Standardization, Paris.

ISO, 2003. Standard 13299: Sensory Analysis – Methodology – General Guidance for Establishing a Sensory Profile. International Organization for Standardization, Paris.

ISO, 2004. Standard 4120: Sensory Analysis – Methodology – Triangle Test. International Organization for Standardization, Paris.

ISO, 2005. General Requirements for the Competence of Testing and Calibration Laboratories. International Organization for Standardization, Paris.

Kemp, S.E., Hollowood, T., Hort, J., 2009. Sensory Evaluation. A Practical Handbook. Wiley-Blackwell, 196 pp.

Kilcast, D. (Ed.), 2010. Sensory Analysis for Food and Beverage Quality Control. A Practical Guide. Woodhead Publishing, 373 pp.

Kim, K.-O., O'Mahony, M., 1998. A new approach to category scales of intensity I: traditional versus rank-rating. Journal of Sensory Studies 13, 441–449.

Larsen, O.V., Simpson, B., Williams, I., Jorgensen, C.W., 2005. Improving flavor panel performance using structure training and validation. MBAA Technical Quarterly 42, 339–341.

Lawless, H.T., 2013. Quantitative Sensory Analysis. Wiley Blackwell, 404 pp.

Lawless, H.T., Heymann, H., 2010. Sensory Evaluation of Food. Principles and Practices, second ed. Springer. 596 pp.

Lyon, D.H. (Ed.), 2002. Guideline No 37. Guidelines for Selection and Training of Assessors for Descriptive Sensory Analysis. Chipping Campden, CCFRA Limited, 64 pp.

McCusker, M.T., Basquille, J., Khwaja, M., Murray-Lyon, I.M., Catalan, J., 2002. Hazardous and harmful drinking: a comparison of the AUDIT and CAGE screening questionnaires. Quarterly Journal of Medicine 95, 591–595.

Meilgaard, M.C., Dalgliesh, C.E., Clapperton, J.F., 1979. Beer flavor terminology. Journal of the American Society of Brewing Chemists 37, 47–52.

Meilgaard, M.C., Reid, D.S., Wyborski, K.A., 1982. Reference standards for beer flavor terminology system. Journal of the American Society Brewing Chemists 40, 119–128.

Meilgaard, M., Civille, G.V., Carr, B.T., 2007. Sensory Evaluation Techniques, fourth ed. CRC Press. 448 pp.

Naes, T., 1990. Handling individual differences between assessors in sensory profiling. Food Quality and Preference 2, 187–199.

Naes, T., Solheim, R., 1991. Detection and interpretation of variation within and between assessors in sensory profiling. Journal of Sensory Studies 6, 159–177.

Piggott, J.R., Hunter, E.A., 1999. Evaluation of assessor performance in sensory analysis. Italian Journal of Food Science 11, 289–303.

Simpson, W.J., 2006. Brewing control systems: sensory evaluation. In: Bamforth, C.W. (Ed.), Brewing: New Technologies. Woodhead Publishing, Cambridge, pp. 427–460.

Stone, H., Sidel, J.L., 1993. Sensory Evaluation Practices. Academic Press, New York, 338 pp.

Tabakoof, B., Mariniz, L., Hoffman, P., 2004. Genetic Diagnosis of Alcoholism Subtype. Patent Number: WO2004067763.

Thompson, S., 1999. Difference-from-control sensory test (ICM). Journal of the American Society of Brewing Chemists 57, 168–171.

CHAPTER

Microbiology

14

A.E. Hill

Heriot-Watt University, Edinburgh, Scotland, United Kingdom

INTRODUCTION

A proactive approach to microbial spoilage, such as implementation of good manufacturing practices and HACCP process control systems, has made a significant impact in reducing incidences of waste batches and product recall. However, as shown in Fig. 14.1, opportunities for microbiological contaminants to enter the

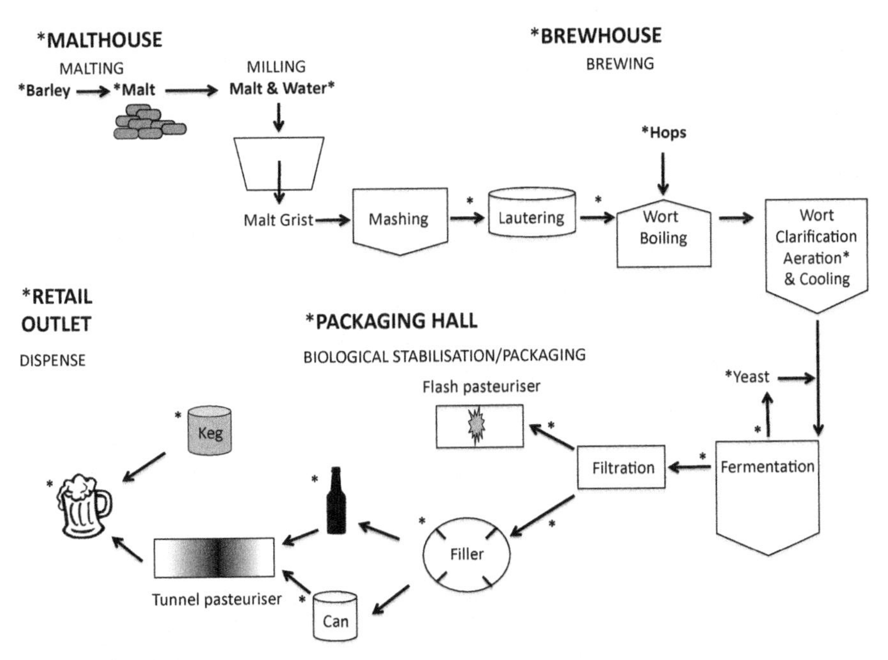

FIGURE 14.1

Schematic of the Brewing Process. Potential Sources of Microbiological Contamination Are Indicated by *.

Redrawn from Vaughan, A., O'Sullivan, T., van Sinderen, D., 2005. Enhancing the microbiological stability of malt and beer—a review. Journal of the Institute of Brewing 111, 355–371.

Brewing Materials and Processes. http://dx.doi.org/10.1016/B978-0-12-799954-8.00014-9

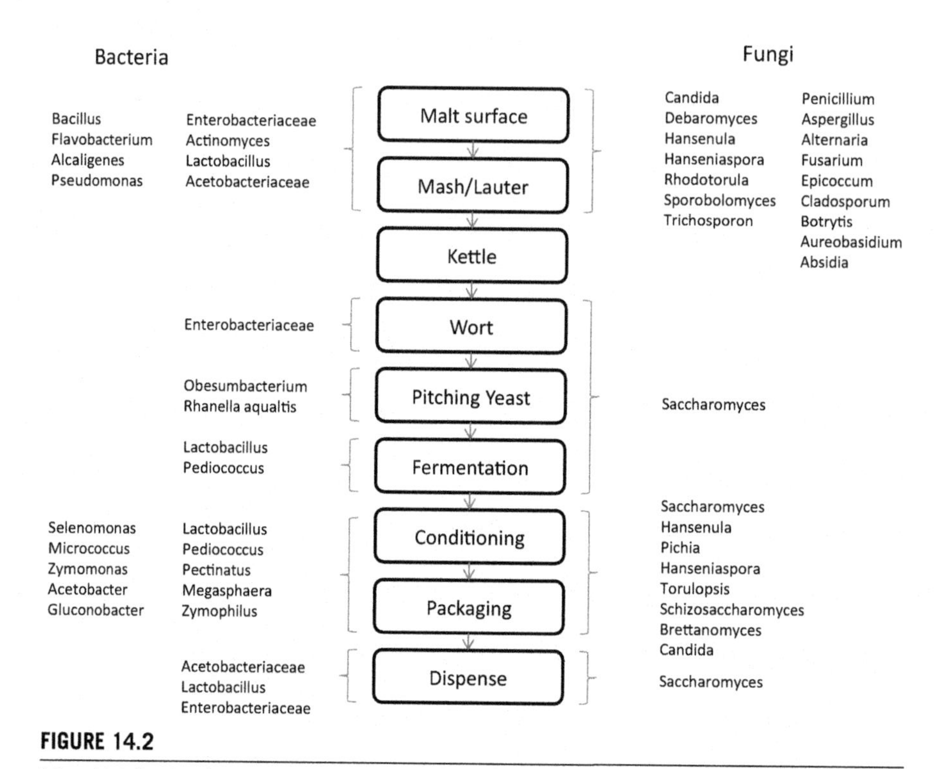

FIGURE 14.2

Microbial Contaminants Within the Brewing Process.

brewing process are available at all stages and their amazing ability to adapt to seemingly hostile conditions makes them a tenacious threat.

The most commonly encountered spoilage microbes are detailed in Fig. 14.2 (Bokulich and Bamforth, 2013; Quain and Storgårds, 2009). Methods employed by breweries to detect and/or identify both yeast and bacteria are constantly changing and vary depending on the scale of operations. Most routinely use adenosine triphosphate (ATP) testing and traditional methods of plating, but as costs for rapid detection methods have decreased, they are increasingly becoming affordable for medium to even very small plants. In this chapter, the practical methods available to detect and identify microbes at each stage of the brewing process are detailed.

CEREALS

Fungi from the genera *Alternaria*, *Cladosporium*, *Epicoccum*, and *Fusarium* are the main hazards in terms of barley and malt infection. Tolerance of fungal growth on cereals is less than 10 colony-forming units (CFU) per gram and zero tolerance of wild yeast. Malt should also be free of mold. Typically, malt is sampled at each intake and at regular intervals during storage and tested for bacteria, fungi, wild yeast, and mycotoxins. Air samples may also be taken.

The simplest way to evaluate the internal microflora of grain is by direct plating. Grain is immersed in full strength or 50% household bleach for 1 min to kill surface microflora then rinsed in sterile distilled water. Either grind the grain before adding to molten agar (express per gram) or place individual grains on the surface of the agar in a Petri dish (express per grain) and incubate at 25–30°C to allow microbes located in the interior of the grain to grow out. The level of internal infection is an indicator of quality and storability of the grain. If *Aspergillus flavus* and *A. parasiticus* or Czapek-Dox Iprodione Dichloran Agar (see Table 14.1) is used, this technique can also give some information about the safety of the grain by indicating whether potentially toxic *Aspergillus flavus*, *A. parasiticus*, or *Fusarium* species are present.

Grain is susceptible to mycotoxins produced either while the crop is growing by *Fusarium* species or during storage by *Penicillium* species. High-quality malting barley should be free from deoxynivalenol (DON) from *Fusarium* head blight and free from disease. For disease, there is no compromise; diseased crops will be rejected and reducing the price will not make them acceptable. Barley from areas with conditions conducive to *Fusarium* head blight is routinely screened for DON, and barley with DON levels over 0.5 ppm will normally be rejected for

Table 14.1 Detection Media for Grain Analysis

Growth Medium	Microorganisms	Incubation Conditions
Aspergillus flavus and *parasiticus* agar (AFPA)	*Aspergillus flavus* and *A. parasiticus*	28°C. 7 days (morphological analysis), 10 days (toxin production).
Czapek peptone yeast extract agar (CZPYA)	*Actinomycetes*	
Czapek-dox iprodione Dichloran agar (CZID)	*Fusarium* species	
Dichloran chloramphenicol peptone agar (DCPA)	*Fusarium* species	25°C. 4 days (Initial examination), 6 days (morphological analysis).
Dichloran glycerol agar (DG18)	*Fusarium* species	
Dichloran-rose bengal-chlortetracycline agar (DRBC)	*Aspergillus flavus* and *A. parasiticus*	
Pentachloronitrobenzene peptone agar (PPA)	*Fusarium* species	
Potato dextrose agar (PDA)	Fungi	

Hocking, A.D., Pitt, J.I., 1980. Dichloran-glycerol medium for enumeration of xerophilic fungi from low-moisture foods. Applied and Environmental Microbiology 39, 488; Mostafa, M.E., Barakat, A., El-Shanawany, A.A., 2005. Relationship between aflatoxin synthesis and Aspergillus flavus *development. African Journal of Mycology and Biotechnology 13, 35–51; Thrane, U., 1996. Comparison of three selective media for detecting* Fusarium *species in foods: a collaborative study. International Journal of Food Microbiology 29, 149–156.*

malting purposes (tolerance is 1 ppm if for human consumption). Unprocessed common wheat and barley are also usually screened for Zearalenone (ZON) (tolerance is up to 100 ppb). DON and ZON are both included in the Home Grown Cereals Authority (HGCA) Grain Passport.

Methods for detecting mycotoxins are summarized in Table 14.2. Chromatographic methods are frequently used for detecting, quantifying, and confirming the presence of mycotoxins. These methods include thin-layer chromatography, high-performance liquid chromatography, liquid chromatography combined with mass spectrometry, gas chromatography, and gas chromatography combined with mass spectrometry.

Various detection methods, such as fluorescence, ultraviolet absorption, and others have been combined with chromatographic methods. New methods based on the production of antibodies specific for individual mycotoxins have also been developed and include enzyme-linked immunosorbent assays and immunoaffinity columns. These methods allow for specific and precise detection and quantification of specific mycotoxins. This has led to the development of test kits for mycotoxins, such as VICAM, which are rapid and simple to use and can be used in the field and throughout the processing stages.

WATER

As the main ingredient of beer and a utility in the production process, water quality is central to brewing. Algae, protozoa, fungi, yeasts, and bacteria may all be present in water, but fortunately, very few waterborne microbes are able to cause serious problems to brewers. Typically, water from boreholes contains fewer microorganisms than surface water, ie, rivers, ponds, and tanks, and public water supplies are, of course, rigorously tested. Microbiological tests on water predominantly involve detection of an indicator organism. The primary fecal indicator organism is *Escherichia coli* which is an ideal indicator because it is abundant and has similar survival qualities to *Salmonella*. Water used for human consumption can have no more than one positive sample (>1 coliform/100 mL) in 40 samples

Table 14.2 Methods for Detection of Mycotoxins (Mirocha and Christensen, 1986)

Mycotoxin(s)	Method(s)
Aflatoxins	TLC, HPLC, ELISA, immunoaffinity column
Deoxynivalenol	GC, HPLC, ELISA, immunoaffinity column
Fumonisins	HPLC, ELISA, immunoaffinity column
Moniliformin	HPLC
Ochratoxin	TLC, HPLC, ELISA, immunoaffinity column
Zearalenone	TLC, HPLC, ELISA, immunoaffinity column

Mirocha, T.J., Christensen, T.M., 1986. Mycotoxins and the fungi that produce them. Proceedings of the American Phytopathological Society 3, 110–125.

tested in a month, and the concentration of fecal coliforms must be zero. It is good practice to monitor nonpublic supplies (borehole, spring, etc.) seasonally.

Brewing water should be tested before entering the hot-liquor tank or any mashing vessels. Similarly, water for dilution should be tested prior to use. Clean in place (CIP) and rinse waters should be checked every cycle. Generally, however, most supplies are checked weekly or upon encountering unstable wort. Sampling points should be uniformly distributed throughout a piped distribution system, and the number of sampling points should be proportional to the number of links or branches. The points chosen should generally yield samples that are representative of the system as a whole and of its main components.

The traditional method for detection of waterborne microbes is direct plating. Samples may also be filtered either on- or offline and filters placed directly on the surface of an agar plate. A range of media for the detection of coliforms is available (Table 14.3) and confirmation of thermotolerant *E. coli* is possible by incubation at 44°C.

Table 14.3 Microbiological Media for Water Analysis

Medium	Target Microorganism(s)	Incubation Conditions
Azide	Intestinal enterococci	40–48 h at 36 ± 2°C
Bismuth Sulfite NPS	*Salmonella typhi* and other salmonellae	40–48 h at 36 ± 2°C
Chromocult NPS	Total coliforms and *Escherichia coli*	20–28 h at 36 ± 2°C
ECD NPS	*E. coli*	16–18 h at 44 ± 2°C
ENDO NPS	*E. coli* and coliform bacteria	18–24 h at 36 ± 2°C
Heterotrophic plate count		
Lactose-multi-concentrated (LMC) broth	Coliforms and *E. coli*	1–2 days at 30–35°C
Lysine	Wild yeast	3–5 days at 30–35°C
MacConkey	Coliform bacteria and other enterobacteriaceae	18–72 h at 30–35°C
Malt extract	Yeasts and molds	3–5 days at 20–25°C or at 30–35°C depending on the target of the investigation
Meat extract-peptone	Total count	<5 days at 30–35°C
Fecal coliform medium (mFC)	*E. coli* and fecal coliform bacteria	18–24 h at 36 ± 2°C
MLGA (membrane lactose glucuronide agar)	Coliforms and *E. coli*	30°C for 4 h, then 37°C for 14 h
Reasoner's 2A agar (R2A)	Heterophilic organisms	>5 days at 30–35°C

Continued

Table 14.3 Microbiological Media for Water Analysis—cont'd

Medium	Target Microorganism(s)	Incubation Conditions
Rainbow agar	*E. coli*	18 h 37°C
Sabouraud	Yeast and molds	<5 days at 20–25°C
Schaufus Pottinger (m Green yeast and mold)	Yeast and molds	2–5 days At 20–25°C or at 30–35°C depending on the target of the investigation
Soybean–Casein digest medium (Caso)	Total count	Bacteria: <3 days at 30–35°C Yeasts and molds: <5 days at 30–35°C
Standard 2,3,5-Triphenyltetrazoliumchloride (TTC)	Total count	<5 days at 30–35°C
Teepol (lauryl sulfate medium)	*E. coli* and fecal coliform bacteria	18–24 h at 36 ± 2°C
Tergitol TTC nutrient pad sets (NPS)	Coliform bacteria and *E. coli*	18–24 h at 36 ± 2°C
Tryptone glucose extract	Total count	<5 days at 30–35°C
Wallerstein (WL nutrient)	Microbiological flora of brewing and fermentation processes	2–5 days at 30–35°C aerobic or anaerobic depending on the target of the investigation
Wort	Yeast and molds	3–5 days at 20–25°C or at 30–35°C depending on the target of the investigation
Yeast extract	Aerobic bacteria	44 ± 4 h at 36 ± 2°C; 68 ± 4 h at 22 ± 2°C

Techniques available to rapidly detect bacteria include fluorescence microscopic methods (eg, Epifluorescence microscopy using acridine), detection of specific metabolites, antibody methods, and deoxyribonucleic acid (DNA)-based methods. However, many of these methods are expensive, require an enrichment step, sophisticated equipment, and expertise, and/or are not suitable for routine analysis. The determination of ATP with a bioluminescence assay has emerged as the main method for rapid detection of viable bacteria in breweries. The main function of the assay is to quantify ATP, a compound in metabolism that is found within all living cells. The assay is based on the reaction between luciferase (enzyme), luciferin (substrate), and ATP. Light is emitted during the reaction and can be measured quantitatively and correlated with the quantity of ATP extracted from bacteria. Several commercial kits are available eg, AquaSnap (Hygiena) and Clean-Trace

(3M). The test systems typically consist of a sampling stick to collect a known volume of water. The sample is then mixed with luciferin and luciferase and placed in a luminometer to determine the relative light units (RLU) emitted. The devices are often used for monitoring CIP systems, and for industrial water treatment systems such as cooling towers and closed water systems.

Despite ATP being considered a robust monitoring parameter for microbial drinking water quality, a significant increase in ATP should be accompanied with methods for detection of specific bacteria to validate whether or not contamination has occurred. Therefore, the best approach for monitoring microbial drinking water quality, to enhance water security and safety, is by combining rapid methods with methods targeted for specific bacterial detection (Vang et al., 2013).

YEAST

Of all raw materials, the most likely source of contamination is from yeast because it is added after wort boiling. Yeast-handling plants also tend to be very complex and difficult to clean (Briggs et al., 2004). Acid washing can be used to reduce or remove bacterial contaminants, but this process does not remove wild yeast. The most common contaminants are the lactic acid bacteria (LAB) lactobacilli and pediococci, and the microbiological media employed to check pitching yeast reflects their nutritional requirements. Yeast should be checked before pitching (preferably 2–4 days prior to brewing), and monthly checks should be carried out to test for nonbrewing strains that may populate over time. Tolerance is less than 10 CFU/mL for bacteria and zero tolerance of wild yeast.

Media employed in traditional plate checks typically include inhibitors and/or stimulators (Table 14.4). Such chemicals might include lysine, a nitrogen source that brewing yeast cannot utilize but wild yeast can; copper sulfate, which is also inhibitory to strains of *Saccharomyces*; or plates containing actidione, which selectively promotes bacterial growth. Such techniques typically require a two-day 25 degree incubation period ["European Brewery Convention (EBC) Analytica Microbiologica: Part II Continued", 1984].

When attempting to identify particularly hard-to-culture LAB strains, Suzuki (2011) maintains that "advanced beer-spoiling detection" media is the quickest and most effective media, primarily because of the low pH levels which they employ. When seeking to identify LAB, anaerobic incubation at around 28°C is typically employed (Suzuki, 2011). Supplementing de Man, Rogosa, Sharpe (MRS) with catalase can potentially speed up growth of LAB (Deng et al., 2014).

For breweries using flow cytometry to determine yeast count and viability, it is possible to extend use of this method to detect beer spoilers such as *ZygoSaccharomyces*, *Dekkera* (*Brettanomyces*), and *Lactobacillus* (Bouix et al., 1999; Donhauser et al., 1993; Jespersen et al., 1993). The principle of flow cytometry is based on fluorescence staining or labeling, and the cells are brought in a fluid stream within a thin capillary in which the fluorescence molecules are excited by a laser and the emission is detected. The laser is also used to count the particles and determine the size.

Table 14.4 Microbiological Media for Pitching Yeast Analysis

Medium	Target Microorganism(s)	Incubation Conditions
$CuSO_4$	Wild yeast	48 h, 25°C
Lysine	Enteric, acetic, and lactic bacteria, wild yeast	48 h, 25°C
MacConkey + actidione	Enteric bacteria	48 h, 25°C
MRS (de Man, Rogosa, Sharpe)	Enteric, acetic, and lactic bacteria, wild yeast	48 h, 25°C
MYGP (malt yeast extract glucose peptone)	Wild yeast	48–72 h at 25–30°C. Combined anaerobic (3 days) followed by aerobic (2 days)
Nocive Brewers Bacteria (NBB)-broth	Lactic acid bacteria	72 h, 25–30°C; aerobic and anaerobic
Nutrient agar/broth	General purpose medium for bacteria although many yeasts will grow	48 h, 25°C
WLN (Wallerstein laboratory differential)	General purpose medium for bacteria	48 h, 25°C
Yeast morphology agar	Assessment of yeast colony morphology	48 h, 25°C

All data are collected and software analyses the data to generate a report with the result of live/dead cells or detection of beer spoilers.

Yeast screening kits for detection and differentiation of *Saccharomyces* species (including differentiation between *S. pastorianus* and S. *bayanus*), *Dekkera* species, and *Pichia anomala* using Multiplex-Fluorescence resonance energy transfer polymerase chain reaction (FRET PCR) are available as are methods for qualitative analysis and detection of *Candida*, *Dekkera*, and *Pichia* species and for top/bottom-fermenting *Saccharomyces* species [Institute of Applied Laboratory Analyses (GEN-IAL GmBH)]. Such methods represent a significant step forward in rapid detection (further details on PCR methodology later in this chapter).

HOPS

The antibacterial properties of hops are one of the main reasons for their use in brewing. Hops are dried down to 8–10% moisture to prevent spoilage, but nonetheless remain susceptible, and, as with other raw materials, checks should be made for each batch as a matter of quality control. A number of breweries carry out dry hopping postbrewhouse which increases the risk of introducing contaminants. For hops, the most likely microbes are fungi, molds, and mildew, and the tolerance is less than 10 CFU/g, with zero tolerance of wild yeast.

Table 14.5 Microbiological Media for Hops Analysis

Medium	Target Microorganism(s)	Incubation Conditions
Lysine	"Wild" yeast	3–5 days at 30–35°C
Malt extract agar	Fungi eg, *Podosphaera castagnei*, mold, mildew	48 h, 25°C
Sabouraud	Yeast and molds	5 days at 20–25°C
Schaufus pottinger	Yeast and molds	2–5 days at 20–25°C or at 30–35°C
WLD (Wallerstein differential broth)	Flora of brewing and fermentation processes	2–5 days at 30–35°C; aerobic or anaerobic
Wort agar	Yeast and molds	3–5 days at 20–25°C or at 30–35°C

For traditional plating, a weighed sample is rinsed in sterile water, and the rinse water either mixed with molten agar or spread on the surface of an agar plate. Media for the detection of common spoilage organisms is given in Table 14.5.

A range of molecular-based techniques are available to detect major fungal crop pathogens including PCR and DNA macroarray methods. Multiple species of *Phytophthora* may be detected directly in infected plant tissues using *Phytophthora*-specific primers and denaturing gradient gel electrophoresis (PCR–DGGE) (Rytkönen et al., 2012). A padlock probe (PLP)-based multiplex method of detection and identification for many *Phytophthora* spp. simultaneously is available (Sikora et al., 2012). A generic PCR reaction is carried out first followed by a species-specific padlock probe ligation. The padlock probes are long oligonucleotides containing complementary sequence regions that can be ligated (joined) onto the target DNA to create a circular molecule. The point at which the circular molecule is joined is mutation specific and can be detected on a microarray. Njambere et al. (2011) state that use of dimeric oligonucleotide probes can improve the process. Recently, a chromogenic detection method built on DNA microarray has been developed. The protocol uses digoxigenin (DIG)-labeled targets. Species-specific oligonucleotides on the array are first hybridized to a labeled target, and the hybridized array is incubated with anti-DIG conjugated alkaline phosphatase substrates to give visible target signals. Samples may be taken from pure cultures, soil, or plants (Wong and Smart, 2012).

SUGARS AND SYRUPS

Noncereal adjuncts and priming sugars are commonly employed in brewing. Irrespective of the point of addition, any materials added to the process should be checked as a matter of quality control. For sugars and syrups, the low water potential

Table 14.6 Microbiological Media for Analysis of Syrups and Sugars

Medium	Target Microorganism(s)	Incubation Conditions
BACARA: chromogenic media	*Bacillus* species	48–72 h at 25–30°C. Combined anaerobic (3 days) followed by aerobic (2 days)
Bacillus differentiation agar	Differentiation between *Bacillus cereus (colorless)* and *Bacillus subtilis (yellow)*	35–37°C for 18–24 h
Malt extract agar (MXA)	All microorganisms	22–25°C for 3 days
Mannitol–Egg Yolk–Polymyxin (MYP) agar	*Bacillus cereus*	30°C. 24 h incubation

prevents growth of contaminants, but the main threat is survival of spores from *Bacillus* species. A range of media is available to selectively culture *Bacillus* species (Table 14.6). A sugar solution may be mixed with molten agar or spread directly on the surface of an agar plate.

Rapid methods of analysis tend not to be used for detection of contaminants in sugars and syrups due to the difficulties in extracting genetic material from the complex medium, but PCR primers for *Bacillus* species are readily available.

BREWING PROCESS

As we move toward fermentation it is essential that the integrity of the system is maintained and that all raw materials and adjuncts are appropriately stored and transferred under sterile conditions. The first challenge for the microbiologist is to ensure that all vessels and pipework are tested.

BREWERY SURFACES

As the beer progresses through the brewery, it comes into contact with vessel surfaces, pipes, and fittings, all of which can harbor infection, particularly on the cold side of the brewery after wort clarification. Any surface that is exposed to wort, beer, or yeast should be thoroughly cleaned and sterilized ie, vessels, piping, and implements. Soiled surfaces can support a microbiological growth which can be introduced into the beer. Any recurrent contamination may indicate the presence of a biofilm. Biofilms are particularly difficult to clean as they can bind strongly to the vessel or pipe.

Almost ubiquitously, ATP tests are employed to check plant hygiene. This includes both swab tests of vessels, pipework, and liquid tests of CIP final rinse water. Typical sample locations and types are shown in Table 14.7.

Table 14.7 Brewery ATP Testing Schedule

Location	Sample Type	Frequency	Target Organism(s)
Hot liquor tank	Liquid water sample Final rinse CIP	Weekly Each CIP	All microorganisms
Cereal cooker	Swab Liquid (CIP final rinse)	Each use/cleaning cycle	All microorganisms
Mash tun	Swab Liquid (CIP final rinse)	Each use/cleaning cycle	All microorganisms
Lauter tun	Swab Liquid (CIP final rinse)	Each use/cleaning cycle	All microorganisms
Mash filter	Liquid (CIP final rinse)	Each use/cleaning cycle	All microorganisms
Kettle	Swab Liquid (CIP final rinse)	Each use/cleaning cycle	All microorganisms
Paraflow/chiller	Liquid (CIP final rinse) Liquid (wort)	Each use/cleaning cycle Each batch	All microorganisms

There are a number of ATP systems currently available for surface hygiene testing including UltraSnap (Hygiena), PocketSwab Plus (Charm Sciences), Hy-Lite (VWR), and Clean-Trace (3M). The UltraSnap system contains a premoistened swab in a tube and the reaction reagent in a cap on the top. A certain area of surface is wiped over with the swab (usually a minimum of 10×10 cm). Once the cap is closed, the reagent moves down to the swab and the tube is shaken a few times. ATP present on the swab can then be detected in a luminometer as described earlier.

An alternative system to ATP uses nicotinamide adenine dinucleotides (NAD/NADH) and nicotinamide adenine dinucleotide phosphates (NADP/NADPH) which are compounds used for energy transfer in the metabolism in living cells or compounds found in food debris. HY-RiSE (Merck) uses a strip or a card, and the result is visible by a color reaction. Once the strip has been wiped across the test surface, two further reagents are added. After 5 min incubation in the dark, the reaction zone will appear yellow if negative or pink/purple to bluish if positive.

Secondary testing (either traditional plating or rapid methods) tend to only be used if ATP levels are consistently breaching tolerance limits and/or are not reduced by additional cleaning. Malt extract agar or similar may be used for the detection of acetic and lactic bacteria and wild yeast.

AIR AND PROCESS GASES

Microorganisms are ever-present in the air, often in association with dust particles or airborne moisture droplets. They can also be introduced to the environment by insects and other pests. Every effort must be made to keep the brewing environment as clean as possible and to minimize the ingress of outside contamination. Wherever possible, all vessels should be covered to reduce the risk of aerial contamination.

Process gases such as oxygen used during pitching and carbon dioxide applied during packaging may provide a route for contaminants to enter the system. Any hosing should be inspected for leaks and all gases and air checked. The method involves either exposing an agar plate to the air for a set period or filtration of the air through a sterile filter which is then placed on the surface of an agar plate. Typically, a nonspecific medium is used, such as malt extract agar, and incubated for 48 h at 25°C.

WORT

As the temperature falls following wort boiling, any bacteria or wild yeast present on surfaces will multiply in the nutrient-rich medium. Problems can also arise if wort is saved and allowed to stand (such as with weak-wort recycling). Contaminated wort can result in decreased fermentation rate, off-flavors/odors, and haze, and therefore should be checked in each batch. Enterobacteria are the most common contaminants, and the tolerance is less than 10 CFU/mL, with zero tolerance of wild yeast.

Following boiling, the likely prevalence of spoilage organisms is low, and the wort will therefore need to be filtered through a sterile filter and placed on an agar plate. Typical media are given in Table 14.8.

It is from this stage forward that rapid methods of analysis are increasingly likely to be used, simply due to the increased cost of spoilage as we move toward final product. Such methods include:

Microcolony method: The microcolony method employs microscopy to detect growing cells that have not yet reached visibly discernible colonies. Several systems are available including Rapid Micro Biosystem's Growth Direct test which uses digital imaging technology to automatically enumerate microcolonies. The system captures the native fluorescence (autofluorescence) that is emitted by all living cells. More advanced systems use 96-well microplate formats and automated plate-handling systems. The μFinder Inspection System (Asahi Breweries, Tokyo, Japan) involves trapping cells onto 0.4-μm pore-size polycarbonate-membrane filters followed by anaerobic incubation on Advanced Beer Detection (ABD) agar medium at 25°C. The incubated filters are then soaked with carboxyfluorescein diacetate (CFDA) staining buffer for 30 min at 30°C. Fluorescent-stained cells may be discriminated from other particles based on their morphological characteristics and fluorescence intensities. In general, the microcolony-CFDA method lacks the selectivity for spoilage strains, and largely depends on the selectivity of media

Table 14.8 Microbiological Media for Wort Analysis

Medium	Target Microorganism(s)	Incubation Conditions
Carr's bromocresol green medium	Gram negative bacteria	27°C for 1 day
Hsu's *Lactobacillus/ Pediococcus* medium (HLP)	*Lactobacillus, Pediococcus*	25°C for 2 days
Lee's multi-differential agar (LMDA)/Schwarz differential agar (SDA)	Lactic acid bacteria	25°C for 2 days
Lin's cupric sulfate medium (LCSM)	Wild yeast	25°C for 1–3 days
Lins wild yeast media (LWYM)	Wild yeast	25°C for 1–3 days
Malt extract agar (MXA)	All microorganisms	22–25°C for 3 days
de Man, Rogosa, Sharpe (MRS)	Enteric, acetic, and lactic bacteria, wild yeast	25°C for 2 days
MRS + Actidone	Enteric, acetic, and lactic bacteria	25°C for 2 days
Sabouraud dextrose agar	Yeast	22–25°C for 3 days

used for microcolony formation, but it can discriminate beer-spoilage strains from nonspoilage strains upon detection of microcolonies on ABD, and enables the intra-species differentiation of beer-spoilage ability of LAB species, such as *Lactobacillus brevis*, *L. lindneri*, and *L. paracollinoides* (Asano et al., 2009).

Antibody-Direct Epifluorescent Filter Technique (Ab-DEFT): Direct epi-fluorescent filter technique (DEFT) employs dyes to enable distinction between viable and nonviable cells. Ab-DEFT is an extension of this method combining membrane filtration and pathogen-specific fluorescent antibodies can enumerate spoilage bacteria. Antibodies are available to distinguish between different lactic acid bacteria such as several spoiling species of *Lactobacillus* and those innocuous or necessary for brewing (Priest and Stewart, 2006). Ab-DEFT may be performed in less than 1 h (Tortorello and Gendel, 1993; Tortorello et al., 1997; Tortorello and Stewart, 1994). A preceding enrichment of the cells, which allows cells to proliferate initially, could significantly improve the sensitivity of the Ab-DEFT method and result in a detection limit of 0.1 CFU/g and only 10-h assay duration (Raybourne and Tortorello, 2003; Restaino et al., 1996).

Oligonucleotide-Direct Epifluorescent Filter Technique (Oligo-DEFT): Oligonucleotide-direct epifluorescent filter technique (Oligo-DEFT) represents a further, more sensitive and specific detection method for the direct enumeration of microorganisms. It has been demonstrated for *E. coli* in water and beverages that the oligo-DEFT method can achieve a detection limit of 1 CFU/mL (Tortorello and Reineke, 2000). Fluorescent-labeled oligonucleotides complementary to 16S ribosomal ribonucleic acid (rRNA) were combined with DEFT. Because of the high abundance in cells, the ribosomal RNA represents an optimal target for fluorescence microscopy analysis. After membrane filtration and a short 2-h hybridization step, it is possible to distinguish between either species or groups of microorganisms (Raybourne and Tortorello, 2003). The range of probes is wide, and the success of this technology relies on finding probes which are highly specific and have an excellent hybridization rate. Specificity is achieved by targeting conserved or unique rRNA sequences. A probe is normally composed of an oligonucleotide with about 20 nucleotides (Kempf et al., 2000), but some tests are carried out with peptide nucleic acid (PNA) which can have some advantages as the molecule is more stable, specific, and sensitive (Almeida et al., 2010). For pathogens and beer-spoiling organisms, 16S rRNA, 23S rRNA, or the corresponding DNA are usually selected as the target (Almeida et al., 2010; Frischer et al., 1996; Fuchs et al., 2001). Different probes are available including detection probes labeled with a fluorescence marker or digoxigenin which is used for further linkage with an antibody–enzyme complex and then a later colorimetric reaction with a chromogenic substrate (such as nitro blue tetrazolium for alkaline phosphatase) (Kempf et al., 2000; Helentjaris and McCreery, 1996). Capture probes are used to bind the target sequence (RNA or DNA) to a plate or another surface. In most cases, the probes are labeled with biotin to react with avidin which is coated on a plate.

Polymerase Chain Reaction (PCR): The Polymerase Chain Reaction (PCR) is a process used to make a large copy number of a specific DNA fragment from genetic material (DNA) in a relatively short time. Target DNA may be extracted from bacterial/yeast cells using simple kits eg, the Simplex Easy DNA Kit. The other basic components needed to perform a PCR include two specific oligonucleotides (15–25 nt), called primers, which are derived from both strands of the target sequence. Thus, they determine the size and specificity of the resulting PCR product. The other components are a thermostable DNA polymerase, deoxyribonucleotides, and a defined reaction buffer (Saiki et al., 1988). PCR is a cycle reaction and is carried out in a thermocycler. Each cycle consists of three steps. In the first step, denaturation, the initial DNA and primers are heated at 95°C and denatured into single strands. In the second step, annealing, the primers hybridize to their opposite sequence on both DNA strands. The annealing temperature is usually 3–5°C below the melting temperature of the primer. If the annealing temperature is too low, the primers may also bind at positions which are not 100% complementary leading to nonspecific products. If the temperature is too high, the primers do not bind or bind incompletely and little/no product is formed. During the third step, elongation, the new DNA strands are synthesized from the 5′ end to the 3′ end by the polymerase. The elongation temperature depends on the working optimum of the DNA polymerase (68–72°C). Thus, the new DNA fragment, which results after the first cycle is further multiplied in the following cycles. Ideally, each newly emerging DNA segment is duplicated in 20–50 cycles (Saiki et al., 1988).

Based on the method of product detection, PCR can be categorized into endpoint PCR and time-point (real-time) PCR. In endpoint PCR, the product is visualized by agarose or polyacrylamide gel electrophoresis followed by staining with fluorescent dyes, for instance ethidium bromide or SYBR Green. Several primer sets have been designed for the group-specific detection of Lactic acid bacteria in brewery samples, in wine, in food, or in the gut (Stewart and Dowhanick, 1996; Heilig et al., 2002; Lopez et al., 2003; Neeley et al., 2005; Walter et al., 2001). Endpoint PCR may be carried out in 3–4 h (Juvonen et al., 2008). The presence of *Lactobacillus brevis*, *L. lindneri*, and *P. damnosus* in beer samples (50 mL) was detected after 30–40 h incubation in Nocive Brewers Bacteria (NBB)-C broth (Bischoff et al., 2001). PCR kits are also available for the detection of *Megasphaera cerevisiae*, *Oenococcus oeni*, *Pectinatus cerevisiiphilus*, *Pectinatus frisingensis*, *Pediococcus acidilactici*, *Pediococcus damnosus*, *Pediococcus pentosaceus*, *Selenomonas lacticifex*, and *Zymomonas mobilis* (Genial PCR).

Real-time PCR: Real-time PCR is currently used in many breweries for the detection of Lactobacilli, *Pectinatus*, *Megasphaera*, and pediococci (Hutzler et al., 2008). Real-time monitoring is possible through used of fluorescent dyes which are incorporated into the PCR product through Fluorochrome-labeled oligonucleotides. The fluorescence signal increases in proportion to the amount of amplified DNA. Dyes such as ethidium bromide and SYBR Green I, which intercalate into the DNA, are the simplest way to follow at real time. The disadvantage of this method is that distinguishing between different PCR products is not possible. A further method to monitor PCR product formation uses Förster resonance energy transfer (FRET) between a donor and an acceptor molecule. The donor fluorochrome is stimulated by a light source and transfers energy to the acceptor fluorochrome that emits a fluorescent signal, which is detected. The FRET signal of the acceptor increases only with the incorporation of the donor fluorochrome into PCR product. This method provides a high specificity, but is very expensive. The FRET principle is also applied in TaqMan probes and molecular beacon probes.

A range of systems is currently available for real-time PCR with the Lightcycler the most popular (Espy et al., 2006). They are normally composed of a fluorescence measuring thermocycler, a computer, and software for operation and data analysis. A LightCycler was the first instrument based on rapid-cycle PCR. It has the capability to run 30 cycles in 10–15 min (Wittwer et al., 1997). Options for Multi-Component Analysis (MCA) and three to four fluorescence channels are standard features in modern instruments. The configuration with 384-well blocks essentially enables a low-density array setup, but nanoplate systems can accommodate up to 3072 reactions within the size of a standard microscope slide (Brenan et al., 2009). Downscaling of a PCR thermocycler on a microchip shortens the run time to a few minutes (Pipper et al., 2008; Juvonen, 2009).

The aim of real-time PCR is to detect and identify exactly the spoiling microorganisms and reduce the analysis time compared to traditional cultivation methods. However, molecular biological detection only works in combination with a quick and easy nucleic acid extraction method and precultivation to achieve the detection limit of the PCR. The detection limit for a PCR should lie in a range from 1 to 10 viable cells per sample, which can contain 10^6–10^9 yeast cells. Using a Lightcycler system, the total procedure takes 1–2 days including precultivation and DNA extraction (Kiehne et al., 2005).

Another automated platform is the GeneDisc Rapid Microbiology System. This system consists of two components: the DNA extractor and the Gene Disc Plate. The GeneDisc DNA Extractor is used to prepare samples for analysis, based on four simple steps (filtration, sonication, heating, and DNA purification). The Gene Disc Plate is a two-part molded device with the same diameter as a DVD, and enables the detection of a range of microorganisms simultaneously within the same sample DNA extract. The Cycler performs gene amplification in the plate, and each plate can be used to test 6, 9, or 12 samples in parallel. When all subunits are in use, the Cycler can analyze up to 96 samples in an hour. All known beer-spoiling bacteria species and genera can be analyzed by the system.

Multiplex-PCR: A relatively recent development in PCR technology is multiplex PCR in which genes are amplified from two or more microbes within the same reaction mix. The sequence of each gene must be sufficiently different to prevent nonspecific binding of primers, and the amplified fragments should ideally differ in size by at least 100 base pairs to discriminate between fragments easily. Beer-spoilage Lactic acid bacteria, *Pectinatus* spp. and *Megasphaera* spp. can be detected by Multiplex-PCR (Asano et al., 2008; Haakensen et al., 2007, 2008a,b; Paradh et al., 2011).

Nested PCR: In situations in which there is only a small amount of target DNA in comparison to the total sample amount of DNA, nested PCR is an option. In nested PCR, two reactions are performed consecutively. In the first PCR, the target gene is amplified in addition to unwanted sequence regions because of nonspecific binding of the primers. The products of the first reaction are then used as template for a second round of PCR with other primers that bind within the first target region and generate a product with very high specificity. Because the DNA region of choice is amplified a second time, it produces sufficient DNA for further procedures (Busch, 2010). Maugueret and Walker (2002) and Koivula et al. (2006) developed nested PCR primers for the detection of *Obesumbacterium proteus* Biotyp I and II in beer, wort, and yeast slurry. The detection limit varies in the individual matrices between 10^2 and 10^7 cells/sample.

Reverse-Transcriptase (RT)-PCR: The reactions described above all use DNA as a template for PCR. However, it is possible to convert RNA by a reverse transcriptase into complementary DNA (cDNA). This RT-PCR followed by PCR or real time-PCR is a powerful technique for the qualitative and quantitative detection of messenger RNA. Bergsveinson et al. (2012) investigated the expression level of *horA* and *horC* in *Lactobacillus brevis* and *P. clausenii* during growth in beer. A deoxyribonuclease (DNase) pretreatment has been shown to be very effective to eliminate DNA contamination when applied prior to RT-PCR analysis, resolving one of the concerns related to this technique, that of ribosomal RNA or pre-rRNA in living cells.

Loop mediated isothermal amplification: Loop mediated isothermal amplification (LAMP) is a one-step amplification reaction that amplifies a target DNA sequence with high sensitivity and specificity under isothermal conditions (about 65°C). The reaction takes place in three steps: the initial step, a cycling amplification step, and an elongation step by a DNA polymerase with strand displacement activity. For the amplification, a set of two outer and two inner primers is required, which are derived from six regions of the target sequence. It provides high amplification efficiency, with DNA being amplified 10^9–10^{10} times in 15–60 min. The increase in reaction product can be monitored by turbidity (Mori and Notomi, 2009). The method can also be combined with an RT-PCR. It should be able to amplify a few target copies and be less sensitive to nontarget DNA than PCR. A LAMP-based application for the identification of *Lactobacillus brevis*, *L. lindneri*, *P. damnosus*, and *Pectinatus* from isolated colonies in 1.5 h has been developed (Tsuchiya et al., 2007). The advantage of this technology is that significant investments in equipment are unnecessary.

Matrix-Assisted Laser Desorption Ionization Time of Flight Mass Spectrometry (MALDI-TOF): MALDI-TOF is a mass spectroscopy requiring at least 10^3 to 10^6 cells for a determination. The basic principle of MALDI is to determine the mass spectrum of the protein profile of the cells and the so-called fingerprints are unique for each microbe. After cultivation on an agar plate, a colony can be picked and then the crude cells or the extracted proteins spotted on a special slide and covered with α-cyano-4-hydroxycinnamic acid-saturated solution. The drop is dried for a few minutes. Extraction of proteins is carried out by mixing the colony with 80% ethanol followed by centrifugation for 2 min. The pellet is resuspended in 70% formic acid and the same volume of acetonitrile and again centrifuged for 2 min; 1 μL of the supernatant is then covered with α-cyano-4-hydroxycinnamic acid-saturated solution and spotted on a template and allowed to dry. Within MALDI, a laser is shone through the dried droplets propelling positively charged proteins toward the electrode. Based on the weight and the electrical charge, they fly faster or slower, and with the detector the time of flight is measured and then converted into a mass (Krishnamurthy et al., 1996). A full spectrum of proteins present is generated giving a specific "fingerprint" for the microbe. Newer research studies have demonstrated that it is possible to differentiate *Lactobacillus brevis* strains based on their spoilage potential (Kern et al., 2014).

FERMENTATION

Fermentation conditions are ideal for bacterial growth and contamination can retard or extend fermentation and cause off-flavors and odors. Typically, specific gravity, pH, and flavor are checked while brewing, and microbiological analysis is only carried out if issues arise during fermentation. Lactic and acetic bacteria and wild yeast are the main threats with a tolerance of less than 10 CFU/mL and zero count, respectively. Media for their detection is given in Table 14.9.

All rapid methods described above may be employed for the detection of microbes in fermentation samples.

Table 14.9 Microbiological Media for Fermentation Analysis

Medium	Target Microorganism(s)	Incubation Conditions
Hsu's *Lactobacillus/ Pediococcus* medium (HLP)	*Lactobacillus*, *Pediococcus*	25°C for 2 days
Lee's multi differential agar (LMDA)/Schwarz differential agar (SDA)	Lactic acid bacteria	25°C for 2 days
Lin's cupric sulfate medium (LCSM)	Wild yeast	25°C for 1–3 days
Lin's wild yeast media (LWYM)	Wild yeast	25°C for 1–3 days
Malt extract agar (MXA)	All microorganisms	22–25°C for 3 days
MRS	Enteric, acetic, and lactic bacteria, wild yeast	25°C for 2 days
MRS + Actidone	Enteric, acetic, and lactic bacteria	25°C for 2 days

PRODUCT

Once we reach the final stages of the brewing process, the sample volume is very low in relation to the batch volume (typically, 250 mL from 1000 hL) and levels of beer-spoiling microbes are extremely low. Filtration is therefore needed to improve the likelihood of detecting contaminants, and this can be carried out either in- or offline.

BRIGHT BEER

Lactic and acetic bacteria present in bright beer cause vinegary, sour astringent off-flavor and odor, excessive gassing, and strong head retention. Every batch should be tested, and the tolerance is less than 10 CFU/mL. Commonly used media for analysis of bright beer are given in Table 14.10.

The complex nature of beer results in difficulties in sensitivity/interference for many rapid methods; however, the following kits may be used in addition to PCR methods described earlier:

FISH (fluorescent in situ hybridization): Three classical FISH detection kits are currently available (VIT-beer from Vermicon) which detect the beer-spoiling organisms *Lactobacillus brevis*, *Megasphaera cerevisiae*, and *Pectinatus.* Obligate and potentially fermentative spoilage yeasts in beer and beer-based drinks may be detected. The kits can be used directly for isolates or for beer samples after an enrichment step. The target for the detection probes is the rRNA of the spoiling organisms. First, the cells are fixed on a slide and then the cell membrane has to be made permeable with an enzyme mix. A drop of reagent containing a fluorescent-labeled probe is added which can penetrate into the cell. During incubation at

Table 14.10 Microbiological Media for Bright Beer Analysis

Medium	Target Microorganism(s)	Incubation Conditions
Beer agar	All microorganisms	22–25°C for 3 days
Malt extract agar (MXA)	All microorganisms	22–25°C for 3 days
MRS	Enteric, acetic, and lactic bacteria, wild yeast	25°C for 2 days
MRS + Actidone	Enteric, acetic, and lactic bacteria	25°C for 2 days
NBB-broth	Lactic acid bacteria, *Pectinatus* and *Megasphaera*	25–30°C for 3 days; aerobic and anaerobic
Raka-Ray	Lactic acid bacteria, *Pectinatus* and *Megasphaera*	30°C for 3 days; anaerobic

46°C, hybridization of rRNA and the probes is performed. Then the slide is washed and examined under a fluorescence microscope; positive identification is made as fluorescent glowing cells (Thelen et al., 2001).

HybriScan: HybriScan is a relatively new quantitative and qualitative method also known as sandwich hybridization. The HybriScan system (Sigma–Aldrich/Scanbec) is based on the detection of rRNA via hybridization events and specific capture and detection probes. The sandwich hybridization is very sensitive, detecting attomoles of the target rRNA molecules. Firstly, the cell walls are destroyed enzymatically, and the rRNA is extracted. The method is highly specific as it uses two probes for the hybridization: capture probes, which are used to immobilize the bacteria on a streptavidin-coated microplate, and detection probes, which are used for the detection reaction. The capture probe is biotin labeled, whereas the detection probe is digoxigenin labeled. After the hybridization at 50°C, the probes and the targets are fixed on the microplate. Horseradish peroxidase is linked with the detection probe using an anti-DIG-horseradish peroxidase Fab fragment. Following a washing step, the bound complex is visualized by horseradish peroxidase substrate TMB (3,3′,5,5′-tetramethylbenzidine). Photometric data is measured at 450 nm giving a positive/negative result. It is possible to calibrate the reaction to calculate the number of colony-forming units.

The HybriScan ***D*** Beer kit detects all beer-spoiling bacteria of the genera *Lactobacillus*, *Pediococcus*, *Pectinatus*, and *Megasphaera*. The sensitivity is 1–10 CFU/L after 24–30 h preenrichment in NBB broth or isolates can directly be used.

The HybriScan ***D*** Yeast kit is used for the detection of yeasts in filterable, nonalcoholic drinks. The specificity covers yeasts including the genera *ZygoSaccharomyces*, *Saccharomyces*, *Candida*, *Dekkera*, *Torulaspora*, and *Pichia*. For direct detection and quantification at least 500 CFU/mL is recommended, after an enrichment step detection of 1–10 CFU/L is possible.

The test can be performed in 2–2.5 h and, as it is in a microtiter plate, it is quite economical and can be automated. The workflow is very similar to the enzyme-linked immunosorbent assay (ELISA) test (Taskila et al., 2010).

PACKAGING

The packaging process exposes bright beer to a range of new surfaces from the buffer tank through the filling machine and the eventual container. The potential for secondary contamination is high, with this stage representing one of the most common points for entry of spoilage organisms, and as such strict attention should be paid to both hygienic design and cleaning regimes. Process quality control parameters for should include turbidity (haze), dissolved oxygen, CO_2 content, original extract or alcohol, and the presence of Acetic and Lactic acid bacteria. A range of bacteria and yeasts may be present in biofilms, such as *Pseudomonas* and *Enterobacteria* species, *Rhodotorula* and *Cryptococcus*, and molds including *Geotrichum* and *Aureobasidium* (Back, 2003). However, the most significant spoilage organisms for brewers are *Pectinatus* and *Megasphaera*; improvements in packaging to reduce oxygen in the headspace of bottled beers has led to an increase in incidence of anaerobic beer-spoilage bacteria which are able to survive within filler heads and CO_2 recovery systems (Paradh et al., 2011).

As with other brewing surfaces, ATP testing is the most common method for detection of microbial activity within the packaging area. Media used for traditional plating tests are also identical to those described for bright beer with Raka Ray and NBB the most common media used within the UK brewing industry (Paradh et al., 2011). The range of rapid methods described above is also available to check swab and rinse samples. A recent development in rapid detection of *Pectinatus* and *Megasphaera* species has been made by Paradh et al. (2014). Acridinium ester-labeled DNA probes specific for *Pectinatus* and *Megasphaera* rRNA have been designed for use in a Hybridization Protection Assay. The assay can detect bacterial contaminants at levels of 5×10^2 to 1×10^3 CFU. Enrichment with MRS or NBB greatly enhances the detection speed (Paradh et al., 2014).

DISPENSE

At the dispense stage, product security is usually outside the control of the brewery. The most common spoilage organisms are Acetic and Lactic acid bacteria which cause surface film, haze, and vinegary flavor/odor. Wild yeasts may also proliferate in dispense lines in which biofilms are a common hazard. Cask dispensing introduces oxygen into beer adding further risk, and therefore line cleaning should be carried out at least biweekly. Tolerance is less than 10 CFU bacteria and 0 CFU wild yeast.

As with packaging, ATP testing is commonly employed. Detection media for traditional plating are detailed in Table 14.11.

Table 14.11 Media for the Detection of Dispense Spoilage Organisms

Medium	Target Microorganism(s)	Incubation Conditions
MRS	Enteric, acetic, and lactic bacteria, wild yeast	25°C for 2 days
Rainbow agar	Enteric, acetic, and lactic bacteria, *E. coli*, *Salmonella*, *Shigella*, and *Aeromonas*	18 h 37°C
WLN (Wallerstein laboratory differential)	Enteric, acetic, and lactic bacteria, yeast	2–5 days at 30–35°C; aerobic or anaerobic depending on the target of the investigation

PCR primers for the detection of Acetic Acid bacteria are available for both real-time and nested PCR (Gonzalez et al., 2006), as are primers for a range of Lactic Acid bacteria, but it is rare for companies to employ such methods at the dispense stage.

SUMMARY

The combination of hygienic plant design, effective CIP, and quality assurance of raw materials represents a sensible strategy for minimizing the risk of microbial contamination during the brewing process. However, "reactive" testing throughout is also essential in quality control. Plate counting and enrichment remain the principal methods for detection of microbes in breweries during the brewing process and in final products. In recent years, various new methods have been adopted in the brewing industry based on cell and microcolony visualization and extensive analysis of cellular and genetic content. PCR-based methods have been widely evaluated in brewing laboratories in recent years.

The wide range of media and methods available may seem overwhelming, but there are common media and tests that may be used for several/all stages, such as ATP testing, making effective control and maintenance reachable for all regardless of budget.

REFERENCES

Almeida, C., Azevedo, N.F., Fernandes, R.M., Keevil, C.W., Vieira, M.J., 2010. Fluorescence in situ hybridization method using a peptide nucleic acid probe for identification of *Salmonella* spp. in a broad spectrum of samples. Applied and Environmental Microbiology 76 (13), 4476–4485.

Asano, S., Iijima, K., Suzuki, K., Motoyama, Y., Ogata, T., Kitagawa, Y., 2009. Rapid detection and identification of beer-spoilage lactic acid bacteria by microcolony method. Journal of Bioscience and Bioengineering 108, 124–129.

Asano, S., Suzuki, K., Ozaki, K., Kuriyama, H., Yamashita, H., Kitagawa, Y., 2008. Application of multiplex PCR to the detection of beer-spoilage bacteria. Journal of the American Society of Brewing Chemists 66, 37–42.

Back, W., 2003. Biofilme in der Brauerei und getrankeindustrie. Brauwelt 24/25, 766–777.

Bergsveinson, J., Pittet, V., Ziola, B., 2012. RT-qPCR analysis of putative beer-spoilage gene expression during growth of *Lactobacillus brevis* BSO 464 and *Pediococcus claussenii* ATCC BAA-344T in beer. Applied Microbiology and Biotechnology 96, 461–470.

Bischoff, E., Bohak, I., Back, W., Leibhard, S., 2001. Schnellnachweis von bierschädlichen Bakterien mit PCR und universellen Primern. Monatsschrift fur Brauwissenschaft 54, 4–8.

Bokulich, N.A., Bamforth, C.W., 2013. The microbiology of malting and brewing. Microbiology and Molecular Biology Reviews 77 (2), 157–172.

Bouix, M., Grabowski, A., Charpentier, M., Leveau, J., Duteurtre, B., 1999. Rapid detection of microbial contamination in grape juice by flow cytometrie. Journal International des Sciences de la Vigne et du Vin 33, 25–32.

Brenan, C.J., Roberts, D., Hurley, J., 2009. Nanoliter high-throughput PCR for DNA and RNA profiling. Methods in Molecular Biology 496, 161–174.

Briggs, D.E., Brookes, P.A., Stevens, R., Boulton, C.A., 2004. Brewing: Science and Practice. Elsevier.

Busch, U., 2010. Molekularbiologische Methoden in der Lebensmittelanalytik: Grundlegende Methoden und Anwendungen. Springer Verlag, Berlin, ISBN: 978-3-642-10715-3, p. 113.

Deng, Y., Liu, J., Li, H., Li, L., Tu, J., Fang, H., Chen, J., Qian, F., 2014. An improved plate culture procedure for the rapid detection of beer-spoilage lactic acid bacteria. Journal of the Institute of Brewing 120, 127–132.

Donhauser, S., Eger, C., Hubl, T., Schmidt, U., Winnewisser, W., 1993. Tests to determine the vitality of yeasts using flow cytometrie. Brauwelt International 3, 221–224.

Espy, M.J., Uhl, J.R., Sloan, L.M., Buckwalter, S.P., Jones, M.F., Vetter, E.A., Yao, J.D.C., Wengenack, N.L., Rosenblatt, J.E., Cockerill III, F.R., Smith, T.F., 2006. Real-time PCR in clinical microbiology: applications for routine microbiological testing. Clinical Microbiology Reviews 19, 165–256.

Frischer, M.E., Floriani, P.J., Nierzwicki-Bauer, S.A., 1996. Differential sensitivity of 16S rRNA targeted oligonucleotide probes used for fluorescence in situ hybridization is a result of ribosomal higher order structure. Canadian Journal of Microbiology 42 (10), 1061–1071.

Fuchs, B.M., Syutsubo, K., Ludwig, W., Amann, R., 2001. In situ accessibility of *Escherichia coli* 23S rRNA to fluorescently labeled oligonucleotide probes. Applied and Environmental Microbiology 67 (2), 961–968.

González, A., Hierro, N., Poblet, M., Mas, A., Guillamón, J.M., 2006. Enumeration and detection of acetic acid bacteria by real-time PCR and nested PCR. FEMS Microbiology Letters 254, 123–128.

Haakensen, M.C., Butt, L., Chaban, B., Deneer, H., Ziola, B., Dowgiert, T., 2007. horA specific real-time PCR for detection of beer-spoilage lactic acid bacteria. Journal of the American Society of Brewing Chemists 65, 157–165.

Haakensen, M., Schubert, A., Ziola, B., 2008a. Multiplex PCR for putative *Lactobacillus* and *Pediococcus* beer-spoilage genes and ability of gene presence to predict growth in beer. Journal of the American Society of Brewing Chemists 66, 63–70.

Haakensen, M., Dobson, C.M., Deneer, H., Ziola, B., 2008b. Real-time PCR detection of bacteria belonging to the *Firmicutes* phylum. International Journal of Food Microbiology 125, 236–241.

Helentjaris, T., McCreery, T., 1996. Utilization of DNA probes with digoxigenin-modified nucleotides in southern hybridizations. Methods in Molecular Biology 58, 41–51.

Heilig, H.G., Zoetendal, E.G., Vaughan, E.E., Marteau, P., Akkermans, A.D., de Vos, W.M., 2002. Molecular diversity of *Lactobacillus* spp. and other lactic acid bacteria in the human intestine as determined by specific amplification of 16S ribosomal DNA. Applied and Environmental Microbiology 68, 114–123.

Hutzler, M., Schuster, E., Stettner, G., 2008. Ein Werkzeug in der Brauereimikrobiologie. Real-time PCR in der Praxis. Brauindustrie 4, 52–55.

Jespersen, L., Lassen, S., Jakobsen, M., 1993. Flow cytometric detection of wild yeast in lager breweries. International Journal of Food Microbiology 17, 321–328.

Juvonen, R., 2009. DNA-Based Detection and Characterisation of Strictly Anaerobic Beer-Spoilage Bacteria. VTT Publications, p. 723.

Juvonen, R., Koivula, T., Haikara, A., 2008. Group-specific PCR-RFLP and real-time PCR methods for detection and tentative discrimination of strictly anaerobic beer-spoilage bacteria of the class *Clostridia*. International Journal of Food Microbiology 125, 162–169.

Kempf, V.A.J., Trebesius, K., Autenrieth, I.B., 2000. Fluorescent in situ hybridization allows rapid identification of microorganisms in blood cultures. Journal of Clinical Microbiology 38 (2), 830–838.

Kern, C.C., Vogel, R.F., Behr, J., 2014. Differentiation of *Lactobacillus brevis* strains using matrix-assisted-laser-desorption-ionization-time-of-flight mass spectrometry with respect to their beer spoilage potential. Food Microbiology 40, 18–24.

Kiehne, M., Grönewald, C., Chevalier, F., 2005. Detection and identification of beer spoilage bacteria using real-time polymerase chain reaction. MBAA TQ 42 (3), 214–218.

Koivula, T., Juvonen, R., Haikara, A., Suihko, M.-L., 2006. Characterization of the brewery spoilage bacterium *Obesumbacterium proteus* by automated ribotyping and development of PCR methods for its biotype 1. Journal of Applied Microbiology 100, 398–406.

Krishnamurthy, T., Ross, P.L., Rajamani, U., 1996. Detection of pathogenic and non-pathogenic bacteria by matrix-assisted laser desorption/ionization time-of-flight mass spectrometry. Rapid Communications in Mass Spectrometry 10 (8), 883–888.

Lopez, I., Ruiz-Larrea, F., Cocolin, L., Orr, E., Phister, T., Marshall, M., Van der Gheynst, J., Mills, D.A., 2003. Design and evaluation of PCR primers for analysis of bacterial populations in wine by denaturing gradient gel electrophoresis. Applied and Environmental Microbiology 69, 6801–6807.

Maugueret, T.M.-J., Walker, S.L., 2002. Rapid detection of *Obesumbacterium proteus* from yeast and wort using polymerase chain reaction. Letters in Applied Microbiology 35, 281–284.

Mirocha, T.J., Christensen, T.M., 1986. Mycotoxins and the fungi that produce them. Proceedings of the American Phytopathological Society 3, 110–125.

Mori, Y., Notomi, T., 2009. Loop-mediated isothermal amplification (LAMP): a rapid, accurate, and cost-effective diagnostic method for infectious diseases. Journal of Infection and Chemotherapy 15 (2), 62–69.

Njambere, E.N., Clarke, B.B., Zhang, N., 2011. Dimeric oligonucleotide probes enhance diagnostic macroarray performance. Journal of Microbiological Methods 86, 52–61.

Neeley, E.T., Phister, T.G., Mills, D.A., 2005. Differential real-time PCR assay for enumeration of lactic acid bacteria in wine. Applied and Environmental Microbiology 71, 8954–8957.

Paradh, A.D., Hill, A.E., Mitchell, W.J., 2014. Detection of beer spoilage bacteria *Pectinatus* and *Megasphaera* with acridinium ester labelled DNA probes using a hybridisation protection assay. Journal of Microbiological Methods 96, 25–34.

Paradh, A.D., Mitchell, W.J., Hill, A.E., 2011. Occurrence of *Pectinatus* and *Megasphaera* in the major UK breweries. Journal of the Institute of Brewing 117 (4), 498–506.

Pipper, J., Zhang, Y., Neuzil, P., Hsieh, T.M., 2008. Clockwork PCR including sample preparation. Angewandte Chemie International Edition in English 47 (21), 3900–3904.

Priest, F.G., Stewart, G.G., 2006. Handbook of brewing. In: Microbiology and Microbiological Control in the Brewery, second ed. CRC Press, pp. 607–627. Print ISBN:978-0-8247-2657-7, eBook ISBN:978-1-4200-1517-1.

Quain, D., Storgårds, E., 2009. The extraordinary world of biofilms. PLoS Biology 5, 2458–2461.

Raybourne, R., Tortorello, M., 2003. Detecting pathogens in food. In: McMeekin, T.A. (Ed.), Microscopy Techniques: DEFT and Flow Cytometry. Woodhead Publishing Limited, Cambridge. Print Book ISBN:9781855736702, eBook ISBN:9781855737044.

Restaino, L., Castillo, H.J., Stewart, D., Tortorello, M.L., 1996. Antibody-direct epifluorescent filter technique and immunomagnetic separation for 10-h screening and 24-h confirmation of *Escherichia coli* O157:H7 in beef. Journal of Food Protection 59, 1072–1075.

Rytkönen, A., Lilja, A., Hantula, J., 2012. PCR–DGGE method for in planta detection and identification of *Phytophthora* species. Forest Pathology 42, 22–27.

Saiki, R.K., Gelfand, D.H., Stoffel, S., Scharf, S.J., Higuchi, R., Horn, G.T., 1988. Primer-directed enzymatic amplification of DNA with a thermostable DNA polymerase. Science 239, 487–491.

Sikora, K., Verstappen, E., Mendes, O., Schoen, C., Ristaino, J., Bonants, P., 2012. A universal microarray detection method for identification of multiple *Phytophthora* spp. using padlock probes. Phytopathology 102, 635–645.

Stewart, R.J., Dowhanick, T.M., 1996. Rapid detection of lactic acid bacteria in fermenter samples using a nested polymerase chain reaction. Journal of the American Society of Brewing Chemists 54, 78–84.

Suzuki, K., 2011. 125th Anniversary review: microbiological instability of beer caused by spoilage bacteria. Journal of the Institute of Brewing 117, 131–155.

Taskila, S., Tuomola, M., Kronlöf, J., Neubauer, P., 2010. Comparison of enrichment media for routine detection of beer spoiling lactic acid bacteria and development of trouble-shooting medium for *Lactobacillus backi*. Journal of the Institute of Brewing 116 (2), 151–156.

Thelen, K., Beimfohr, C., Bohak, I., Back, W., 2001. Spezifischer Schnellnachweis von Bierschädlichen Bakterien mittels fluoreszenzmarkierter Gensonden. Brauwelt 141 (38/01), 1596–1603.

Tortorello, M.L., Gendel, S.M., 1993. Fluorescent antibodies applied to direct epifluorescent filter technique for microscopic enumeration of *Escherichia coli* O157:H7 in milk and juice. Journal of Food Protection 56, 672–677.

Tortorello, M.L., Reineke, K.F., 2000. Direct enumeration of *Escherichia coli* and enteric bacteria in water, beverages and sprouts by 16S rRNA in situ hybridization. Food Microbiology 17, 305–313.

Tortorello, M.L., Stewart, D.S., 1994. Antibody-direct epifluorescent filter technique for rapid, direct enumeration of *Escherichia coli* O157:H7 in beef. Applied and Environmental Microbiology 60, 3553–3559.

Tortorello, M.L., Reineke, K.F., Stewart, D.S., 1997. Comparison of antibody-direct epifluorescent filter technique with the most probable number procedure for rapid enumeration of *Listeria* in fresh vegetables. Journal of AOAC International 80, 1208–1214.

Tsuchiya, Y., Ogawa, M., Nakakita, Y., Nara, Y., Kaneda, H., Watari, J., Minekawa, H., Soejima, T., 2007. Dentification of beer-spoilage microorganisms using the loop-mediated isothermal amplification method. Journal of the American Society of Brewing Chemists 65, 77–80.

Vang, Ó.K., Albrechtsen, H.J., Bentien, A., Smith, C., 2013. ATP Measurements for Monitoring Microbial Drinking Water Quality (Doctoral dissertation, GRUNDFOS Holding A/SGRUNDFOS Holding A/S).

Walter, J., Hertel, C., Tannock, G., Lis, C., Munro, K., Hammes, W., 2001. Detection of *Lactobacillus*, *Pediococcus*, *Leuconostoc*, and *Weissella* species in human feces by using group-specific PCR primers and denaturing gradient gel electrophoresis. Applied and Environmental Microbiology 67, 2578–2585.

Wittwer, C.T., Ririe, K.M., Andrew, R.W., David, D.A., Gundry, R.A., Balis, U.J., 1997. The LightCycler: a microvolume multisample fluorimeter with rapid temperature control. BioTechniques 22, 176–181.

Wong, M., Smart, C.D., 2012. A new application using a chromogenic assay in a plant pathogen DNA macroarray detection system. Plant Disease 96, 1365–1371.

APPENDIX

Practicalities of Achieving Quality

C.W. Bamforth

University of California, Davis, California, CA, United States

At present, early in the 21st century, beers have never been more consistent. The achievement of this speaks not only to an awareness of the vast complexity of factors that impact quality but also an understanding of how to control them. The entire success or failure of a company stands on its ability to consistently deliver product to the satisfaction of the consumer.

DEFINITIONS OF QUALITY

There are many definitions of *quality*:

> *Quality is the achievement of consistency and the elimination of unwanted surprises.*
>
> *The supply of goods that do not come back to customers who do.*
>
> *The extent of correspondence between expectation and realization.*
>
> *The match between what you want and what you get.*

Quality is very much a personal issue. There are innumerable types of beer worldwide, and no drinker is expected to like all of them equally. However, the problem is more subtle than that. For example, a beer leaving a brewery destined for export is expected to be inherently "clean and bright," whereas by the time it has traversed an ocean or two, encountered extremes of temperature, and only emerged on supermarket shelves after prolonged storage in a warehouse, then it will certainly display aged aroma and may even be somewhat turbid. Although the brewer, trained to champion fresh beer that is in most (but not all) cases brilliantly clear, will be horrified to sample that product at retail, the customer may actually conclude that this is what the beer should taste and look like and, indeed, any effort by the brewer to lessen the deterioration may be met with complaint.

RESPONSIBILITY FOR QUALITY

Quality is not the exclusive preserve of a Quality Department. To achieve quality depends on a commitment from all employees—in other words, a holistic quality environment. It is fashionable to speak of *Total Quality Management* (TQM), in which everybody in a company has a commitment to excellence, manifesting itself in superiority in the organization. The provision of a comfortable, attractive, and pleasing working environment, for example, will lead to dividends in quality performance through the enhanced motivation and pride of the workforce.

Quality must be to the forefront in every aspect of a company's operations: purchasing, manufacture, marketing, etc. It can only succeed if driven from the highest level of management.

QUALITY SYSTEMS

TQM is, in part, achieved via the adoption of a quality system. Nowadays, formalized approaches are available, embodied in international standards such as those of the International Organization for Standardization (ISO) 9000 series. These are exercises in focusing, achieving compliance, and ensuring that systems are documented. They represent good discipline. However, it must be realized that they do not necessarily ensure that a product is "good" or "bad." They do not guarantee that process stages are necessarily the correct ones. All they seek to do is ensure that standard procedures are followed. It is up to the company to ensure that best practices prevail. To have such accreditation is really a stamp of approval that a company is paying heed to the need for a quality system.

There are companies that actively seek out suppliers who possess this type of accreditation. It is rather more farsighted for companies to look for suppliers that have genuine quality systems in place, which in effect means the presence of, and unwavering adherence to, a Quality Manual. This should document everything that pertains to the product: specifications for the beer, for the raw materials from which it will be derived, and for key control points in the process; standard operating procedures (SOPs) for everything from materials purchase, storage, handling, and use, all the way through the brewery to beer shipment.

The most common quality management standard is ISO 9001:2015. It specifies the quality system that proves the capability to manufacture and supply a product to an established specification. It defines the activities that must be controlled, but not how they are to be controlled.

The key elements of a quality system are:

1. A quality policy, usually signed by the senior executive, which speaks of the company's commitment to quality and to meeting the needs of customers.
2. A Quality Manual that outlines and maps the scope and elements of the quality management system and how these may be cross-referenced to relevant procedures and work instructions.
3. Written procedures for all key operations that affect product quality.
4. Systems for ensuring uniform use of updated documentation, including specifications, control diagrams and schemes, codes of practice, operating manuals, Hazard Analysis and Critical Control Points (HACCP) (see later).

5. Secure records to demonstrate that the system is working effectively and meets legal and contractual requirements.
6. A schedule of meetings to review the quality system.
7. Training of all staff whose jobs impact product quality.
8. Control of purchasing of all materials and services that can affect product quality; eg, lists of approved suppliers, auditing regimes, etc.
9. Calibration programs for equipment according to recognized standards.
10. Internal audits to ensure that the system is effectively implemented and maintained.
11. Systems for handling nonconforming product.
12. Systems for handling complaints.
13. Systems for adjusting standard procedures so that problem recurrence is avoided.

QUALITY ASSURANCE VERSUS QUALITY CONTROL

Quality Control (QC) is a reactive approach, one which is invariably associated with waste in that it seeks to respond to measurements that are made and effect corrections if the values are outside specification. Much more effective is the Quality Assurance (QA) approach, in which systems are introduced to *ensure* that the product at every stage in its production is within specification. The emphasis is one of *prevention* rather than *detection*. Brewers speak of "*right the first time*."

Analysis and quantification of the process and product are, of course, necessary. However, it is preferable that these measurements should be proactive rather than reactive. For example, measuring and controlling wort gravity, oxygenation, and viable-yeast pitching rate are vastly preferable in the pursuit of controlled fermentation than is the monitoring of specific gravity in fermenter and response to deviations arising because the appropriate control strategies had not been put in place. Wherever possible, measurements are made either automatically by an inline or online device linked to a feedback loop (for example, inline measurement of turbidity to feedback control a filter-aid dosing pump), or manually by the person operating that part of the process. This individual can take ownership of correcting the issue, rather than waiting for analysis to be performed by a laboratory technician with the attendant delays and lack of sensitivity of response.

SPECIFICATIONS

Achievement of consistency demands that meaningful specifications are brought to bear. The need is for specifications that pertain to the finished product, that allow a quantified or at least qualified gauge of how close any batch of product is to perfection. Specifications are needed to allow selection and handling of raw materials.

Moreover, specifications are required at all points from raw materials to product that guarantee that the stream at all stages is fit for use.

Specifications should be realistic and meaningful. They need to relate to what is expected in the product but not set to such a rigor that they are unachievable. For example, if we demand that our small-pack beer display minimum-aged character in six months of optimum storage, then the specification for oxygen needs to be as low as possible but nonetheless not so low that cannot be achieved consistently in the equipment at our disposal. The same considerations apply to the specifications that we set on our raw materials.

THE COST OF QUALITY

We can consider quality-related costs under the headings of internal failure, external failure, appraisal, and prevention.

Internal failure costs are the ones associated with product being out of specification at any stage in the production process and before the beer has left the brewery for retail. They can be subdivided into

- Rework: The cost of correcting matters to return a product to "normal," for example rectifying carbon dioxide levels in a bright-beer tank out of specification.
- Reinspection—The cost of checking reworked material
- Scrap—Product that is beyond repair, for example beer that has become contaminated.
- Analysis of failure—The cost of investigating the cause of the internal failure.

External failure costs pertain to the implication of a product actually getting out into trade with a quality defect.

- Recall—The cost of investigating problems, recovering product, and probably above all else, the cost of lost reputation and market.
- Warranty—The cost of replacing product
- Complaints—The cost of handling customer's objections
- Liability—The implications of litigation

Appraisal costs are the expense of analyses throughout the process, from raw materials to product. They can be subdivided into

- Inspection and test: Analytical methodology, whether in laboratory or online
- Internal auditing
- Auditing of suppliers

Prevention costs are the ones devoted to operating a well-run QA system—See Table 1.

Table 1 Elements of a Quality Assurance Program

1. A defined quality system: what is worth measuring, when, where, how often, and by whom
2. Established quality standards and procedures for raw materials with spot-checking and auditing protocols
3. Audits and surveys of processes and procedures
4. Projects—eg, installation of inline analysis and feedback control systems
5. Training and auditing of operators on QC checks
6. Collaborative exercises—eg, interlaboratory method checking
7. Specialized analysis and procedures within a QA laboratory
8. Complaints handling procedures
9. Quality awareness campaigns
10. Instrument checking and calibration

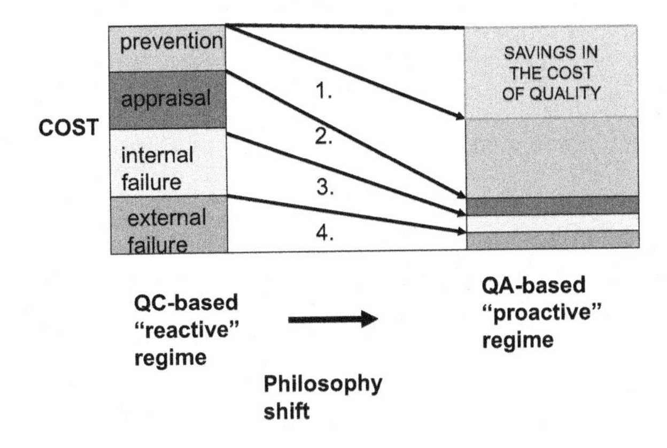

FIGURE 1

The cost impact of moving to a Quality Assurance strategy.

Derived from Bamforth, C.W., 2002. Standards of Brewing. Brewers Publications, Boulder, Colorado.

Fig. 1 illustrates that increased investment in prevention enables a reduction in appraisal costs but much more importantly allows for greatly reduced losses accruing from internal and external failure.

STATISTICAL PROCESS CONTROL

Robust analysis in any production operation depends on the generation of reliable data and an appreciation of its relevance and reliability.

It is important to have tools that "locate" an individual measurement and its status in relation to other values. It is also necessary to quantify how "disperse[sic]" the data are—How broad is the spread?

Most values in an adequate population are distributed as illustrated in Fig. 2. The *mean* (M) is the average of all the values (x_i), obtained by adding them up (Σ) and dividing by the number of measurements (n).

$$M = \frac{1}{n}\sum x_i$$

The extent to which the values are dispersed (ie, the subtraction of the lowest value from the highest value) tells us how widely spread the data set is, but it does not signify where the data is congregated. The deviation of any individual measurement from the mean is given by the expression $x_i - M$. The *standard deviation* (σ) is given by

$$\sigma = \sqrt{\frac{1}{n-1}\sum [x_i - M]^2}$$

σ^2 is called the *variance*.

The coefficient of variation (CV) is a simple way to illustrate variation.

$$\mathrm{CV} = \frac{100\sigma}{M}$$

It allows expression on a percentage basis of the error inherent in a method. The lower is the CV, the more reliable is an analysis.

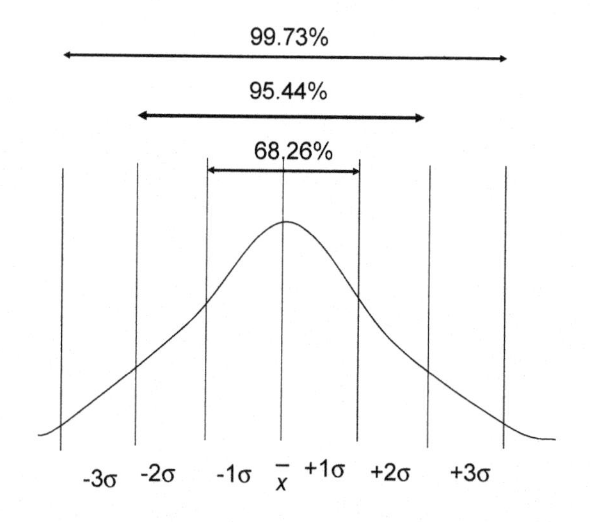

FIGURE 2

Standard distribution.

Derived from Bamforth, C.W., 2002. Standards of Brewing. Brewers Publications, Boulder, Colorado.

The normal distribution, as depicted in Fig. 2, is an illustrative way of describing probability, ie, the likelihood of a value being a certain distance from the mean. Thus, there is a 68.26% chance of the value being within one standard deviation of the mean, 95.44% probability of it being within two standard deviations, and a 99.73% chance of it being within three standard deviations.

PROCESS CAPABILITY

No system is immune from variation and some degree of inconsistency or "*noise*." There are two types of noise:

- Internal noise, examples of which include fluctuations within a batch of raw materials (malt, hops, yeast, water, etc.), sampling inconsistencies, and wear and tear in machinery
- External noise, such as human factors (different operators) and differences between sources of raw materials (eg, harvest year)

If a process generates a product that has a stable mean value, then it is said to be "in control." There may be much variation about the mean, but only within clearly understood limits. When a process deviates more than this, leading to a shift in the mean or an increase in the variability, it is then "out of control."

Process capability is given by

$$\frac{\text{Upper limit of measurement} - \text{lower limit of measurement}}{6\sigma}$$

The denominator 6σ speaks to the breadth of a normal distribution (see Fig. 2) and has in recent years been used as the name for a well-known approach to quality.

CONTROL CHARTS

Charting data makes it easier to spot when a process is moving out of control. Properly done, these charts allow the operator to determine whether any changes are within or outside normal variation. If a change is observed in a parameter, but nonetheless it was entirely within normal fluctuation, it would be foolhardy to make adjustments. Only those parameters the value for which will lead to a go/no-go decision are measured on every batch.

The simplest type of control diagram to use is to plot data within a framework of action (reject) and warning lines (Fig. 3). Convention has it that these lines should be positioned at three standard errors ($3\sigma/\sqrt{n}$) above and below. This explains 99.73% of the variation in data for a normal distribution. It is often the case that two other "warning lines" within the action lines are used. These are set at two standard errors ($2\sigma/\sqrt{n}$). The action and warning lines are sometimes referred to as the Reject Quality Limits (RQL) and Acceptable Quality Limits (AQL), respectively.

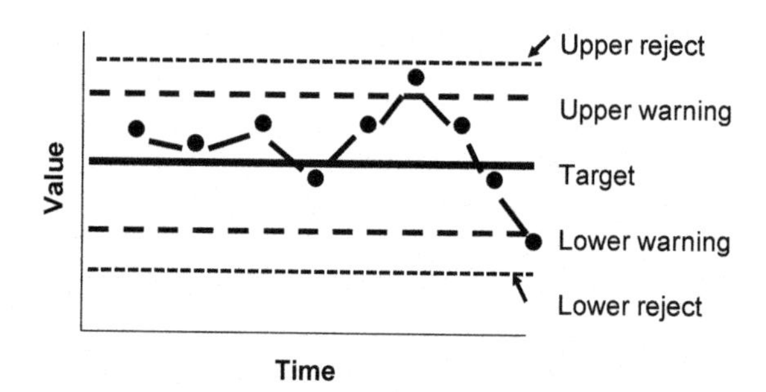

FIGURE 3

Trend plotting.

Derived from Bamforth, C.W., 2002. Standards of Brewing. Brewers Publications, Boulder, Colorado.

Another type of plot frequently used is the cumulative sum (CUSUM) plot

$$\mathrm{CUSUM} = \sum_i (x_i - T)$$

in which x_i is the ith measurement of a quantity and T is the target value (what the measurement ideally should be, or *specification*). Therefore, the CUSUM is the summation of all the deviations from normality.

CUSUM plots are valuable for highlighting where substantial changes are occurring in a parameter. When the slope of these plots trends upward, this indicates that the average is somewhat above the target value. When the slope trends downward, this indicates that the average is somewhat below the target value. The more pronounced the slope, the more does the value deviate from target.

STANDARD METHODS OF ANALYSIS

Worldwide, there are several sets of methods, with many overlaps but some significant differences, with origins lying in brewing practices in separate nations. The standard methods of the Institute of Brewing (now called the Institute of Brewing and Distilling, IBD) were originally developed in England for the analysis of ale-type beers and so, for example, use small-scale infusion mashes at constant temperature. They have now been subsumed into the methods of the European Brewery Convention (EBC), which are based on the production of lager-style products: their mashes (Congress mashes) have a rising temperature regime. There have been major efforts to merge the two different compendia. The methods of analysis of the American Society of Brewing Chemists (ASBC) owe more to the EBC than the IBD, but they form the analytical reference point for all North American brewers, whether producing lagers or ales. The box offers a crib of the contents of the ASBC methods. Mention is also made in the present book of the Central European Brewing Analysis Commission (MEBAK) methods, which are the ones routinely used in Germany and a few other places.

Methods are written in a standard format that offers the best opportunity for whosoever is pursuing it to be consistent in their approach. The method is then distributed to different laboratories, together with a set of samples that have been produced in a single location. Following analysis, the values obtained are returned to a central data coordinator.

There may be several sources of variance (defined earlier) in analytical data. Although not applicable in the establishment of standard procedures in which carefully regulated samples are distributed from a central location, for routine samples one of the main risks is associated with unrepresentative sampling. The other sources of variance (replication error and systematic error) have diverse origins: different batches of chemicals, unconscious deviations from the laid-down method, contaminating species in glassware or water, atmospheric conditions, human imperfections, and so on. The extent to which each of these matters depends on method robustness: the more robust the procedure, the less the scatter.

The usual approach to testing for errors was developed by Youden and is used not only in setting up new recommended methods but also routinely for screening and comparing performance on existing methods between laboratories, eg, the different laboratories across locations within a major brewing company. In this technique, the various labs are sent a pair of samples, representing two different levels of the analyte under examination. Each lab would be asked to make their measurements on this parameter in the two beers, the values for beer one being the *X* series and those for beer two being *Y* values. The collated data is then plotted as shown in Fig. 4. The circle has a radius that is determined by multiplying the standard deviations of the replication error. Essentially, there is a 95% probability

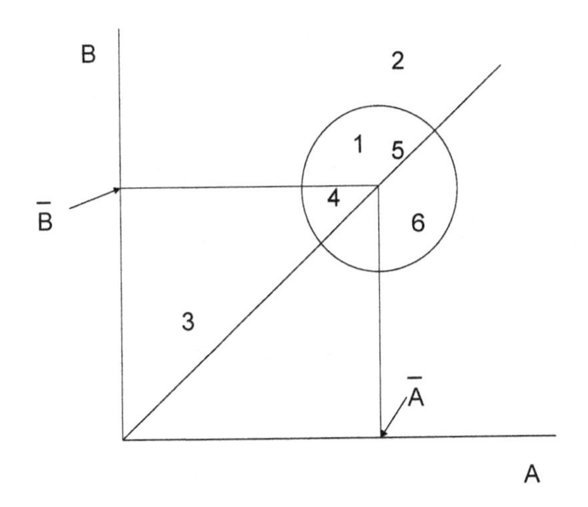

FIGURE 4

Youden plot.

Derived from Bamforth, C.W., 2002. Standards of Brewing. Brewers Publications, Boulder, Colorado.

Table 2 Procedure for Calculating Errors Using the Youden Method

	Laboratory 1	Laboratory 2	Laboratory *n*	Total	Average
Result A	A_1	A_2	A_n	$\sum A$	$\overline{A}$
Result B	B_1	B_2	B_n	$\sum B$	$\overline{B}$
A − B	X_1	X_2	X_n	$\sum X$	$\overline{X}$
A + B	Y_1	Y_2	Y_n	$\sum Y$	$\overline{Y}$

$$\text{Standard deviation of the total error} = \frac{\sum (Y-\overline{Y})^2}{2(n-1)}$$

$$\text{Standard deviation of the replication error} = \frac{\sum (X-\overline{X})^2}{2(n-1)}$$

of a result falling within the circle. The digits on the plot indicate the data generated by different laboratories, in this example revealing that lab 2 is reliable for measure A though not B, and lab 3 underestimates in both instances. Table 2 illustrates how the data are handled.

Two other values are important. The first of these is *repeatability* (*r*), which is an index of how consistently data can be generated by a single operator in a single location using a standard method. The difference between two single results found on identical test material by one operator using the same apparatus within the shortest feasible time interval will exceed *r* on average not more than once in 20 cases in the normal and correct operation of the method. The second parameter is *reproducibility* (*R*), this being the maximum-permitted differences between values reported by different labs using a standard method. This is the value obtained in collaborative trials between labs. Thus, the lower are the *r* and *R* values, the more reliable is the procedure.

SETTING SPECIFICATIONS AND MONITORING PERFORMANCE

The information gleaned in this type of intercollaborative exercise helps the brewer achieve the goal of setting realistic and meaningful specifications for the product, raw materials, and process samples.

When setting specifications, there are two key requirements:

1. Knowledge of the *r* and *R* values, which indicate how reliable a method is, and
2. Appreciating the true range over which a parameter can vary before a change is observed in quality.

WHAT IS MEASURED

Tables 3–6 give typical analytical elements for malt and hops in process and final product, respectively. Incoming water will have a depth of analysis associated

Table 3 Components of a Malt Specification

Parameter	Rationale
Moisture	Excess moisture reduces stability. Brewer does not wish to buy water.
Hot water extract	Indicates potential extractable material obtainable in the brewhouse
Saccharification time	Indicates sufficiency of starch-degrading enzymes.
Color	Correct color for finished beer style. Indication of kilning severity.
Protein	Excess protein means less carbohydrate extract, plus increased haze risk
Kolbach index (also known as soluble nitrogen ratio or soluble/Total protein ratio)	Measure of protein modification—relevant re: colloidal stability
Diastatic power	Total starch-degrading enzyme activity
Wort pH	Impacts extraction and stability
β-Glucan in wort	Predictor of wort separation and beer filtration problems
Friability	Assessment of extent of malt modification and predictor of wort separation problems
Partly unmodified grains ("homogeneity")	Predictor of wort separation and beer filtration problems
Dust and extraneous materials	Index of unusable and potential problematic material that would need to be screened out
Dimethyl sulfide precursor	Potential dimethyl sulfide levels in lager beers
Arsenic, lead, nitrosamines, ochratoxin A, deoxynivalenol	Food safety issues
Gushing	Predictor of overfoaming in end product

Additionally, the contract is likely to specify variety, minimum storage period, extent of pesticide usage, freedom from genetically modified material.

Table 4 Components of a Hop Specification (Pellets)

Parameter	Rationale
Moisture	Stability
α-Acid	Potential bitterness
Pesticides	Food safety issue
Heavy metals	Food safety and product quality issue
Hop storage index	Extent of deterioration of hop

Table 5 Key Inprocess Checks

Check	Frequency	Rationale
(a) Milled grist		
Particle size distribution	Weekly	Confirmation of correct mill gaps
(b) Mash		
Iodine test	Every mash	Indicator of starch conversion
(c) Sweet Wort		
Specific gravity	Every batch	Monitoring extract recovery, cut-off and strength of wort to kettle
pH	Every batch	Cut-off point for wort collection. Check for cleaning agent contamination
Clarity	Varies	Some brewers believe it relates to product quality
(d) Boiling and hopped wort		
Specific gravity	Every batch	Strength of wort to next stage and per cent evaporation
pH	Every batch	No cleaning agent contamination
Bitterness	Varies	Hop utilization
Color	Varies	Predictor of color in final product
Clarity	Varies	Some brewers believe it relates to product quality
Oxygen in cooled wort	Every batch	Needed for yeast
(e) Yeast		
Viability	Every batch	To refine pitching rate and monitor deterioration in yeast performance
Cell count	Every batch	To define pitching rate
(f) Fermentation and "green beer"		
Specific gravity	Every batch	Progress of fermentation
pH	Every batch	Progress of fermentation
Total vicinal diketones (ie, diacetyl and pentanedione and their precursors)	Every batch	Avoidance of flavor problems

Monitoring of temperature at all stages is a given, as too is tasting and smelling. The efficiency of inplace cleaning systems should be evaluated by monitoring the strength of detergent and by screening swabs of tanks of postrinse samples using traditional or rapid microbiological tests.

Table 6 Quality Control Checks on Beer

Parameter	Frequency
Clarity	Every batch
CO_2	Every batch
Foam	Pour—every batch Instrumental method—monthly
Haze breakdown	Small pack—monthly
Color	Every batch
Ethanol	Every batch
Apparent extract (ergo original extract)	Monthly
Fermentable extract	Monthly
Bitterness	Every batch
Free amino nitrogen	Monthly
pH	Every batch
Volatile compounds	Monthly
Inorganics (notably iron, copper, sulfate, chloride, nitrate)	Monthly
Sulfur dioxide	Every batch if there are legal requirements to label if above a certain level (eg, 10 ppm in United States); otherwise monthly
Polyphenols	Monthly
O_2	Every batch
N_2 (if used)	Every batch
Microbiological status	Monthly
Taste clearance	Every batch
Flavor profile (or trueness to type)	Monthly

From Bamforth, C.W., 2002. Standards of Brewing. Brewers Publications, Boulder CO.

with it that far exceeds any of these, but almost without exception, it is the raw material the analysis for which is left in its entirety to the supplier (Tables 7 and 8). The brewer should, however, constantly be checking its organoleptic quality and its pH on arrival into the brewery and at all places where it is processed (carbon filtration, ion exchange, and deaeration).

Table 7 Extract from the National Primary Drinking Water Regulations

Component	Maximum Contaminant Level Goal	Maximum Contaminant Level (mg/L Unless Stated)	Potential Health Effects	Sources of Contaminant
Cryptosporidium or *Giardia*	Zero	99–99.9% removal/ inactivation	Diarrhea, vomiting, cramps	Fecal waste
Legionella	Zero	Deemed to be controlled if *Giardia* is defeated	Legionnaire's Disease	Multiplies in water heating systems

Continued

Table 7 Extract from the National Primary Drinking Water Regulations—cont'd

Component	Maximum Contaminant Level Goal	Maximum Contaminant Level (mg/L Unless Stated)	Potential Health Effects	Sources of Contaminant
Coliforms (including *Escherichia coli*)	Zero	No more than 5% samples positive within a month	Indicator of presence of other potentially harmful bacteria	Coliforms naturally present in the environment; *E. coli* comes from fecal waste
Turbidity	n/a	<1 nephelometric turbidity unit.	General indicator of contamination, including by microbes	Soil runoff
Bromate	Zero	0.01	Risk of cancer	Byproduct of disinfection
Chlorine	4	4	Eye/nose irritation; stomach discomfort	Additive to control microbes
Chlorine dioxide	0.8	0.8	Anemia; nervous system effects	Additive to control microbes
Haloacetic acids (eg, trichloracetic)		0.06	Risk of cancer	Byproduct of disinfection
Trihalomethanes		0.08	Liver, kidney, or central nervous system ills, risk of cancer	Byproduct of disinfection
Arsenic		0.05	Skin damage, circulation problems, risk of cancer	Erosion of natural deposits; runoff from glass and electronics production wastes
Asbestos	7 million fibers per liter	7 million fibers per liter	Benign intestinal polyps	Decay of asbestos cement in water mains; erosion of natural deposits
Copper	1.3	1.3	Gastrointestinal distress, liver, or kidney damage	Corrosion of household plumbing systems; erosion of natural deposits
Fluoride	4	4	Bone disease	Additive to promote strong teeth; erosion of natural deposits
Lead	Zero	0.015	Kidney problems; high blood pressure	Corrosion of household plumbing systems; erosion of natural deposits

Table 7 Extract from the National Primary Drinking Water Regulations—cont'd

Component	Maximum Contaminant Level Goal	Maximum Contaminant Level (mg/L Unless Stated)	Potential Health Effects	Sources of Contaminant
Nitrate	10	10	Blue Baby Syndrome	Runoff from fertilizer use, leaching from septic tanks, sewage, erosion of natural deposits
Nitrite	1	1	Blue Baby Syndrome	Runoff from fertilizer use, leaching from septic tanks, sewage, erosion of natural deposits
Selenium	0.05	0.05	Hair or fingernail loss, circulatory problems, numbness in fingers and toes	Discharge from petroleum refineries, erosion of natural deposits, discharge from mines
Benzene	Zero	0.005	Anemia; decrease in blood platelets; risk of cancer	Discharge from factories; leaching from gas storage tanks and landfills
Carbon tetrachloride	Zero	0.005	Liver problems; risk of cancer	Discharge from chemical plants and other industrial activities
Dinoseb	0.007	0.007	Reproductive difficulties	Runoff from herbicide use
Dioxin	Zero	0.00000003	Reproductive difficulties, risk of cancer	Emissions from waste incineration and other combustion; discharge from chemical factories
Alpha particles	Zero (as of 12/8/03)	15 pCi per liter	Risk of cancer	Erosion of natural deposits
Beta particles and photon emitters	Zero (as of 12/8/03)	4 mrem per year	Risk of cancer	Decay of natural and humanmade deposits

Reproduced from Lewis, M.J., Bamforth, C.W., 2006. Essays in Brewing Science. Springer, New York. (The full table can be found at http://www.epa.gov/safewater/mcl.html).

Table 8 National Secondary Drinking Water Regulations

Contaminant	Secondary standard
Aluminum	0.05–0.2 mg/L
Chloride	250 mg/L
Color	15 color units
Copper	1 mg/L
Corrosivity	Noncorrosive
Fluoride	2 mg/L
Foaming agents	0.5 mg/L
Iron	0.3 mg/L
Manganese	0.05 mg/L
Odor	3 threshold odor number
pH	6.5–8.5
Silver	0.1 mg/L
Sulfate	250 mg/L
Total dissolved solids	500 mg/L
Zinc	5 mg/L

These are nonenforceable guidelines regulating contaminants that may cause cosmetic effects (eg, skin or tooth discoloration) or aesthetic effects (taste, odor, color). States may choose to adopt them as enforceable standards.
Reproduced from Lewis, M.J., Bamforth, C.W., 2006. Essays in Brewing Science. Springer, New York.

HAZARD ANALYSIS CRITICAL CONTROL POINTS (HACCP)

HACCP is a management system designed to assure the safety of food products.

Several countries have legal requirements for HACCP.

Prerequisites for implementing HACCP are:

1. A broadly articulated policy declaring the company's commitment to making safe products.
2. "Securing of the borders," meaning confidence that there is no threat from surrounding industries and from accidental or malicious contamination.
3. Identification of those elements of the process that present the greatest risk, for example packaging locales. In the latter, there needs to be rigorous policies regarding the presence of glass.
4. Rigor in buildings and equipment, ensuring that they are properly cleaned and maintained. Consideration should be given inter alia to restroom facilities and pest control.
5. Rigor in all suppliers of materials, with checking procedures at receipt.
6. Precautions regarding the inviolability of vehicles employed for raw materials arriving at the brewery and beer and co-products leaving it.
7. Proper training programs for all employees.
8. Existence of a robust product-recall system.

Customary steps in implementing HACCP are

1. Expression and realization of the commitment to the policy and procedures of senior management.
2. Definition of scope and extent, from raw material delivery to shipment of product from the brewery.
3. Establishment of HACCP team(s).
4. Generation of a process flow chart that includes all raw materials and process steps within the scope of HACCP.
5. Identification by the HACCP team(s) of all potential hazards and determination of appropriate control measures. Hazards include chemical and microbiological contamination (eg, deoxynivalenol in malt, caustic contamination from faulty inplace cleaning, glycol from leaking wort coolers) and physical risks (eg, broken glass or filling tubes in packaging).
6. Identification of the Critical Control Points (CCP), viz. the stages in which controls must be introduced to minimize hazards.
7. Establishment of what the critical limits are to be at each CCP—ie, realistic specifications bases on ready measurement capabilities.
8. Establishment of response procedures for each CCP, ie, corrective actions and records.
9. Training of personnel
10. Updating whenever there are changes in raw materials, equipment, or declared process procedures.

A SUMMARY OF THE ASBC METHODS OF ANALYSIS

This list offers an indication of the breadth of methods available through the American Society of Brewing Chemists. All methods and supplementary material (including training videos) is available on line to members of ASBC (http://methods.asbcnet.org/).

1 Barley Methods

1. Barley-1. Importance of representative sampling. Triers for static beds (eg, boxcars) or "spout samplers" or "belt samplers" for grain under transfer. Grading on basis of malting vs. feed quality and on other attributes, such as damage.
2. Barley-2. Physical tests.
   Variety by physical characteristics: appearance.
   Test weight per bushel (US) is "the weight in pounds of the volume of grain required to fill level full a Winchester bushel measure of 2150.42 cubic inches capacity."
   Assortment by sieving
   Thousand kernel weight—Weight of a 1000 kernels
   Texture of endosperm—Mealy or steely?
   Skinned and broken kernels
   Weathering and kernel damage
3. Barley-3 Germination
   Germinative Energy (GE)—Assessment of dormancy
   Germinative Capacity (GC)—Viability
   GE/GC/Water sensitivity simultaneously
4. Barley-4. Milling for analysis: Obtaining a representative powder
5. Barley-5. (A) Moisture by drying. (B) Alternate oven method. (C) NIR

Continued

A SUMMARY OF THE ASBC METHODS OF ANALYSIS—cont'd

6. Barley-6. Extract—Try to predict what you will get after malting using enzymes
7. Barley-7. Protein (A) Kjeldahl and N $\times$ 6.25; (B) NIR; (C) Combustion and N $\times$ 6.25; (D) Protein in whole grain by NIR
8. Barley-8. Potential diastatic power
9. Barley-9. Kernel brightness by instrument
10. Barley-10. Pregermination by fluorescein dibutyrate
11. Barley-11. Deoxynivalenol by GC
12. Sprout damage

2 Malt methods

1. Standard sampling, directly comparable with methods for barley.
2. Physical tests, very similar to those for barley (test weight, assortment by sieving, 1000 kernel weight, foreign seeds, and skinned and broken kernels, mealiness). Additional method is acrospire length—longer acrospires mean more modification.
3. Moisture (cf. barley)
4. Extract: specific gravity from a small-scale mash; rate of filtration through filter paper is a loose predictor of wort separation
5. Soluble protein (Kjeldahl) and free amino N (ninhydrin) levels in wort
6. Diastatic power: Titration (by thiosulfate) of iodide not binding to starch. Rapid method: Measurement of reducing sugars by 4-hydroxybenzoic acid hydrazide (PAHBAH). Automated flow analysis.
7. Alpha-amylase. Use of excess beta-amylase, measurement of residual starch by iodine
8. Use of Kjeldahl or combustion to measure protein (N $\times$ 6.25) (cf. barley)
9. Colored malts; mash with standard malt and measure extract and color
10. Nitrosamines by Gas Liquid Chromatography (GLC)
11. Sulfur dioxide by color reaction with *para*-rosaniline hydrochloride
12. Modification using the friabilimeter; second sieving for gross undermodification
13. Deoxynivalenol by GLC (see barley)
14. Dimethyl sulfide (DMS) precursor: Caustic hydrolysis of S-methyl methionine and GLC measurement of DMS
15. Grist analysis

3 Adjunct methods

Cereals vs. syrups
Cereals-1 Sampling (cf. barley and malt)
Cereals-2 Physical characteristics—Color, odor, husks, germs and foreign seeds, mold, infestation, assortment of grits
Cereals-3 Moisture by oven drying
Cereals-4 Oil by extraction with petroleum ether
Cereals-5 Extract: Use of malt (method A) or added enzymes (method B)
Cereals-6 Nitrogen—Either Kjeldahl or by combustion
Cereals-7 Ash—By ignition at 550°C
Sugars and Syrups-1 Sampling of containers—Square root of total number of containers
Sugars and Syrups-2 Color, odor, taste
Sugars and Syrups-3 Clarity of 10% solution
Sugars and Syrups-4 Color of 10% solution
Sugars and Syrups-5 Extract: Direct measurement of specific gravity
Sugars and Syrups-6 Moisture = 100 − Extract
Sugars and Syrups-7 (A) Fermentable extract by fermentation or (B) rapid fermentation
Measure change in Extract

A SUMMARY OF THE ASBC METHODS OF ANALYSIS—cont'd

Sugars and Syrups-8 Iodine reaction for residual starch (blue color) or dextrins (reddish)
Sugars and Syrups-9 Acidity by either potentiometric titration or using phenolphthalein as an indicator
Sugars and Syrups-10 pH
Sugars and Syrups-11 Protein (Kjeldahl)
Sugars and Syrups-12 Ash by ignition
Sugars and Syrups-13 Diastatic Power for malt syrups—Same method as for malt
Sugars and Syrups-14 Total reducing sugars—Fehlings
Sugars and Syrups-15 Sucrose—Measure reducing sugars before and after inversion using either acid or invertase
Sugars and Syrups-16 Dextrose in presence of other reducing sugars
Sugars and Syrups-17 Fermentable saccharides by chromatography—GLC or high-performance liquid chromatography (HPLC)
Sugars and Syrups-18 Fermentable carbohydrates by cation exchange HPLC

4 Brewers Grains Methods

1. Sampling—Wet grains and dry grains
2. Preparation of sample
3. Moisture—Wet samples, preliminary dried samples, dry grains, wet samples (rapid)
4. Available extract—Wet grains and dry grains
5. Soluble extract—Wet grains and dry grains
6. Feed analysis
7. Protein by combustion

5 Hops Methods

1. Sampling "Oregon sampler" to take cylindrical core; statistical sampling. Grinding of cones or pellets
2. Physical Examination—Leaves and stems, color, luster, size of cones, condition of cones, lupulin (amount, color, stickiness), aroma, seeds
3. Aphids
4. Moisture—By distillation, vacuum drying, oven drying
5. Resins
6. Alpha and Beta acids: Extraction in toluene and measurement of absorbance or lead acetate conductimetric method
7. Alpha acids in hops and hop products by ion-exchange chromatography
8. Nonisomerized hop extracts. Isopropyl ether to extract prior to spectrophotometric or conductimetric analysis
9. Isomerized hop extracts—Ion exchange and HPLC analysis of iso-alpha-acids
10. Hop bitter acids in nonisomerized extracts by stepwise ion-exchange chromatography
11. Gradient elution of hop constituents using ion-exchange chromatography
12. Hop storage index. Increase in absorbance at 275 nm relative to 325 nm.
13. Total essential oils by steam distillation
14. cf. method 6, but by HPLC
15. Iso-α-acids in isomerized pellets by HPLC
16. Iso- α, α and β acids in hop extracts and isomerized hop extracts by HPLC
17. Hop essential oils by capillary gas chromatography-flame ionization detection

6 Wort Methods

1. Sampling. Risk of infection, insolubilization, color change, etc. Best to analyze fresh—Otherwise, define storage conditions and attempt to standardize between labs. Any filtration through paper, not kieselguhr (but see 8).

Continued

A SUMMARY OF THE ASBC METHODS OF ANALYSIS—cont'd

2. Specific gravity. (A) by pycnometer—Now archived. (B) By digital density meter—Wort should be bright.
3. Extract. By calculation from specific gravity.
4. Apparent extract by hydrometer.
5. Fermentable extract. Use of yeast—Either (A) over 48 h (or till fermentation complete), or (B) rapidly (5 h)
6. Iodine reaction—Archived
7. Total acidity by potentiometry or indicator method
8. pH
9. Color. Must be bright, but filter paper insufficient. Standardized kieselguhr filtration, then A_{430}.
10. Protein by Kjeldahl or combustion
11. Reducing sugars
12. Free amino nitrogen–ninhydrin reaction
13. Viscosity—Dynamic viscosity (cP) which is resistance to shear flow within a liquid; kinematic viscosity (cS) is a measure of time taken for a liquid to flow through an orifice under gravity.
14. Fermentable saccharides by chromatography. (A) GC, (B) HPLC
15. Magnesium by Atomic Absorption Spectrophotometry (AAS)
16. Zinc by AAS
17. Protein in unhopped wort by spectrophotometry—Measured at 215 and 225 nm
18. β-Glucan in Congress wort by fluorescence—Calcofluor
19. Fermentable carbohydrates by cation exchange HPLC
20. Elemental analysis by Inductively Coupled Plasma-Atomic Emission
21. Thiobarbituric acid index
22. Wort and beer fermentable and total carbohydrates by HPLC
23. Wort bitterness: bitterness units by spectrophotometry; bitterness units by segmented flow analysis . Iso-α-acids in wort by HPLC
24. International Bitterness Units in Wort by Spectrophotometer

7 Beer methods

1. Sampling. Preparation: attemperation to 15–20°C—decarbonation by either manual swirling in Erlenmeyer or on a rotary shaker; procedures for aseptic sampling from tank, line, and finished package.
2. Specific gravity—Pycnometers—Archived, digital density meter
3. Apparent extract—Hydrometer
4. Alcohol—Distillation, refractometry, gas chromatography, Servo Chem Automatic Beer Analyzer (SCABA) (alcohol by catalytic combustion, density by densitometer), enzymic (low-alcohol beers)
5. Real extract—Density of residue from distillation, by refractometry
6. Calculations—Original extract, real degree of fermentation, apparent degree of fermentation, carbohydrate content
7. Iodine reaction—Archived
8. Total acidity—Titration with sodium hydroxide—Endpoint by potentiometry (pH meter) or indicator (phenolphthalein)
9. pH
10. Color—Spectrophotometry at 430 nm, photometer
11. Protein—Kjeldahl, combustion
12. Reducing sugars—Fehling's
13. Dissolved carbon dioxide—Pressure methods, manometry/volumetry.
14. Ash—ignition
15. Total phosphorus—archived
16. End fermentation—Yeast to ferment away any residual sugars
17. Dextrins—Archived

A SUMMARY OF THE ASBC METHODS OF ANALYSIS—cont'd

18. Iron—Colorimetry with bipyridine or *o*-phenanthroline; atomic absorption spectrophotometry; color formation with Ferrozine
19. Copper—Colorimetry with zinc dibenzyl dithiocarbamate (ZDBT) or Cuprethol; AAS
20. Calcium—Titrimetry, AAS
21. Total SO_2—Monier-Williams (reaction with *p*-rosaniline hydrochloride)
22. Foam collapse—Sigma value—or foam flashing (for bottles)
23. Bitterness—A_{275} of iso-octane extracts (including automated flow); HPLC
24. Heptyl hydroxybenzoate—archived
25. Diacetyl—Reaction with α-naphthol or dimethylglyoxime; GLC
26. Formazin standards for haze
27. Physical stability—Haze after chilling—Elevated-temperature storage
28. Chill-proofing enzymes—Archived
29. Lower boiling volatiles—GLC for esters and alcohols
30. Triangular taste test
31. FAN–ninhydrin
32. Viscosity
33. Calories by calculation—From alcohol, real extract, and ash
34. Dissolved oxygen—Colorimetry using indigo carmine
35. Total polyphenols—Colorimetry with Fe(III) in alkali
36. Sodium by AAS
37. Potassium by AAS
38. Magnesium and calcium—AAS, sequential titration with ethylene diamine tetraacetic acid (EDTA)
39. Chloride—conductometric or mercurimetric titration
40. Nitrosamines distillation or adsorption prior to GLC
41. Total carbohydrate by spectrophotometry (phenol reaction) or HPLC
42. Aluminum by graphite furnace AAS
43. Chloride, phosphate, and sulfate by ion chromatography
44. DMS by GLC and chemiluminescence detection
45. Elemental Analysis by Inductively Coupled Plasma-Atomic Emission Spectroscopy
46. Measurement of oxidative resistance in beer by electron paramagnetic resonance
47. Iso-α-acids in beer and wort by HPLC
48. Headspace GLC–flame ionization detection analysis of beer volatiles
49. Determination of gluten using the R5 competitive enzyme-linked immunosorbent assay (ELISA) method

Not numbered; beer inclusions; common causes of elevated turbidity.

8 Microbiological methods

8.1 Yeast

1. Representative sampling important—Avoidance of stratification—taking samples at different stages in pumping. Rousing of holding tank contents. Taking samples from different places in compressed cake. Must hold cold 0–2°C and analyze as soon as possible (cf. autolysis, bug growth). Sample in lab jar must be well mixed.
2. Examination. Use of your senses—Color, flocculence, odor, taste. Microscope—cell shape, condition, cleanliness, cell abnormalities, damaged cells, contaminants
3. Viability—Methylene blue, methylene violet. Latter thought more reliable. Spores—Grow yeast in acetate as main C source and then count spores after staining with malachite green. Respiratory deficient yeasts using a triphenyl tetrazolium chloride overlay technique—If respiratory deficient, do not reduce dye.
4. Counting—Hemocytometer and microscope
5. Yeast solids. Dry weight will give contribution from other solids. Spin down can allow differentiation between yeast and trub.

Continued

A SUMMARY OF THE ASBC METHODS OF ANALYSIS—cont'd

6. Viability by slide culture: only living organisms give colony on microscope slide. Compare with hemocytometer.
7. Yeast sporulation—Good for identifying wild yeasts because brewing yeast sporulates poorly if at all—Use of media to induce aerobic growth on acetate
8. Killer yeast identification plate test in which an agar plate seeded with a sensitive strain and then inoculated with killer strain. Clear zone of inhibition. Can use to test whether a yeast produces killer toxin.
9. Giant colony morphology. Growth on wort gelatin medium: production of characteristic morphologies.
10. Differentiation of ale and lager yeast. Growth on 5-bromo-4-chloro-3-indolyl-α-D-galactoside (X-α-gal)—Agar plates—melibiose (lager yeasts) produce α-galactosidase which can break this down. Release indole which oxidatively polymerizes to give an insoluble blue-green dye that does not diffuse into agar; or growth on Bacto yeast mold (YM) agar at 37°C—ale yeasts can grow; or growth in a broth containing melibiose—monitor pH drop by a pH indicator
11. Flocculation. Helm assay—Observe yeast sedimentation in a calcium sulfate solution at pH 4.5. Or Absorbance Method—Similar, but measure yeast in suspension using absorbance at 600 nm
12. Vitality by fluorescence
13. Differentiation of brewing yeast strains by PCR fingerprinting
14. Miniature fermentation assay
15. Differentiation of ale and lager yeast strains by rapid X-α-GAL analysis

8.2 Microbiological

1. Aseptic sampling—Beer in process, packaged beer, process water, brewery air, equipment, and surfaces. Important to get a representative sample.
2. Detection Agar. Incubate separately for aerobic/anaerobic. Organisms collected by membrane filtration for beer, rinse water, etc. Cellulose acetate or cell nitrate membranes 0.45 μM. Put filter directly onto plate.
3. Differential staining. Gram positive/negative
4. General culture media universal beer agar medium and brewers' tomato juice medium. For all types of organism including brewing yeast. Nystatin suppresses growth of brewing yeast
5. Differential culture media
   - Nystatin (formerly Cycloheximide) medium (for bacteria).
   - Lee's Multi-Differential Agar (LMDA)—General bacteria.
   - Raka Ray—Lactic acid bacteria.
   - Lysine medium—Wild yeasts can use lysine but brewers yeast cannot.
   - Lin's Wild yeast Differential Medium—Fuchsin-sulfite and crystal violet.
   - Barney–Miller Brewery Medium: *Lactobacillus* and *Pediococcus*
   - DeMan Rogosa Sharpe (MRS)—Detects *Lactobacillus* and *Pediococcus*.
   - Malt extract, yeast extract, glucose, peptone (MYGP) + Cu—Detects wild yeast
   - Cadaverine, lysine, ethylamine, and nitrate (CLEN)—Some wild yeast. Selective medium for *Megasphaera* and *Pectinatus*.

   Selective Medium for *Megasphaera* and *Pectinatus* (SMMP)
   Nystatin in selective media.
   Hsu's *Lactobacillus* and *Pediococcus* Medium.

9 Filter aids

1. Sampling
2. pH of a suspension of the filter aid in water
3. Impact of filter aid on odor and taste
4. Iron pickup by beer

A SUMMARY OF THE ASBC METHODS OF ANALYSIS—cont'd

10 Packages and Packaging Materials

10.1 Bottles

1. Dimensions (height, outside diameter, out-of-perpendicular, identification marks, glass distribution, weight, locking ring diameter, reinforcing ring diameter, width of locking ring, throat diameter, finish)
2. Defects
3. Color (amber, redness ratio)
4. Capacity (overflow, fill point)
5. Surface protective coatings (lubricity, coating quality, rub tests)

10.2 Bottle Closures

1. Defects glossary and classification
2. Test pressure (for crowns, pilfer-proof closures)
3. Gas retention capacity of crowns
4. Resistance to pasteurization
5. Removal torque procedures
6. Crimp determination for crowns

10.3 Cans

1. Defects glossary and classification
2. Rusting tendency
3. Dimensions (metal gauge thickness, flange width, filled-can countersink depth)
4. Ends (curl opening, seaming-chuck fit, ring-pull end pop-and-pull tests)
5. Capacity (overflow, headspace)
6. Enamel rater for evaluating metal exposure
7. Beverage can terminology. Defines the standard terminology used to describe different parts of a beverage can and can end
8. Copper sulfate test. Determines whether there are any exposed metal surfaces on the interior of containers

10.4 Fills

1. Total contents of bottles and cans by calculation from measured net weight
2. Total contents of cans of known tare weight

11 Sensory Analysis

1. Terms and definitions—Some key ones: ascending method of limits test, bias, category scaling, descriptive analysis, detection threshold, difference threshold, directional difference test, duo–trio test, hedonic, paired comparison test, paired preference test, ranking test, rating test, recognition threshold, reference standard, synergism, terminal threshold, threshold, triangular test
2. Test room, equipment, and conduct of test
3. How to choose the appropriate method
4. Selecting and training assessors
5. Reporting data
6. Paired comparison test—Presentation of two samples for discussion of attributes
7. Triangular test—Pick out the different beer from three samples (two of one, one of another beer)
8. Duo–trio test—Presentation of a reference sample, followed by a pair of beers; subject asked to pick out the nonreference beer from the pair
9. Threshold of added substances—Ascending method of limits test; assessors presented with a series of triangles of beers, with progressively increasing amounts of a given flavor substance. Aim is to identify where the flavor threshold lies.
10. Descriptive analysis
11. Ranking test—Rank attributes of a range of beers
12. Flavor terminology and reference standards
13. Difference-from-control—Score magnitude of difference of attributes in a beer from those in a presented control.

Continued

A SUMMARY OF THE ASBC METHODS OF ANALYSIS—cont'd

Identification Guides

The ASBC methods also include these guides:

Beer Inclusions: Common Causes of Elevated Turbidity.
Common Brewery-Related Microorganisms.
Flavor Standard Spiking Calculator.
Beer Flavor Database.
Hop Flavor Database.

Index

'*Note:* Page numbers followed by "f" indicate figures and "t" indicate tables.'

I

K

L

M

N

O

P

Q

R

S

T

U

V

W

Y

Printed in the United States
By Bookmasters